The Word of God
and the
Languages of Man

SCIENCE AND LITERATURE

A series edited by George Levine

The Word of God and the Languages of Man

*Interpreting Nature in
Early Modern Science and Medicine*

*Volume 1
Ficino to Descartes*

JAMES J. BONO

The University of Wisconsin Press

The University of Wisconsin Press
114 North Murray Street
Madison, Wisconsin 53715

3 Henrietta Street
London WC2E 8LU, England

Printed in the United States of America

Library of Congress Cataloging-in-Publication Data

Bono, James J. (James Joseph)
    The word of God and the languages of man: interpreting nature in
early modern science and medicine / James J. Bono.
    332 p.   cm. — (Science and literature)
    Includes bibliographical references and index.
    Contents: v. 1. Ficino to Descartes —
    ISBN 0-299-14790-8 (hc: alk. paper). — ISBN 0-299-14794-0 (pbk.: alk. paper)
    1. Science, Renaissance—History.   2. Science—History—17th
century.   3. Science—History—18th century.   4. Medicine—15th–18th
centuries.   5. Literature and science.   I. Title.   II. Series.
Q125.2.B66   1995
509'.03—dc20   95-15652

To Barbara, Joseph, and John

Contents

Acknowledgments

Without generous support provided by a number of public and private institutions this book would not have been possible. Grants from the National Science Foundation enabled me to write an essay that was the precursor to Chapter 4 (Grant No. SES-8218405) and to complete most of my research and begin drafting the book as a whole (Grant No. DIR-8720745). Research and writing were completed under the ideal circumstances provided by the Institute for Advanced Study at Princeton during the 1990–91 academic year and summer of 1992, where my membership in the School of Social Science was made possible by the generous support of the Andrew W. Mellon Foundation. I am grateful to the School of Social Science, to its permanent faculty and staff, and to the visiting members of the School of Social Science and the School of Historical Studies for the precious gifts of time and intellectual community. Work on seventeenth-century materials was facilitated by a grant from the William Andrews Clark Memorial Library; a New Faculty Development Award from the New York State/ UUP Joint Labor-Management Program enabled me to consult materials at Harvard University; a research award from the School of Medicine and Biological Sciences at the State University of New York at Buffalo facilitated earlier stages of my research. I would also like to thank, from the School of Medicine, Dean John Naughton; from the Department of History, my former chair, William S. Allen, and present chair, Jonathan Dewald; and from the Faculty of Social Sciences, Dean Ross MacKinnon, for their granting of leaves and for their continued support of my scholarly work.

I am indebted, and offer my sincerest thanks, to numerous research libraries here and abroad and to their generous and knowledgeable librarians and staff, including Widener, Houghton, and Andover Theological Libraries of Harvard University; Princeton University's Firestone Library; the Library of the Institute for Advanced Study, Princeton; Cornell University's Olin Library and its History of Science Special

Collections; the National Library of Medicine and its Division of the History of Medicine; the William Andrews Clark Memorial Library of UCLA; the Huntington Library of San Marino, California; the Bibliothèque Nationale, Paris; the British Library; the Library of the Wellcome Institute for the History of Medicine, London; and, last, the Library of the Warburg Institute, University of London. From my home institution, SUNY at Buffalo, I would like to thank, in addition, the staff of Lockwood Library (and especially its interlibrary loan department), and Lilli Sentz of the Health Sciences Library's History of Medicine Collection.

Chapter 4 represents a slightly revised version of my article "Reform and the Languages of Renaissance Theoretical Medicine: Harvey versus Fernel," *Journal of the History of Biology* 23 (1990): 341–87. Copyright © 1990 Kluwer Academic Publishers. Reprinted by permission of Kluwer Academic Publishers.

Numerous friends and colleagues contributed to this book through their generous conversations, advice, and encouragement. I would especially like to thank Hans Aarsleff, Mario Biagioli, Barbara Bono, Roger Chickering, Harold Cook, Brian Copenhaver, Luke Demaitre, Julia Douthwaite, William Eamon, Judith Farquhar, Moti Feingold, Paula Findlen, Clifford Geertz, Faye Getz, Pamela Gossin, Hugh Grady, David Hollinger, Jim Holstun, Kenneth Knoespel, Ray Kondratus, David Lux, Robert Markley, Bruce Moran, Nancy Nersessian, Cornelius O'Boyle, Martin Pernick, Suzanne Pucci, Karen Reeds, Barbara Reeves, Lissa Roberts, Shirley Roe, George Rousseau, Joan Scott, Michael Shank, Nancy Siraisi, Julie Solomon, Nick Steneck, Charles Stinger, David Vampola, Sarah Van den Berg, Albert Van Helden, and Susan Wells. For providing a creative and supportive context for the pursuit of critical interdisciplinary approaches to the history of science, without which this study would not be what it is, I would like to acknowledge the Society for Literature and Science and its members, especially (in addition to those cited above) Mark Greenberg, N. Katherine Hayles, Stuart Peterfreund, and Stephen Weininger. To my teachers Charles Lohr, Patrick Heelan, John Murdoch, and Everett Mendelsohn, and to the late Charles Schmitt and D. P. Walker, I owe a special debt that I cannot hope to repay. All errors remain mine.

I would like to thank George Levine of Rutgers University, editor of Wisconsin's Literature and Science series, for his continuing support and for including this book in his series. I would also like to thank Barbara Hanrahan and Allen Fitchen for encouraging me to write this book and publish it with the University of Wisconsin Press.

For enlivening and giving purpose to this *corpus*, for continually lightening my burden and refreshing my soul as this work journeyed toward completion, for cheerfully understanding my absences and tolerating divided attentions, and wondrously, for never letting me forget that which most retains ineffaceable value and meaning beyond all books and words, I thank Barbara and my sons, Joseph and John—my animating *spiritus*—and dedicate this book to them.

The Word of God
and the
Languages of Man

1

Introduction

The Word, the Text, and the Narrative

In principio erat verbum, et verbum erat apud Deum, et Deus erat verbum.
—The Gospel According to Saint John

The *Word* occupies a special place in the West. Not only do we find entire discourses dominated by the Word, even spaces created within the very interstices of contesting discourses have been haunted by its presence—or absence. The Word also functions as source of legitimating myths of origins. Premodern Europe, for example, witnessed a powerfully unifying culture of the Book behind which stood the authorizing presence of the "Word." This "bookish culture" pursued the search for truth, figured as eternal and essential, through a variety of strategies. Through dialectic, commentary, the question, literary and rhetorical analysis, exegesis, etymology, explication, allegory, philology, textual criticism, and the disputation, to name a few prominent ones, such strategies focused their efforts on texts. Cutting across disciplines—across, for example, the various and distinct theoretical and practical sciences comprising the scholastic division of the sciences, or across the *trivium* and *quadrivium*—the pursuit of knowledge and search for truths proceeded through the mediation of texts. Behind such pathways toward truth lay the conviction that the knowledge and truths sought by mankind were housed, maintained, and sheltered from corruption, change, even temporality itself, by their stable containment within the Book.

The Book *was* eternal and stable, the product of the divine, not of mortal men. It was God's Word that gave force and validity to the many texts that became the object of intense scrutiny, and that gave urgency

3

to the effortful practices used to probe them and unlock their secrets. Behind each text, however limited or corrupt it might be, stood this ideal of a perfect, comprehensive, consistent, and edifying Book authorized by God's Word and waiting to be uncovered, discovered, or reconstructed by mankind. Thus, "bookish culture" meant more than the mere habit of lavishing attention upon texts, or even of placing technologies of reading the text at the center of the culture's search for truth. At the heart of bookish culture was the authority of the Book and the presence of the Word.

Europeans, however, were not alone among premodern civilizations in fashioning a cultural identity that placed texts at the center of the search for meaning and stability. As John Henderson (1991) persuasively argues, virtually all of the major premodern civilizations had at their core "sacred" texts around which developed complex, but uncannily similar, exegetical and commentarial traditions. Moreover, these traditions soon occupied central places in the ongoing cultural life and evolution of these major world civilizations. Among Confucian, Vedantic, Qur'anic, rabbinic, classical epic, and Christian traditions commentators and exegetes surrounded their "sacred" texts with dense layers of interpretation, while sharing common assumptions about the universality, comprehensiveness, coherence, and self-consistency of such texts that had far-reaching implications for the very cultural identities and foundational beliefs and practices of their respective civilizations.

While all major cultures in world history seem thus to share certain common and deeply rooted features, differences exist as well. For one, the exact nature of the "sacred" corpus of texts varies from one premodern civilization to the next; in some cases, in Christian Europe or medieval Islam, central cultural practices are in fact defined by the necessity of reconciling, or otherwise accommodating, a "sacred" text—the Qur'an or the Bible—with other texts that assumed a canonical status and generated layers of commentary often intertwined with traditions of "sacred" commentary. For both medieval Europe and Islam, the textual inheritance of classical and Hellenistic Greek writings provided such an alternative canon. Then, too, one can point to the obvious fact that in the West the tradition of commentary and exegesis—as *central* and *privileged* cultural practice—seemingly came to lose much of its cultural authority and paradigmatic status in the early modern period, while other civilizations remained very actively traditions of the text.

This is not the place to take on the monumental world-historical question of how, and why, the West followed a different trajectory from that of the rest of the world. But we can and should note that the question, and more specifically the observation, that the West largely "abandoned" exegetical and commentarial traditions in the seventeenth century is misleading and fraught with opportunities for invidious comparisons. Here science and a supposed "scientific mentality" are at the center of historiographical, not to say ideological, controversy. For the observation that premodern European "bookish culture" was replaced by modern European scientific culture seems to suggest a discursive rupture of such proportions that it calls for the construction of a new mentality—one that so-called "Hellenocentric" historiography traces to ancient Greek origins.[1]

While not explicitly entering this larger debate, the present study takes rather a different view of this transition from bookish to scientific culture. Most baldly put, scientific culture[2]—at least in early modern Europe—does not represent a break from enchantment with the figure of the Book, nor from the authorizing presence of the Word. Rather, it represents the transformation of such enchantment and presence into new cultural practices that remain "textual," though no longer wedded to the exegetical and commentarial traditions of the old, bookish culture. While the new "scientific culture" of the seventeenth century does break from the particular kind of attention "bookish culture" paid to texts and even discards many of its technologies of reading, it nonetheless shares a foundational tropological deployment of the metaphor of the Book. Each "culture" deploys that metaphor in markedly different ways, to be sure; each enacts its own "topical contextures"[3] and

1. I do not mean to suggest that the "scientific culture" of the seventeenth century simply represents the recreation of an ancient Greek "mentality." Instead, I merely wish to point out that the question of the unique character of the West and its special genius in giving birth to science often assumes a special Western "mentality": one that is rational, scientific, nonsuperstitious and that is assumed to have deep roots in ancient Greek civilization. For "Hellenocentrism" and its critique, see Bernal (1987, 1991). For a critique of the notion of mentalities and a corrective to such Hellenocentrism, see Lloyd (1990); in addition, see Vernant (1982); Staden (1991, 1992, 1993).

2. I use the phrase "scientific culture" with great concern and only as a shorthand for the emerging culture of seventeenth-century Europe that decried the exegetical and commentarial practices of its predecessors and contemporaries. This is *not* to say that this new culture is not concerned with texts, nor that it does not rely upon the critical textual and philological methods of humanism and the new classical scholarship. Nor do I mean to suggest that the seventeenth century constructed a scientific culture in a late-nineteenth- or twentieth-century sense. Quite obviously, the term "scientific" is an anachronism. See Grafton (1991).

3. For the notion of topical contextures, see the important article by Lynch (1991). Lynch's topical contextures provide a corrective to the global force of Foucault's discourses. By invoking Lynch's term here I intend to signal the local and hybrid, even at

discursive practices. But each as well establishes itself as a discourse of the Book.

Thus, while the West breaks from the rest of the world in the seventeenth century by initiating a turn away from textual traditions of commentary and exegesis, from "bookish culture" in the narrow sense, within its own cultural arena the West experienced a discursive shift very much rooted in attempts to rework and reconstitute the meaning of the Book. Old practices waned or found themselves strategically redeployed within the regime of new practices. Newly defined relationships of man, society, and nature to a divinely authored Book produced new practices, including technologies of "reading" central to the emergence of modern forms of science. The "natural and experimental history" of Bacon, the probabilism of Mersenne's mathematical and observational methods and mechanism, the Cartesian "mathesis, ou mathématique universelle," and Boyle's "experimentalm life" with its instrumental and literary technologies all constitute new practices incorporating new technologies for reading—and reconstructing—the divine Book. Each, of course, had its limits and strengths. But all depended upon the central trope of the Book; all purported to read that Book. All were therefore authorized by the Word, access to whose "meaning" legitimized their respective discursive practices.

This understanding of critical transformations within early modern European civilization suggests, I argue, a rather different view of its dynamics. Henderson (1991:201–2), for example, stresses the singular importance of the sheer proliferation of texts—and alternatives made possible by diverse texts, including the possibility of contradiction and error—as the engine driving cultural change in the European tradition. This argument places great weight on the new technology of printing. To be sure, printing did have an enormous impact upon early modern culture; it did foster and even accelerate the development of critical and historical approaches to texts—first developed within humanist textual criticism and philology in fifteenth-century Italy before the advent of the printing press—that challenged the old exegetical and commentarial tradition (Eisenstein 1979). But printing and the sheer multiplicity of books alone did not drown the old "bookish culture" in a sea of conflict, contradiction, error, and texts! Rather, critical transforma-

---

times self-contesting, nature of these discursive complexes. For more on these points see Bono (1990a).

tions within early modern culture owe their origin to the steady reworking of formative cultural narratives within the late Renaissance of the
sixteenth and seventeenth centuries. Indeed, the very metaphors of
the Book and the Word so central to Latin Christian culture became the
battleground over which an astonishing variety of contesting narratives fought for the privilege of establishing an authoritative interpretation. The new science—better understood as a set of cultural practices
and technologies for reading and manipulating the Book of Nature—
owes its very formation as a complex and hybrid discursive regime to
the fabricating power of newly established metaphoric "contextures"
produced by narrative situation of key metaphors in new, local networks of meaning.[4]

Thus, I argue that the dynamic of cultural and scientific change in
early modern Europe does not come from outside, whether from the
impact of printing or from a "social context" figured as external and
prior to science. Rather, we must look to a complex process whereby
cultural actors continually negotiate meanings produced by the contestation of new narratives vying for interpretive authority over formative tropes of the Book and Word.

Looking to such a process of negotiation, contestation, and change
may strike the historian of science as odd, or even prove unsettling.
By displacing familiar practices, it may evoke fear of the abandonment
of "real history" for "theory." Yet all historical practice is simultaneously theoretical. Indeed, contesting a sharp dichotomy between
"theory" and "practice," I would point to this study as an example of
their interdependence and even mutual construction. Nonetheless,
both the subject matter and the analytic approach of my book are
sufficiently unconventional—at least within the history of science—
as to call for some attempt to articulate what is at stake theoretically.
In recent years, the history of science has begun to gaze upon a radically transformed panorama, one in which the foreground of discoveries and the middle ground of scientific theories have become situated
against, and within, a background of social, cultural, and ideological
contexts. Exploring that new and rough terrain has proven exhilarat-
ing. But mapping passages and tracing exchanges among these diverse, contiguous territories have proven far more resistant. I believe
that attention to the role of language, metaphor, narrative, and scientific practice can help us begin to carve out useful paths through the

4. Bono (1990a: 1995). Bono (in progress) will discuss connections between metaphor,
narrative, and scientific change.

landscape that the history of science must traverse in the task of exploring this terrain.[5]

To do so we must, however, think of the history of science as part of a larger enterprise, that of science studies—better still, of the cultural studies of science. Without privileging any one component in the usual litany of "internal" and "external" factors that contesting parties invoke to stake their claims about the "nature" of science, we should, in addition, start from the premise that science is not "a natural kind" (Rorty 1991; Rouse 1993) and cull from literary theory, from literature and science, and from theoretical discussions in history viewpoints and practices that can help us dismantle troublesome obstacles. By reexamining assumed dichotomies between literature and science, theory can permit us to articulate useful understandings of the textuality of science, its inherent culturality, and the discursive meaning of all practices, including scientific practices.[6]

By problematizing the very status of the historian as interpreter of the past, theory in contemporary history can help us grasp the evidence we use as multivalent "texts." The very "documents" once seen by historians as providing evidence for or against historical narratives and explanations now present themselves as complex "texts" that are themselves the site of discursive conflicts and self-contesting motives and "interests." The appeal to explanatory "contexts" to legitimate readings of "documents" or "complex texts" has also proven to be far more troublesome than past practitioners of history may have understood. While the call by some to reject context as a legitimate historical concept appears unnecessary, undesirable, and perhaps simplistic, historians do need to rethink the notion of context. In particular, we must contest the binary opposition between text and context and, with it, the notion that stable, universal contexts—such as "culture" or "society"—can be unproblematically invoked to explain the meaning of a text or cultural practice. Contextual arguments are important, but how contexts are formed and the role of texts and textuality in constructing contexts must enter into framing of such arguments by

5. This topographical metaphor has its limitations. One of them, which I hope will become quite obvious, is that the elements of the scientific landscape that I have mentioned are not as sharply separated from one another as the metaphor suggests. The scientific and the sociocultural are not different terrains; each defines the other. See the review essay by Latour (1990).

6. For an overview of recent trends in literature and science, see Rousseau (1986); Levine (1987); Amrine (1989); Peterfreund (1990); Batt and Pierssens (1993); and, finally, work appearing in the new journal *Configurations: A Journal of Literature, Science, and Technology* 1 (1993) and later issues.

historians. With respect to scientific discourse, then, the subject position of the historian entails that the interpretation of "documents" and "contexts" constitutes a thicket that requires vigilance and theoretical wariness to penetrate.[7]

Wandering through the landscape traversed by the history of science, theory can suggest to the historian a more highly problematized relationship between language and science than has typically been assumed. If not ignored, language is often treated simply as a landmark signaling the fact that one has entered the terrain of a new scientific paradigm or theoretical framework. Language becomes, in a sense, an epiphenomenon of some larger, totalizing, paradigmatic perspectival shift. While the features of such a new scientific language—the semantic import of theoretical terms, the relationship between words and things in scientific representation—may attract the attention of scientists and later historians as intrinsically interesting, the role of language as agent or site of scientific change is dismissed or, at the very least, reduced to the level of the inconsequential. In short, if language is part of scientific change, it is either the consequence of such change or the passive impediment that enlightened change seeks to overcome. Whether mere landmark, or formidable barrier to progress in the journey of science, language, in this view, provides neither passage nor medium of exchange among the foreground, middle ground, and background of the historical terrain mapped by historians of science.

This view fails, however, to problematize scientific language in a way which recent theoretical notions of discourse, context, and the subject position of the historian demand. Scientific language does not constitute the mere discursive reflection of some totalizing worldview. Rather, one can instead regard scientific language as itself the site of intensive and extensive discursive exchanges. Here scientific language, rather than simply monolithic and stable, is itself complex, self-contesting, and a source alternately of temporary discursive stability and of a tensely unstable, complexly negotiated exchange.

Problematized in this fashion, language can serve a crucial role as an object of scrutiny for the historian of science. Language may provide a focus for historical analysis that can expose the "textual" paths and processes linking the practices of science and scientific theory with sociocultural "contexts." Such terrain can open itself up to exploration

7. On the linguistic turn and history, see LaCapra (1983); Orr (1986); Scott (1986); Toews (1987); Kloppenberg (1987); LaCapra (1988); Berkhofer (1988); Novick (1988); Harlan (1989); Hollinger (1989); Appleby (1989); and Hunt (1989).

by historians of science who join the hunt for the role language plays in facilitating historically situated discursive exchanges. Metaphoric language, often regarded as opposed in principle to the precise, literal language of modern science, pervades scientific discourse and, as I have argued elsewhere, provides the very site for exchanges between scientific and other discourses (Bono 1990a). Such metaphoric exchanges can also facilitate the adoption of narrative strategies that structure the articulation of theoretical explanations rendering the work of science into a rhetorically cohesive scientific discourse.

This book offers to illustrate a number of ways in which attention to "language" presents fertile ground for the history of science and science studies. Clearly, its chief focus is upon language theory and its role in the Scientific Revolution. The book does not, however, assume an "interaction" between two fully articulated and separate domains of "language theory" and "science." Quite to the contrary, my intention has been to treat both language and science and complex and mutually dependent cultural practices. Although we may in some sense retrospectively identify a set of core ideas that we associate with each, as *practices* language and science do not respect boundaries that exclude one from the other. Renaissance magic enacts Neoplatonic language "theory," just as sixteenth-century natural history as a practice orders and articulates the natural world as a complexly differentiated, but fundamentally unified, system of linguistic signs and symbols.

As mutually dependent cultural practices, language theory and science constitute intersecting discourses linked by common metaphors. Two dominant metaphors in particular foster such links and exchanges: the metaphors of the Book and of the Word. Yet these metaphors *also* link language and science to other discursive regimes, particularly to religious discourse. As a result, the metaphors of the Book and the Word facilitate discursive exchanges between religion on the one hand and both language and science on the other. Through such exchanges (and we must not forget the ways in which religious discourse intersects with the discourses of society and politics in the sixteenth and seventeenth centuries), the metaphors of the Book and the Word become subject to competing religious-cultural narratives: stories about the origins and nature of language, for example. Such narratives, then, attempt to fix, or stabilize, the meanings associated with these metaphors by providing a specific social and cultural location within the overall *ecology* of contesting sociocultural discourses and values (Bono 1995).

The result of the narrative reinscription and location of the metaphors of the Book and the Word is that the very practices of language and science take on new meaning. In this processs, the recast cultural narratives fundamentally alter the metaphoric relations through which natural things ("animals," "plants," "heart," "blood," "matter") and language ("words," "names," "grammar") are discursively related to the "Book of Nature" and the "Word of God." Consequently, such narrative reinscriptions authorize new hermeneutic strategies for interpreting and articulating the relations of things and words to nature and the divine, and to each other. We get, in short, new practices and new ways of engaging the world linguistically and scientifically, new technologies for mapping nature and knowledge.

The title of my book suggests that just such narrative contexts generated contestation and change in the understanding of texts and nature in early modern Europe. The mythos surrounding the Western notion of the Word lent profound meaning to the trope of the Book. The Word as *verbum Dei* signified the Bible; the Word also authored, through its originative creative power, God's Works—the world and creatures. Behind the literal surface of words and creatures found respectively in the text of the Bible and in the text of nature therefore stood the presence of the Word. The Word sustained each text as a living and comprehensive Book, pregnant with divine meaning for all humans potentially to read. But *how* could humans read those twin books?

The text of the Bible, although authored by God, presented His Word to humans in the wrappings of ordinary human languages. To penetrate God's Word—to unlock the secret wisdom of His written Book—tools such as exegesis must be applied to the language of the Bible. Humans must approach divine wisdom by removing the dead husk of mere fallen and corrupt human language to expose the living kernel of the Word within. Exegesis and commentary—indeed, the entire array of technologies of reading developed by Latin Christian culture—were then applied to the text of the Bible in order to uncover the meaning and message of the Book. That meaning and message was itself the living Word that mortal humans hoped to grasp in their quest for salvation and release from the finality of death.

The search for the *logos* implanted by God within things in the Book of Nature also entailed the mediation of human language. The Judeo-Christian metaphorics of creation with its emphasis upon the creative power of God's speech—His Word—transferred to human apprehension of nature the characteristics of textuality associated with the Word and authorship, thus making nature a book whose secrets could best be contained in words properly framed in human languages. Then, too, the close association of knowledge of nature with the categories and vocabulary of Aristotelian philosophy in the Middle Ages, and consequently with syllogistic or propositional reasoning, ensured that interpretation of the Book of Nature would closely align itself with careful analysis of terms framed within prevailing human, if scholarly, languages. Knowledge of the Book of Nature was, in short, embedded in linguistic mediations. Hence, the same techniques used to read God's Book of Scriptures could be, and were, transferred to reading His other Book, nature.

For both biblical criticism and natural philosophy access to the rich significance of these two divinely authored books required intimate engagement with language and with an entire toolkit of technologies of reading. Thus the metaphoric associations of the Word and the Book laid foundations for a set of assumptions about the divine and nature that generated specific technologies of reading and authorized hermeneutic practices for the production of theological and philosophical, or natural, knowledge. The result was the construction of a "bookish" culture whose activities centered upon texts, language, and their interpretation.

This bookish culture was, however, far from static. While the focus on the text remained a constant and the full array of technologies of reading constituted a common resource, the interpretive strategies and specific hermeneutic practices deployed by actors in this bookish culture were responsive to local variation. More precisely, the interpretation of nature and texts so central to bookish culture came to be crucially dependent upon the construal of the relationship between the "Word of God" and the "languages of man." Here cultural narratives come to play a decisive role.

For most exegetes of the Middle Ages, reigning narrative schemes cast human languages as thoroughly implicated in fallen human history, as inherently corrupt, darkened by the shadow of the ruined Tower of Babel, and but distantly related to the Word of God. While human languages were therefore largely divested of any special transcendent meaning, they nonetheless served an important function as vehicle for the articulation of divine and natural knowledge. For better or worse, the textual focus through which such divine and natural knowledge come to be reflected within "man's" mind required the mediation of human languages, imperfect as they might be. Thus, Saint Augustine would, in his *De doctrina christiana* and other works, authorize the use of pagan learning and texts and justify the human's ability to fathom the divine Word locked within the confines of postlapsarian and post-Babylonic fallen human languages. Although producing an inevitable density, human language nonetheless provides mankind with its only access and prod to knowledge of the Word. Through careful comparison and explication of texts, through techniques of reading such as allegory, mankind can strive to lift the veils of ordinary language obscuring the Word.[8] To Augustine's limited, but accommodating, approval of the use of the "languages of man" in the search for divine truths in the text of the Bible, the Scholastics of the High Middle Ages added an ambivalent Aristotelian perspective on language. Not only were the languages of man fallen, according to Aristotelians they were thoroughly social and conventional in origin and character. Yet, despite the conventionality of language, Aristotle's example fostered a system of Scholastic inquiry in natural philosophy (as well as other areas of knowledge) that enshrined textual practices that were part and parcel of the bookish culture of the late Middle Ages and Renaissance and that relied upon an elaborate word-based logic and semantics.[9]

Indeed, as I argue in this book, those who would link the emergence of modern science in the seventeeth century with adoption of a conventionalist theory of human languages simplify what is a far more complex and historically interesting cultural phenomenon. Many Aristotelians of the Renaissance and early modern period continue to subscribe to the master's conventionalist view of language while participating in what we may broadly call "bookish culture." Julius Caesar Scaliger even produced an elaborate investigation of language and language theory that enshrined this Aristotelian perspective in opposition to proponents of more "mystical" views (Scaliger 1540; Jensen 1985, 1990). The transition from bookish to scientific culture *was* critically dependent upon shifting theories of language in the sixteenth and seventeenth centuries, but to understand this link we must understand how both language and "science" were mutually implicated cultural phenomena.

Earlier I stated that the metaphor of the Book was foundational to both bookish and scientific cultures of the early modern period. Rather than an abrupt *rupture* dissociating the "two cultures," I contend that the transition between the two involves attempts to contain and negotiate new meanings that try to attach themselves to the metaphor of the Book in the sixteenth and seventeenth centuries. These meanings were

8. On Augustine see especially the illuminating discussion in Cave (1979:80–83), and the chapter in Colish (1968 [1983]).
9. On Scholasticism, language, and logic, see E. J. Ashworth (1988).

produced by the adaptation of formative master narratives, specifically biblical narratives that concern themselves with the origins and nature of language. The contestation of narratives and narratively defined meanings surrounding language in this period represents a relatively unstudied cultural phenomenon that operated with enormous implications both within and across discourses of natural philosophy, medicine, magic, and science, and within and across bookish and scientific cultures.

Within the boundaries of bookish culture, the Neoplatonism revived by Ficino's Florentine Academy later provoked seminally important reworkings of biblical narratives of the Fall and the confusion of tongues associated with the building of the Tower of Babel. Combined with the Renaissance search for origins (Quint 1983), such efforts reopened the question of a postlapsarian and post-Babylonic survival of the originative divine Word within the Adamic language. As a result, the entire relationship between the "Word of God" and the "languages of man" became subject to redefinition. If the languages of man were in some fashion inherently linked to the Word of God, as some claimed, then attention to language itself held open the prospect of access to divine knowledge and wisdom. Furthermore, since the Adamic language embodied Adam's true and perfect knowledge of the natures of all earthly creatures in the Garden of Eden, access to the pristine Adamic language promised to unlock the secrets of the very Book of Nature for its sixteenth-century readers.

Narratives extolling the presence of the Word in the languages of man assumed many forms in the sixteenth and seventeenth centuries. In turn, such narratives provoked a variety of counter narratives that modified, supplemented, fundamentally altered, or rejected the vision of the Word of God and the languages of man, of Adamic knowledge and postlapsarian "science," associated with them. Within the borders of bookish culture, these new narratives fostered new awareness of the content of and interconnections between the two divine books of the Scriptures and nature and stimulated new interpretive strategies and hermeneutic practices for reading these newly constituted books. By contrast, new counternarratives—often drawing upon discursive resources generated in the wake of Reformation and humanist understandings of the Book and textual traditions—inscribed profoundly different visions of the Book of Nature and its connection to the Word of God. Indeed, the very contest over narrative authority to interpret Scriptures and the Book of Nature ultimately, I claim, led to radically new relationships between the "Word" and the "Book of

Nature," between the languages of man and the Word of God, between God and nature, and between words and things. Within the new discourses defined by these shifting relationships, a new regime of practices developed that came to focus not explicitly upon texts, but upon things inscribed by God in His Book of Nature; not explicitly upon meaning and symbols, but upon descriptions of structures, relations, and things in nature. The agonistic struggle of narratives within the discourses of late Renaissance "bookish culture"—a struggle that was not "just" about ideas and "discourses," but about situating and valorizing discourses as cultural practices and social institutions—and attempts to negotiate among them thus created in the very rifts and fissures opened up by its seismic cultural activities a space for new discourses, and thus new practices, of a purportedly postbookish, "scientific" culture. The present study charts how the narrative reconfigurations of "science" open just such spaces; its companion study will trace the networks of social, religious, and political sites that produce (and are constituted through the production of) alternative narrative configurations in one specific example—that of mid-seventeenth-century England.

I have framed *The Word of God and the Languages of Man* as a two-part study of the cultural transformation of "science" during the sixteenth and seventeenth centuries. This volume constitutes the first part of my study: *"Ficino to Descartes."* The second part, not yet complete, will attempt to analyze in much greater detail how attention to the sort of narratives about language found within the texts and cultures of early modern science explored in the present book can help us grasp the complex *sociocultural dynamics* of change. More specifically, by focusing upon the concrete example of language theory and science in England from 1640 to 1670, the second volume will dissect the networks whereby the scientific, medical, social, political, cultural, and religious simultaneously implicate and fashion each other. Such networks are (at the least) material, institutional, and social; but also (though less obviously) literary, tropological, and narrative. The "solidity" of the former appear to establish "contexts" for the sciences; certainly, material artifacts, institutional arrangements, and social boundaries permit the very work of the sciences to be defined and placed in motion. Yet, to invoke material, institutional, and social networks as *the* "contexts" for the sciences obscures the very operation and circulation of *meanings* such networks put into play. For these networks do not simply constitute and circulate already canonized meanings, thus establishing themselves as the solid and stable *contexts* of meaningful scientific discourse and practice; rather, they themselves are also *"texts"* fully implicated in the exchanges of meaning that construct "sciences" as cultures and practices. Material, institutional, and social networks, and the artifacts, arrangements, and boundaries that they produce, remain enmeshed in literary, tropological, and narrative networks that rework them as texts that construct contexts for the sciences. As a result, the active historical agent—standing at a node of such intersecting networks—receives them as "texts" to be reworked into "contexts" for the production of scientific practices and knowledge. Where the first volume displays the narrative reworking and (re)configuration of "science," medicine, occultism, and natural philosophy in the circulation and adaptation of formative Renaissance cultural narratives, the second volume will turn to the necessarily *local* construction of intermeshing networks themselves and their fashioning of a culture and sociopolitical polity of science in mid-seventeenth-century England.

Given the scope of this second volume, I have deliberately excluded certain topics from the first. Thus, serious consideration of such figures as Jacob Boehme,[10] John Wilkins, and Robert Boyle will figure prominently only in the story I plan to tell in Volume 2. While accomplishments of the newly founded Jesuits and their protégés—from the polymath and occultist Guillaume Postel, to the Coimbra commentaries and Iberian Scholastics such as Benedictus Pererius, to the anti-occultist Martin Del Rio, to, finally, the prodigious Athanasius Kircher—receive some attention, I defer to the later volume discussion of the role of Jesuit, and more generally Catholic, science for two reasons. First, recent scholarship clearly demonstrates the formidable growth of Jesuit scientific traditions starting in the late sixteenth century and mounting steadily higher through the seventeenth. With its emphasis upon "industry," Jesuit and Catholic science will serve as a comparative backdrop against which one may better assess the "Baconianism" of mid-century and of the Restoration Royal Society. Second, the prominent role played by Jesuit missions and scholars in the study of non-European languages clearly connects with our continuing concern with language theory, cultural narratives, and science and with specific developments concerning their interrelations in seventeenth-century En-

10. Boehme plays such an important role in England, particularly among the radical sectarians and those inclining toward a Paracelsian or chemical philosophy, that I have chosen to consider his work in the context of Volume 2. Many of Boehme's works were indeed translated into English during the period I shall cover; for example, Boehme (1651, 1965). See also Koyré (1929); Hutin (1960); Konopacki (1979); and Ormsby-Lennon (1988).

gland.[11] The second volume will also address an issue of pressing importance: what significance do the cultural narratives informing early modern science have for women? In the context of Restoration England, I plan to reexamine the work of Margaret Cavendish, and perhaps (if space and chronological limits permit) Anne Conway.[12]

The argument of Volume 1, "*Ficino to Descartes*," proceeds through seven subsequent chapters. In Chapter 2 ("Ficino and Neoplatonic Theories of Language"), I begin with an examination of the roots of what might be called a Neoplatonic theory of language in the work of the late-fifteenth-century Florentine philosopher, physician, and *magus* Marsilio Ficino. Ficino not only revives the so-called Hermetic tradition of magic stemming from the alleged writings of the legendary Hermes Trismegistus but also re-presents, through his translations and commentaries, both the Plato of such dialogues as the *Philebus* and *Cratylus* and the Platonic theory of language for subsequent generations of Renaissance students of nature. Drawing upon ancient Neoplatonic readings of Plato, Ficino frames ideas fundamental to the Neoplatonic theory of language as it was articulated in the Renaissance that would have such a central and transformative impact on the story told in the rest of Volume 1. More than that, Ficino tentatively and cautiously incorporates such views of language into his own system of natural philosophy, medicine, and magic. In doing so, Ficino unleashes a host of theoretical and practical problems concerning nature, God, the *verbum Dei*, and the languages of man for subsequent early modern thinkers to confront. His own speculations, as I suggest, also provided a model that made it imperative to articulate a specific narrative framework in order to invest such theories of language with potential meaning and use that humans could access and performatively enact.

The framing of narrative contexts for a theory of language and for justifying humankind's access to the divine secrets implanted in nature forms the central concern of Chapter 3 ("The 'Word of God' and the 'Languages of Man': Renaissance Cultural Narratives, the Reformation, and Theories of Language"). I argue that the need for understanding such narrative contexts has been obscured, and is made all the more urgent, by the misleading habit of labeling early modern theories of language as either "natural" or "conventional." To be sure, classical

11. On Catholic and Jesuit science and related matters, see W. B. Ashworth (1986, 1989); Dear (1987); Feldhay (1987); Harris (1989); N. Jardine (1979); Mungello (1985).

12. On Cavendish, see Grant (1957); Sarasohn (1984); and Schiebinger (1989). On Conway, see Merchant (1979, 1980); Gabbey (1977); and Nicolson (1930).

precedent can be found in Plato and Aristotle for such a division; this division, and a vocabulary for expressing it, were also used by Renaissance authors. Nonetheless, two objections remain to its continued use by historians: first, the distinction as we use it today carries too many modern connotations and is freighted with too many ideological implications and with a restrictive finality that many, if not most, sixteenth-century thinkers would not have felt; second, as I illustrate in this chapter, an enormous variety of options regarding the nature and origins of language existed in the early modern period that can not fit neatly into this procrustean bed.[13]

This is not to say that a certain disenchantment with human languages did not occur in the Renaissance. Indeed, a number of fine books have charted a growing recognition from the late Middle Ages through the sixteenth century of the flawed, imperfect nature of language and of "man's" consequent alienation from the ideals of perfect knowledge figured by the tropes of the Book and the Word.[14] These views, I would nonetheless argue, often couple with others to make peculiarly hybrid, to us even monstrous, offspring.[15] For sixteenth-century thinkers, at least, such monstrous progeny combining fundamentally flawed human languages with the prospect of an ideal, universal—if not unified—language was but a marvelous consequence of God's active agency in the world.[16]

Thus, Chapter 3 insists upon the important narrative contexts for language theory found in the biblical episodes of Adam's naming of the creatures in the Garden of Eden and of the destruction of the Tower of Babel and consequent confusion of tongues. Such master cultural narratives became the focus of intense contestation as they were recast with various implications for the understanding of the relationship between the "Word of God" and the "languages of man." I further argue in this chapter that such recasting of Renaissance cultural narratives—often

13. I am indebted to Professor Milad Doueihi of the Johns Hopkins University for bringing to my attention a new book by Marie-Luce Demonet (1992). Word of this book reached me too late to take it into consideration in the present work.

14. For example, Cave (1979); Gellrich (1985); Quint (1983); Waswo (1987). For Commonwealth and Restoration England, see Kroll (1991), which does not pay much attention to these tropes. See review of Kroll by Bono (1993).

15. For tensions and ambivalences that can be found in such sixteenth-century figures as Erasmus who confront the limited, imperfect, and fallen nature of human languages, see the learned, theoretically acute, and subtle analyses offered by Cave (1979).

16. See Céard (1980) for a fine discussion of tensions within sixteenth-century language theory. Céard (1977) has also captured wonderfully the sixteenth-century capacity to entertain the monstrous within the Book of Nature. See Chapter 6, below.

responding to the changed cultural climate produced by the Reformation—produced assumptions about language and "man's" access to the perfect understanding contained in the *verbum Dei* that profoundly shaped how humankind could read the Book of Nature. Fundamentally, what we witness in the sixteenth and seventeenth centuries is the narrative construction of a variety of interpretive strategies or hermeneutic practices available to postlapsarian humans. These range the spectrum from what I call (textual) exegetical to symbolic exegetical to "deinscriptive" practices.

The remaining chapters of my book center upon particular examples of figures who adopt specific hermeneutic strategies for reading the Book of Nature during the Scientific Revolution. Chapters 4 and 5 treat examples of various exegetical and symbolic strategies of interpretation, while Chapters 7 and 8 focus instead upon figures like Bacon who turn to nature as a book written in the language of things and construct "deinscriptive" practices for describing things or their relations. Chapter 6 attempts to pull together a number of intermediate examples while discussing the transition from one set of strategies and practices to another. (This transition, I might add, is not, strictly speaking, linear or irreversible.)

Chapter 4 ("The Priority of the Text: Bookish Culture and the Exegetical Search for Divine Truth") contrasts the hermeneutic practices of two major practitioners of theoretical medicine in the sixteenth and seventeenth centuries. Focusing on the particular role of the blood, spirits, and the theory of matter in physiology, the chapter underscores the significance of language theory for understanding the differences between Harvey and Fernel. Fernel's attitude toward language was shaped by Neoplatonic theories as he adopted Ficino's notion of an ancient tradition of pagan wisdom stemming indirectly from the Mosaic revelation of the *verbum Dei*—what was known as the *prisca theologia*. Consequently, Fernel regarded the language of ancient medical and philosophical texts as containing divine wisdom in a veiled form that could be uncovered through a kind of Christianized etymology and philology associated with exegesis. Harvey, by contrast, rejected such views of language and ancient texts and instead articulated a view of language that deeply affected his efforts to de-in-scribe God's Book of Nature through the adoption of experimental and observational practices. Such practices led Harvey to articulate a complex and subtle understanding of the vital activities of the body's tissues and organs grounded in a theory of vital, active matter.

Chapter 5 ("Reading God's Signatures in the Book of Nature:

Paracelsian Medicine and Occult Natural Philosophy") considers variants of sixteenth-century exegetical hermeneutics that turn away from written texts to the visible "text" of nature. Paracelsus, I argue, leans upon a narrative of postlapsarian man cut off from the pristine, Adamic language. Rejecting fallen human languages and the traditions of textual exegesis, disputation, and commentary centering on language, Paracelsus views nature itself as a book bearing the traces of God's own image stamped upon things, His creatures. Drawing upon the model of Adam-*signator*, practitioner of the signatory art (*kunst signata*), Paracelsus constructed an elaborate practice for the symbolic exegesis of the text of nature itself. Later Paracelsians adapted Paracelsus' hermeneutical practices to the constraints imposed by the local sociopolitical conditions they experienced in a post-Reformation ecology of cultural discourses. One such figure, Oswald Croll, represents a telling example in which the presence and Protestant metaphorics of the Word transmute Paracelsus' signatory art to fit a "radical" Protestant narrative of redemption via the agency of the Word. The consequences of these metaphoric and narrative shifts are analyzed and illustrated in this chapter.

Chapter 6 ("From Symbolic Exegesis to Deinscriptive Hermeneutics") is transitional. Its subject is precisely the making and unmaking of links between words, symbols, and things on the one hand, and the "Word of God," as both archetypal model of the world and authorizing ground for a hermeneutics of nature, on the other. I first consider the case of natural history in the early modern era, which displays a hybrid attention to both written texts and the texts of nature, viewing the latter as symbolic and as participating in a vast intertextual network of words, fables, and things. Fundamentally exegetical in its practice, this form of what has been called the "emblematic world view" (Ashworth 1990) raises questions concerning how, and why, this intertextual view of nature waned in the later seventeenth and eighteenth centuries. Considering the views of Ashworth and others, I follow up suggestions offered by Jean Céard (1980) to propose important links to narrative and language theory in the transition from an exegetical and symbolic to a descriptive natural history. This transition, I argue, involves a shift from a discourse that privileges unity to one that glories in the variety and multiplicity of nature. What seems to be at stake, then, are differing valuations of alterity and the ability of discourses to absorb and contain difference. The model of change that emerges from this discussion clearly questions the metaphorics of rupture and conversion favored by others, suggesting instead a complex process of negotiation involving metaphoric exchanges and the opening up of narrative spaces for new hermeneutic practices. Finally, a highly selective reading of Galileo concludes the chapter with discussion of the implications of his view of the Book of Nature, mathematics, and the role of interpretation.

Chapter 7 ("The Reform of Language and Science: Sir Francis Bacon's Adamic Instauration and the Alphabet of Nature") indicates how Bacon radically reformulates the narrative of Adam's perfect linguistic competence and knowledge of creatures in the Garden of Eden in order to stress postlapsarian man's complete loss of this Adamic inheritance. In its stead, Bacon prescribes a wholly different regimen for man: man must turn directly to the text of nature and, through labor and the methodical use of his senses, recapture Adam's dominion over nature. Emphasizing God's absolute freedom and power, Bacon figures nature as an orderly and coherent text, but one whose very language is but the contingent and arbitrary creation of its autonomous author. It is a language of contingent things, rather than a language of God's symbolic traces stamped upon things. By narratively resituating the foundational tropes of the Book and the Word, Bacon both rejects "bookish culture" and retains a system of traditional tropes while embracing a new hermeneutics of deinscription, or description. Through description, through "experimental and natural histories," Bacon hopes not to recover the original Adamic tongue, but to construct a future, perfect language mirroring things.

Finally, Chapter 8 ("Beyond Babel: Mersenne, Descartes, Language, and the Revolt against Magic") extends consideration of the hermeneutics of deinscription to founders of the mathematical and mechanical approaches to nature in the seventeenth century. Noting the real and important differences these two French luminaries of the new science represent with respect to Bacon, I trace how Mersenne reformulates the narrative of Adam and the Adamic language as a foundational move in his efforts to combat magic and establish the legitimacy of the sciences and his new vision of man. Mathematics and science provide Mersenne and Descartes with keys for unlocking the very "grammar" of the divinely inscribed language of things in which the Book of Nature is written.

The examples analyzed in these chapters are, of course, highly selective and are not intended to be comprehensive. Such a goal would be impossible in a book of this scope. Although I have spent considerable time examining treatises on language by grammarians, philologists, rhetoricians, and those who debate the relative merits of vernacular and classical languages (among others), I have chosen not to attempt to write a survey of the sixteenth- and early-seventeenth-century study of language. A number of studies already exist that treat various aspects of this larger story, and I have tried to acknowledge these studies, where appropriate, in the notes to specific chapters below (see especially Chapter 3). In choosing to highlight certain discourses within the historical record, I have been guided by the goal of examining ways in which study of, and attitudes toward, language have provided resources, models, and narrative configurations for practices and strategies employed by those who wished to read the "Book of Nature." With this goal in mind, I have found the richest sources to be those which explicitly discuss the origins and nature of language on the one hand, and those that incorporate such discussions (or highly coded allusions to such) within scientific, medical, theological, and philosophical texts.

Although I do note the contributions of various "Aristotelianisms" of the late Renaissance, citing figures as diverse as the Iberian Scholastic Benedictus Pererius and the grammarian, "humanist," and neo-Aristotelian Julius Caesar Scaliger, I do not attempt to provide an overview of Aristotelian contributions to the study of language, or even to the relations between language and natural philosophy. Since my own training centered on the Aristotelian tradition, and has been much influenced by such scholars and mentors as Charles Schmitt and Charles Lohr, there is a certain irony in the book's relative neglect of this tradition. My own sense, to hazard a gross generalization, is that much of the impetus for changing views of language, and of the relationship between "words" and "things," come from Neoplatonists, occultists, religious radicals and reformers, and skeptics. One of the challenges, then, is to chart how the provocations staged by such movements were received by those natural philosophers and physicians trained in, and largely committed to, Aristotelian views of nature. In my book, we can glimpse some of the tensions and innovations wrought upon such Aristotelians in, for example, my chapter on Fernel and Harvey.

Readers will quickly note that my choice of figures to analyze is highly selective. Medical texts and authors appear more prominently in the pages that follow than examples from other "sciences." In part, I must confess to my own limited knowledge as dictating some of my choices. Yet, I would also argue that Ficino deserves a prominent place at the beginning of my story, and that medical authors and natural philosophers or occultists like Fernel, Paracelsus (and even his "follower," Croll), and Harvey not only were major figures in their own day, but also exhibit highly telling and significant traces of the mutual implication of language theory, cultural narratives, and "science" and their sociocultural "contexts." For the rest, Galileo, Bacon, Mersenne, and Descartes certainly figure prominently in traditional accounts of the "Scientific Revolution" and play important roles in my own reading of this period as well; I certainly acknowledge, however, that the relative space afforded to each in my book remains arbitrary. Figures like Kepler and Boyle represent major but necessary omissions, although I am confident that the analytic scheme and historical categories advanced in this study could be adapted to these cases and others like them as well.[17] In the case of Kepler, the necessity of thorough familiarity with the pertinent "texts" and "contexts" made me as a nonexpert choose to leave consideration of his place within my historiographical scheme to a more appropriately equipped scholar.

The fine points of the analytic scheme I have outlined above can, of course, be witnessed in the detailed historical analysis I provide in later chapters of this book. I only wish to emphasize three points in conclusion. The first is that my book, as theory *and* practice, insists upon the importance of developing a thoroughgoing cultural and interpretive stance in the variety of interdisciplinary practices that have come to be known as "science studies" and among which I include a relative newcomer, literature and science.[18] "Approach[ing] science," as Clifford Geertz notes, "not as opaque social percipitate but as meaningful social action," the "interpretivist" stance I wish to advocate for science studies neither cordons off the natural from the human sciences, nor reductively explains the scientific in terms of the social.[19] Instead, the work of science *is* the work of constructing meaningful narratives about both human and nonhuman "agents," always understanding that such work and meanings are local in their production and import.[20] They are, as Donna Haraway (1988) stresses, "situated knowledges" or narra-

17. I shall treat Boyle extensively in Volume 2. For now, see Markley (1985, 1993). The recent book by Hallyn (1990) may suggest comparisons for Copernicus and Kepler.

18. The new journal *Configurations*, published by the Society for Literature and Science and the Johns Hopkins University Press, represents a forum for an interpretive approach to science studies.

19. The quotations are from Geertz (forthcoming). I follow Geertz in advocating an "interpretivist approach" that does not take the social as stable, separate from, and prior to the scientific, but rather regards science as meaningful social practice that creates (following Latour) its own local, sociocultural contexts.

20. See the important studies by Rouse (1990, 1993); Latour (1987, 1993). I shall treat these issues more thoroughly in my book-in-progress, *Figuring Science*.

tives and *not* "views from nowhere." Hence to "explain" science, science studies must become a hermeneutical enterprise using the tools of interpretation to trace the paths whereby the human and the natural become narratively configured as agents in a social, cultural, and natural world (Latour 1987, 1993; Rouse 1990, 1993; Traweek 1988).

Thus, second, I deliberately choose not to present a relentlessly developmental and strictly linear, or chronological, story. This book does not chart a decisive transition *from* the "Word of God" *to* the "languages of man," nor *from* the "Book" *to* "Things" or "Nature." With respect to the latter, while there is a gradual drift toward preoccupation with things, I argue that even this "transition" is authorized by the narrative reworking of the very trope of the "Book." Moreover, the marked interconnectedness of these tropes and alternative discourses fostered considerable slippage from one formulation to another, thus making it artificial to speak of a decisive transition. With respect to the "Word of God" and the "languages of man," as the example of Bacon should make clear, many opportunities existed for the recuperation of the ideal of a universal language even where human languages were considered forever cut off from an *originative* divine language. In addition, as numerous examples well into the eighteenth century could attest, even after most Europeans embraced a thoroughly conventional and social understanding of the languages of man, undercurrents of older dreams of a divine, Adamic language persisted. My story, then, is less one of sharp boundaries and unidirectional crossings than of seismic shifts and complex topographical foldings leading to multiple, and sometimes circuitous, paths.

Consequently, my sense of "contextualization" (which I want to defend as necessary to the historian) does not begin and end with the delineation of biographical, social, and intellectual details of "descent." Providing a stronger chronological sense is not, in itself, "contextualization." Indeed, historical practices that rely on notions of stable and unchanging contexts (social, economic, intellectual, psychological, etc.) that are simply described and invoked as explanations beg the question. Obviously, biographical, bibliographical, and historical background are intrinsic to historical analysis. The question is *how* to use such materials in constructing a historical interpretation. I propose to follow a new "cultural" approach to the history of science, to science studies, and to the question of the contextualization of science—and scientific change—as a historical phenomenon. My approach places primacy on the construction of cultural meanings. Such cultural meanings are to be found in objects, practices, language, social and professional habits, arrangements, and protocols as well as in theories, ideas, and texts. They are, I argue, often closely connected with the deployment of significant cultural narratives.[21] In my view, the first step of a contextualist historical analysis ought to be the analysis of cultural meanings and their narrative configuration in a set of historical artifacts such as, in the case of my book, scientific and medical texts and practices. This is largely the task I undertake in Volume 1. That is, the "contextualization" I speak of refers in the first instance to the narrative creation of contexts of cultural meanings that make sense of the texts, theories, and the methodological and hermeneutical practices of the figures whom I study in this book. This is *not* to deny social and political contexts as important to the history of science. But I do insist that such sociopolitical contexts are themselves shaped and reshaped by the process of producing cultural meaning among historical actors. What counts as "social" therefore can be investigated only within a local and highly situated setting for *particular* groups at particular times. For this reason, I have deliberately envisioned the second volume as quite narrow in scope in contrast to the breadth of the first volume. The first volume is capable of analyzing narrative and cultural configurations as changing cultural contexts for science and medicine; the second volume will, I hope, apply such "cultural contexts" to analysis of a highly significant, but restricted, English social, political, and religious scene in order to make much more complex arguments about the relationship between science, religion, society, and politics in the history of science.

Finally, I wish to stress that when I speak about "science" I am referring to a set of "practices" that may be linguistic, rhetorical, literary, conceptual, material, instrumental, experimental, mathematical, and visual (the list is not exclusive). I regard all such practices as discursive, and, indeed, prefer to define discourse itself as meaningful practices, activities, and objects, not, as all too frequently and mistakenly is assumed, as merely "language."[22] This distinction, of course, reinforces my insistence on the interpretive nature of science studies and also suggests how the "turn toward practice" recently taken by this field of study ought to be conceptualized.[23]

21. See Smocovitis (1992: esp. 63–64), for discussion of a similar cultural approach to context as it applies to the history of science.

22. See Woolgar (1986). For a richer notion of discourse and practices as "social activities" along the lines that I advocate here, see Certeau (1986), with the foreword by Wlad Godzich; and Certeau (1988, 1984).

23. On this emerging turn, see Pickering (1992).

2

Ficino and Neoplatonic Theories of Language

Magnum, O Asclepi, miraculum est homo.

A great wonder, Asclepius, is man!

With these words of the legendary Egyptian magus Hermes Trismegistus,[1] Giovanni Pico della Mirandola announced to the world of Renaissance Europe his rapturous vision of the Dignity of Man.[2] Man's dignity, for Pico, lay in his intrinsic ability to fashion his own innermost self, to act, in concert with God, as co-creator of his own nature. This role as self-fashioner, creator, actor set man apart from other creatures. No longer a passive witness to God's creation, man actively manipulates and transforms his world and, in so doing, reshapes himself. Man's destiny rests then for Pico in his willingness to assume the mantle of his divinely endowed birthright: that of Man as Magus.

Pico's vision of the dignity of man, of man's transformative power as magus, owes a profound debt to his fellow Florentine Marsilio Ficino's recuperation of the ancient *hermetica* and of the reputation of their alleged author, Hermes Trismegistus.[3] Together, Ficino and the younger Pico resurrected earlier traditions, some of them highly unor-

---

1. "De hominis dignitate," in Pico (1969:1.313). Translation Pico (1965:3). Following Pico and his translators and commentators, I retain use of the term "man," and of masculine pronouns, rather than substituting gender-neutral language. On the "Egyptian" Hermes, see Festugière (1949–54).

2. For Pico in general, see Anagnine (1937); Baron (1927); Barone (1949); Cassirer (1963, 1942); Dulles (1941); Garin (1937, 1972); Kristeller (1964, 1965); Monnerjahn (1960); and the introduction to Pico (1969:v–xxvi) by Cesare Vasoli.

3. For the texts attributed to the legendary Hermes Trismegistus, see Hermes Trismegistus (1945–54, 1924–36).

thodox, to give birth to a vigorous progeny: Renaissance magic. Yet, apart from the true sources of this movement and the extent of its indebtedness to the Hermetic tradition,[4] Ficino's and Pico's rehabilitation of the legendary Hermes Trismegistus profoundly affected Renaissance understandings of the past and led, as a result, to an increasingly central concern with the powers, origins, and meaning of language. For, embedded in their views of man as magus lay bold assumptions about the relationship of the Divine Word to the languages of man, assumptions which, in their detailed implications, held out the prospect of a thoroughgoing reordering of how man was to read nature and interpret texts.

Two roots, both of which anchor the thought of Ficino and Pico, nourished such later Renaissance readings of nature and texts. A philosophical root gave rise to images of the cosmos, including sublunar nature, as the expression of a divinely wrought architecture in which the individual parts are joined together intrinsically by essential similitudes and systemically through the mediation of pervasive active forces. Such substantive and dynamic links exhibited by natural entities and processes, moreover, themselves reflect the divine ideas after which God modeled his plan for the cosmos. Feeding this philosophical root, a historiographical root promoted the luxuriant growth of narratives about the origins of cultural traditions. Through such narratives one could legitimate attempts to fathom nature and (esoteric) texts through the medium of language. Such roots sanctioned man's ability to grasp the powers of nature by permitting him, through language, special access to the revelations of the Word, the *verbum Dei*.

The philosophical roots of this new vision of mankind arise from the seed of ancient Neoplatonic philosophy and medieval Platonism

---

4. The association of Renaissance magic with Hermes Trismegistus and the *hermetica* is most closely identified with the work of Frances Yates (esp. 1964). Numerous scholars—for example, Schmitt, Clulee, and Westman—have examined such claims cautiously and critically. Copenhaver's detailed studies regard the *hermetica* as a "banal" and inconsequential *philosophical* source for Renaissance magic, including Ficino's *De vita*. See Schmitt (1978); Westman (1977); Clulee (1988: esp. chap. 1); Copenhaver (1978, 1984, 1986, 1988, 1988a, 1990). My stress falls not on the philosophical importance of the *hermetica*, but rather on the significance of Hermes and the "hermetic tradition" in fostering a vision of "history" (through the myth of the *prisca theologia*) that provided a narrative framework for investing Neoplatonic language theory with specific, local cultural meaning within the Renaissance. Perennial conflict between language as "natural" versus "conventional" only carries significance after each "theory" is embedded in a larger cultural narrative, such as that of biblical narratives of Adam, Babel, and Pentecost. See below, Chapter 3.

as reinterpreted by Christian thinkers of the Renaissance.[5] To the extent that these thinkers also adopt some characteristic views of language from such sources, or at any rate utilize them in order to construct their own Christianized variants, we may properly call such views, collectively, Neoplatonic theories of language (Coudert 1978; Demonet 1992; Walker 1954, 1972). Neoplatonic understandings of language were widely current in the sixteenth and seventeenth centuries, and even more widely discussed. We find such Neoplatonic views among occultists—whether physicians, alchemist, or magicians—such as Paracelsus, Cornelius Agrippa von Nettesheim, and John Dee, among theosophists like Jacob Boehme, among religious sectarians like the Puritan John Webster, but also among relatively orthodox physicians, natural philosophers, and "scientists."[6] Such views were still discussed in Robert Boyle's and Newton's day.[7] In fact, as I argue in this book, these views form an important part of the story of the so-called "Scientific Revolution": we cannot, in fact, hope to understand how and why the very strategies that people used to study and understand nature changed during the sixteenth and seventeenth centuries without paying some attention to the role played by views of language in shaping those strategies and practices.

Hence, we need to acquire some sense of Renaissance perspectives on Neoplatonic theories of language if we are to understand such early modern developments. Curiously, and revealingly, however, there was no one "Renaissance view" of this Neoplatonic theory (Coudert 1978; Dubois 1970; and Chapter 3, below). While it has become commonplace to refer to such a theory as the theory of "natural" languages, the fact of the matter is—as we shall see in Chapter 3—that this characterization is misleading (Vickers 1984). It relies on modern categories, or at least on polemically restricted binary oppositions, that fail to capture the richness, subtlety, diversity, and socially situated import that the Neoplatonic theory of language acquired in specific, local situations. Renaissance Europe was not the late classical, Hellenistic world. Neoplatonic theories of language were shaped and defined by those culturally self-fashioning stories that sixteenth- and seventeenth-

5. See Bono (1984); Cassirer (1963); Chenu (1968); Collins (1974); Garfagnini (1986); Hankins (1990); Klibansky (1939); Rossi (1960, 1961); Trinkaus (1970); Walker (1958); Watts (1982); and Yates (1964, 1982: esp. the essays on Lull).

6. For Paracelsus and Agrippa, see Chapter 5, below. For Dee, Chapter 7. I shall discuss Boehme and Webster in Volume 2; see Chapters 1 and 3 for some pertinent bibliography.

7. I plan to discuss Boyle and Newton in Volume 2.

century people desperately tried to expand or constrict to fit new social, cultural, religious, scientific, and "intellectual" situations.[8] Thus a second root—that promoting such luxuriant growth of narratives of origins—was grafted onto the first, giving rise to various hybrid forms of the Neoplatonic theory of knowledge more or less well suited to the soil of late Renaissance and early modern Europe.

But first, before we witness how such new forms challenged and changed the cultural and scientific ecology of Europe,[9] we must try to recapture what this not so new, but still somewhat exotic, foreign seed looked like to those who encountered it just before the dawn of a new century that would alter the European climate decisively, indeed, forever. Marsilio Ficino was there at the beginning.

Marsilio Ficino (1433–1499)

In the case of Ficino, the philosophical roots of a theory of language begin to emerge early in his career. A youthful work written for Michele Mercati da San Miniato suggests that God, in Ficino's view, is the repository of the essential natures, or forms, of all entities found in nature. Man's ability to acquire knowledge of nature entails some measure of similarity with the divine mind.[10] Hence, early on, Ficino draws upon a Christian Platonic idea in which the ideas or forms of created substances mirror the ideas contained in the divine mind (*mens*). Such an idea represents a first step toward a Neoplatonic theory of language. For if one adds to it the notion that God instantiated the divine ideas in corporeal substances through the creative act of his divine Word, one can metaphorically associate a "language of nature" with the "Word of God."

The historiographical roots of Ficino's theory of language reinforce such metaphorical association and transform it into a narrative of the origins of philosophical, and especially esoteric, thought. Simply put, Ficino adds to creation myths, in which the divine brings the natural cosmos into being through the agency of His creative Word, the further element of an ancient revelation. Such a revelation—in Ficino's view transmitted through the *prisca theologia*[11]—directly linked God's Word

8. For "self-fashioning" more broadly, see Greenblatt (1980).

9. See Chapter 3 for this problem, and later chapters for its impact.

10. See the text "Summa philosophie Marsilii Ficini ad Michaelem Miniatensem," edited in Kristeller (1956: 55–97). The relevant passages are on 64–65.

11. The *prisca theologia*, or ancient theology, refers to the Renaissance belief in an esoteric tradition of secret theological wisdom, in most accounts emanating from the Mosaic

with human traditions and discourse. Such a move—which will become one basis for certain Renaissance hermeneutic strategies—
enables Ficino to construct a narrative framework within which he
may embed his theory of language and thus lend a historiographical,
supraphilosophical legitimacy to his reading of nature. Ficino's narrative emplotment of the languages of man, we shall discover, empowers
language, particularly words known to the magus, enabling man to
fathom the language of nature and hence its secret powers.

As we shall also see, late in his career, when writing the *Libri de vita*
(*De vita libri tres*, 1489),[12] Ficino rearticulated in Neoplatonic guise the
idea of the created universe as mirroring the divine ideas in order to
fashion his own system of astrological medicine and spiritual magic. By
that time, Ficino had wedded this idea to a vision of man's power to
harness the hidden forces of nature through talismans, images, and
words. The sources of Ficino's magic have been studied extensively by
scholars such as Paul Oskar Kristeller, D. P. Walker, Frances Yates, and
Brian Copenhaver.[13] In the following discussion I will instead attempt
to recount how Ficino's philosophical and magical concerns converged
in his growing understanding both of the power of words and of the
genetic, historical links between man's use of language and the creative
divine Word.

Ficino's use of talismans and interest in magical images and words
have been traced by some to his use of the *Asclepius* and translation of
the Hermetic treatises for his patron, Cosimo de' Medici. Certainly,
their appearance in the *Hermetica* was critical to Ficino's intellectual development. But, particularly for his understanding of the power of
words, Ficino drew on the rich resources of the Platonic tradition, interpreted through the lenses of Neoplatonism. Ficino's work on Plato and
the Platonic tradition had, however, to await completion of his translation of the *Corpus Hermeticum*, since Cosimo insisted that he first turn to

---

revelation, that formed a link through the *prisci theologi*, or ancient theologians, to pagan
traditions of wisdom in the ancient world. The classic treatment of this Renaissance tradition may be found in Walker (1954), later incorporated into Walker (1972).

12. The first edition is Ficino (1489). The text, in corrupted form, is included in the
standard edition of his collected works, Ficino (1576). I have used the new critical edition and translation Ficino (1989), whose introduction provides an essential review of
its history.

13. Walker (1958, 1986); Yates (1964); Kristeller (1956: esp. 221–47 ["Marsilio Ficino e
Lodovico Lazzarelli: contributo alla diffusione delle idee ermetiche nel Rinascimento"]).
For Copenhaver's studies, see note 4 above. For Ficino see also Kristeller (1943, 1976,
1987); Marcel (1958); and Saitta (1954).

the works of the thrice-great Hermes.[14] Afterward, Ficino produced a number of commentaries on Plato's works.

The *Philebus* commentary[15] contains an important discussion of the power of words and, particularly, of the names of the gods (Ficino 1975: 134–43). Ficino reads into Plato's works a consistent philosophical doctrine:

> As Plato says in the *Cratylus*, a name is some of the power of the thing itself. Initially, it is conceived in the intelligence, then articulated by the voice, finally expressed in writing. But the power of a divine thing is also divine. So we ought to venerate the names of God (since the divine power is present in them) much more than we venerate the shrines and statues of the gods.
>
> . . . God's names are like images or sunbeams of God Himself, penetrating through the heavenly beings, the heroes, the souls of men. However, whoever admires the sun venerates the sun's light too. So you must worship both God and God's sunbeams, the powers, the images lying concealed in the significance of names. (Ficino 1975: 138–41)

Ficino's allusion to Plato's *Cratylus* is, of course, highly selective, drawing only upon an aspect of the dialogue that had become a locus classicus for Neoplatonic views of language as linked essentially to the nature of individual things. Ignoring the counterthesis, that words are mere conventional human constructs with no inherent connection to the inner essence of things in nature, Ficino, like his Neoplatonic predecessors, seizes upon the discussion of divine names to reveal the latent power in words.[16] That power, Ficino suggests, comes from God, insofar as the names of the gods themselves have a divine origin. Ficino's analogy of the sun and its light underscores this point. Indeed, the analogy captures a metaphor central to Neoplatonic language theory, the metaphor of unity.[17] Just as the multiple, scattered rays of light derive their origin and power from a single source, the sun, so, too, do the names uttered by man have their source in the unity of God's being.

For the moment, Ficino's commentary does not capture in its net the wider problem of the contrast between the unity of God's Word and the

14. Yates (1964: 13). For Ficino, the Medici, and the "Platonic Academy of Florence," see Bullard (1990); Field (1988); and Hankins (1990; 1990a; 1991; 1991a).

15. See the critical edition and translation of this text by Michael J. B. Allen (Ficino 1975).

16. The significance of Plato's *Cratylus* and its postclassical reception receive useful review in Vickers (1984a: 97–100).

17. For the metaphor, or myth, of unity and language, see Dubois (1970) and my discussion below, especially in Chapter 6.

diversity of the languages of man. Resolution of that fundamental contrast will require narrative unfolding of the hidden links between human traditions and sacred history. Philosophically, however, Ficino does reveal that the disparate names of the gods participate in the unity and power of God. Invoking (Pseudo-)Dionysius the Areopagite, Origen, the Hebrews, Homer, and Saint Paul, Ficino stresses the living, miraculous power of words, suggesting that some names, because they are closer to the divine Word, enjoy special privilege and power (1975:140, 141). This conception suggests, if only tentatively here, that behind the multiplicity of words and objects lies the single power of the Word: "It is not without great mystery that Paul attributes all the acts of a living being to the divine word," Ficino states, ". . . it's as if God Himself were there in His words even when they're presented through the prophets" (1975:142, 143). Not surprisingly, then, Ficino links the power of "natural objects" and the divine power contained in names, concluding that "so great is the divine force preserved in these names that even men far removed from God and wrong-doers can work miracles by them."[18]

The miracle or "wonder-working" word represents a highly significant theme in the tradition of Renaissance occultism stemming from Ficino (cf. Reuchlin 1964). In its most narrow sense, the wonder-working word refers to just one sacred word, or perhaps a small number of them. But the notion lying behind such a belief—which we might call that of a "Platonic (or Neoplatonic) linguistics"—extends potentially to all of language. That notion would claim that language can, and indeed should, reflect in an intrinsic and essential manner the very nature of the things it names. Such a view of language frequently accompanied magical or mystical beliefs and, hence, underlined the power that lay hidden in the proper understanding and use of the true names of things in nature. The extension of such power to all of language—at least to all properly understood language—required, however, elaborate defense, such as we shall find later authors providing. Ficino's aspirations are perhaps less ambitious, though his desire to incorporate the power of words into his spiritual magic seemed heady enough to many.

The tendency of Ficino's thought with respect to language does, however, point in the direction of a Neoplatonic identification of word and

---

18. Ficino (1975:142; Latin text, 143). The passage continues, "Denique, si fas esset, commemorarem quanta vis insit quinque sacramenti verbis. . . . Per haec confirmatur opinio multa nomina ab ipsa rerum proprietate inventa fuisse."

thing, res and verba.[19] At its very core, language not only represents the nature of individual entities but also embodies and reflects their intrinsic powers. Ficino embraces such views not only in his Philebus Commentary, but in a number of other works as well.[20] Nevertheless, even given such general Neoplatonic views of language, substantial questions remain concerning their import and consequences. For, while the core of language may embody an ideal union of word and thing, of signifier and signified, the question of man's access to that core of linguistic purity and unity, and of the extent to which such access grants humans consequent power over things, remained subject to debate, and often obscure, during the Renaissance.[21]

Ficino himself tacitly provides philosophical grounds for ceding man power over things through words, while his own example of how to harness and control the forces of nature through spiritual magic, especially in his De vita coelitus comparanda (1489), displays a cautious stance toward the power of language. The De vita coelitus comparanda comprises the third and final book of Ficino's influential De vita. In its entirety this work constitutes an important contribution to both Renaissance medicine and philosophy.

Book I deals with preserving the health of scholars, while in Book II Ficino discusses how one might prolong life, once again with the particular temperament and afflictions of the scholar in mind. Much of

19. See Vickers' discussion of res and verba (1984a); and Padley (1976) and Howell (1961) for the rhetorical tradition.

20. See, for example, the Cratylus commentary (next note) and also Coudert (1978: 74–75).

21. The key classical text was, of course, Plato's Cratylus. For Ficino's comments, see In Cratylum, uel de recta nominum ratione, epitome, in Ficino (1576: 1309–14). The text allows Ficino to survey different views of language, including the opinion that words are constituted "naturally" rather than by human convention. For example: "Itaque . . . non tam pro arbitrio nostro, quam pro natura rei id faciemus, ut eo modo & instrumento, quo diuidi possit aptius, diuidamus, similiterque comburamus. At dicere & nominare actio quaedam esse uidetur, ideoque propriam debet habere naturam, ut non ita sicut placet nobis, sed ut rei modus exigit, nominemus" (1311). In addition, see 1311 and 1314. Despite the centrality of Plato's text to the classical tradition, I argue below, in Chapter 3 and later, that the rich variety, subtlety, and authority of various theories of language in the sixteenth and seventeenth centuries were dependent upon how language was integrated into specific Judeo-Christian narratives. Thus, Neoplatonic theories of language, and theories of magic like Ficino's that depended upon a theory of language, acquired specific and local cultural meaning only through their rearticulation and reorientation within such narratives. Such theories of language and magic, in addition, were contested as well by the challenging and recasting of Renaissance cultural narratives. The link between these developments and the construction of hermeneutic strategies for reading the Book of Nature is the major concern of this book.

what Ficino surveys in these books represents standard medical fare
for the late medieval and Renaissance period. Scholars are identified
as particularly prone to *melancholia*, a state which Ficino describes
and seeks to forestall in terms for the most part consonant with the
dietetic medical model of Galenic humoral pathology and therapeutics. Ficino's discussion of the prolongation of life also touches upon
many traditional models of aging associated with ancient medicine
and physiology, laying particular stress upon innate heat and the effects of dryness upon the aged. His suggestions for prolonging life
incorporate time-honored prescriptions for moderating the ravages of
old age through food, drink, and exercise.[22]

Yet Ficino lays special emphasis throughout his work on the role of
spirits in health and aging. Medical spirits were, of course, a standard
component of Galenic medical theory. The vital spirits especially were
associated with life and robust health, playing a significant role in connection with the heart, blood, a life-giving heat of organisms (Bono
1981, 1984). Ficino draws upon both traditional medical sources and
the legacies of Neoplatonic speculation and medieval astrology to associate orthodox medical spirits with the far broader conception of a pervasive *spiritus* linking the macrocosm of the heavens and sublunar
sphere with the microcosm of man. As a result, Ficino's medical preoc-
cupations in the *De vita* unfold to embrace a larger universe of occult
agencies, forces, and entities.

In its root sense, this occultism, for Ficino, supplements the merely
manifest, often palpable, qualities (hot-cold, wet-dry) underlying the
humors and, hence, health and disease. Ficino's occultism, then, constitutes a *scientia* of hidden powers and entities allowing access to the
beneficial influences locked within the very essence of natural and celestial substances.[23] Metaphorical juxtaposition of the manifest (or observable) and occult (or hidden) serves, for Ficino, to raise certain
classes of entities and operations above others as sources of unparalleled power and insight. This same metaphorical process serves, in the
wake of Ficino and Neoplatonic language theory, to raise a kind of hidden, and therefore primitive, language above other manifest, and therefore vulgar, human discourses.

22. See Clark (1986); Gruman (1966); Klibansky et al (1964); Niebyl (1971); Hall (1971);
McVaugh (1974); Aristotle (1964); Galen (1821–33) (for translation see Galen 1971); R.
Bacon (1928). Also Bono (forthcoming).
23. For occult powers, properties of the "whole substance," and Renaissance medicine, see Bono (1981); Nutton (1983); Richardson (1985). Copenhaver (1978) discusses
occult qualities, and has turned his attention to medical authors such as Fernel.

As a *scientia*, a form of knowledge or understanding, Ficino's occultism—his magic—purports to "read" sublunar and celestial nature by lifting the veils of manifest appearances, penetrating to the core of things themselves. In a like manner, as we shall see, by lifting the veils of ordinary language man can hope to uncover the true language of nature, to discover the power of words themselves. But magic, for Ficino, is more than a way of knowing; it is a practice operating on the very nature it purports to read. It functions, in other words, as a transformative "art" enabling man to harness natural forces in order, for instance, to preserve health and transform life.

Ficino's third book, the *De vita coelitus comparanda*, thus dwells upon the possibilities magic permits for actively drawing upon celestial influences to preserve health and prolong life. Insofar as Ficino attempts to achieve such goals through the operation of *spiritus* and its interaction with the vital spirits, Ficino's occultism has been termed "spiritual magic." The basis for Ficino's belief in spiritual magic, for the possibility of "drawing down" the "life" or influences of the heavens in order to fortify natural substances used as medicines or to restore directly man's healthful vigor, rests in his conviction in a world-soul mirroring both nature and the divine mind. This world-soul "possesses by divine power precisely as many seminal reasons of things as there are Ideas in the Divine Mind" (Ficino 1989: 243). In effect, the world-soul operates as a kind of medium between the divine mind and nature. "Seminal reasons" enable the world-soul to produce individual "species," or natural substances of a particular kind, which reflect individual divine ideas.

As a result, then, individual members of a species are not only "linked" to one another by virtue of their origin in the formative power of a "seminal reason," but linked to other material things or natural substances as well. Such "links" result from the replication of divine ideas in specific material forms found in different strata of nature, both sublunar and celestial: in, for example, the material forms of specific mineral, plant, animal, and heavenly bodies. This "network" of material forms constitutes the basis for regenerating the innate power of individual things. For the efficacy of an individual material entity can be enhanced, for example, by drawing upon the power of another natural object to which it is generically "linked" by virtue of the conformity of their seminal reasons to a specific divine idea: "if in the proper manner you bring to bear on a species, or on some individual in it, many things which are dispersed but which conform to the same Idea, into this material thus suitably adapted you will soon draw a particular gift from the Idea, through the seminal reason of the Soul" (1989: 243)

What Ficino envisions, then, is a universe replete with material forms—with species and individuals—which are reflections of divine ideas as mediated through a world-soul and its seminal reasons, and which themselves may act as conduits of divine power. Such "conduits" are highly specific: only those material forms which owe their origins and essential natures to an individual divine idea are capable of attracting, conveying, or receiving restorative divine virtues among themselves. Through the mediation of the world-soul, such virtues may be drawn to material forms precisely because the world-soul "herself has created baits of this kind suitable to herself, to be allured thereby . . ." (1989: 245). Ficino then agrees with the ancient Zoroaster who "called such correspondences of forms to the reasons existing in the World-soul 'divine lures,'" and with Synesius who "corroborated that they are magical baits" (1989: 245).

Thus the possibility of Ficino's spiritual magic rests, in part, upon the assumption of a divine plan through which the ideas in the divine mind endow the universe with an architectonic structure exhibiting patterns of correspondence among its various levels and individual parts, which, in turn, enjoy the potential for sympathetic interaction. The very diversity of material forms necessarily exhibits, on a deeper level, a unity of purpose and a harmony of function. The key to such unity and harmony—and thus to both human knowledge and power—is apprehension of the correspondences linking material forms and understanding of how to make such links operational.

For Ficino the correspondences are not manifest, perceptible qualities of the materials forms, but "occult" or hidden virtues flowing from the conformity of the essences, natures, or "ideas" of certain material forms with those of others. These correspondences constitute "divine lures" and "magical baits" precisely because they draw to them higher, "divine" powers that humans can potentially harness and transmit from one link in the chain of correspondences to another. Ficino's project in the *De vita coelitus comparanda* is precisely to suggest *how* humans may draw upon such "life," upon such power in the universe. His answer invites us to assume the role of *magus*, to turn to astrology and *spiritus* as sources of knowledge, power, and, hence, of magic.

Astrology, for Ficino, enables the Neoplatonic magus to identify potential sources of life-enhancing vitality which he may deem appropriate to the particular temperament and nature of individual men, and especially scholars. His concern is therefore not with the crude judicial

astrology of ill-reputed fame with its emphases upon astral determina-
tion of human nature and fate at birth.[24] Rather, for Ficino, the heavens
are a source of continual restorative virtues into which one can tap
much like one may tap into a hidden stream of underground water.
Unlike a stream of water, however, the streams of astral virtues are
many and subject to manipulation by the magus. Thus man, as magus,
has it within his power to draw down celestial life of a specific nature,
for specific ends.

The key to such a human-centered, nondeterminative astrological
"activism" is the intermediary role played by the heavenly bodies in
Ficino's cosmology. As we have seen, the "species" produced by the
world-soul from seminal reasons mirroring the divine ideas provide for
a kind of interlocking architectonic structure to the created universe
which links different strata of nature, and hence different individual
material forms, together. For Ficino, the heavens are a source of restorative power precisely because they are linked in highly specific ways to
individual species of natural entities here on earth. As Ficino says, "In
the stars, moreover—in their figures, parts and properties—are contained all the species of things below and their properties" (1989: 245).

If, through astrology, the magus may identify specific astral virtues
whose "life" he may attempt to draw down, it is through an all-
pervasive *spiritus* that such power is conveyed to humans. This macrocosmic shift is analogous to the medical spirits operating within the
human body according to Ficino: "Just as the power of our soul is
brought to bear on our members through the spirit, so the force of the
World-soul is spread under the World-soul through all things through
the quintessence, which is active everywhere, as the spirit inside the
World's Body, but that this power is instilled especially into those
things which have absorbed the most of this kind of spirit" (1989: 247).
The function of this quintessence or celestial spirit with respect to man
thus allows him to absorb astral virtues which may prove beneficial.
The beneficial effects of celestial spirits transmit themselves to man via
his own spirit, and man may enhance these effects in at least two ways.
First, man may prepare himself to receive general astral influences by
becoming a better passive vehicle for their absorption: "The world does
wholly life and breathe, and we are permitted to absorb its spirit. This is
absorbed by man in particular through his own spirit which is by its

---

24. On Ficino, astrology, medicine, and magic, see, in addition to works cited earlier,
Couliano (1987); Garin (1969, 1976); Kaske (1986); Vasoli (1980); Zambelli (1973); and Zanier (1977).

own nature similar to it, especially if it is made more akin to it by art, that is, if it becomes in the highest degree celestial. Now it becomes celestial if it is purged of filth, and anything at all inhering in it which is unlike the heavens."[25] One aspect of "art," of Ficino's astrological medicine, enables man to purify himself through an array of practices: through diet, dress, habits, medicines, and the like. Here Ficino offers a wealth of advice concerning those substances, odors, and experiences which can best serve to purify man's spirit, and about how such factors are in accord with the peculiar astral affinities of man's nature.[26]

But man may also enhance the effects of celestial spirits upon his own spirit by attempting, through magic, to draw down actively from the heavens those specific virtues whose properties he deems to be especially beneficial to him individually. Indeed, the real power of the heavens consists in engrafting in things "occult virtues" through the agency of celestial spirits.[27] These occult virtues are especially attracted

25. Ficino (1989: 259). On page 256 Ficino differentiates between the nature of world spirit and human spirit—the one is celestial, the other derives from more mundane elements: "Sed ad mundi spiritum redeamus, per quem mundus generat omnia, quandoquidem et per spiritum proprium omnia generant, quem tum coelum, tum quintam essentiam possumus appellare. Qui talis ferme est in corpore mundi, qualis in nostro noster, hoc imprimis excepto, quod anima mundi hunc non trahit ex quattuor elementis, tanquam humoribus suis, sicut ex nostris nostra, immo hunc proxime (ut Platonice sive Plotinice loquar) ex virtute sua procreat genitali, quasi tumens, et simul cum eo stellas, statimque per eum parit quattuor elementa, quasi in illius spiritus virtute sint omnia." Later in the sixteenth century the physician-philosopher Jean Fernel will further "radicalize" human medical *spiritus* by giving it a celestial origin and denying that it is educed from matter. See Bono (1981) and Chapter 4 below.

26. Ficino (1989) discusses many such life-prolonging practices in book 2, *De vita longa*. Book 3, of course, discusses the role of the heavens and of *spiritus*. See, for example, chapter 11, "The Ways in Which Our Spirit Can Draw the Most from the Spirit and Life of the World; Which Planets Generate and Restore Spirit; and What Things Pertain to Each Planet" (288–97), which, among other topics, discusses odors.

27. Ficino (1989: book 3, chap. 12, "Natural and Even Artificial Things Have Occult Powers from the Stars, through Which They Expose Our Spirit to the Same Stars," esp. 298, 300): "Neque tamen dicimus spiritum nostrum coelestibus duntaxat per qualitates rerum notas sensibus praeparari, sed etiam multoque magis per proprietates quasdam rebus coelitus insitas et sensibus nostris occultas, rationi vix denique notas. Nam cum proprietates eiusmodi earumque effectus elementali virtute constare non possint, consequens est a vita spirituque mundi per ipsos stellarum radios singulariter proficisci, ideoque per eas spiritum affici quam plurimum atque quam primum, coelestibusque influxibus vehementer exponi." See also chapter 16, "On the Power of the Heavens" (322): "Sed quis nesciat virtutes rerum occultas, quae speciales a medicis nominantur, non ab elementali natura fieri, sed coelesti? Possunt itaque (ut aiunt) radii occultas et mirabiles ultra notas imaginibus imprimere vires, sicut et ceteres inserunt. Non enim inanimati sunt sicut lucernae radii, sed vivi sensualesque tanquam per oculos viventium corporum emicantes, dotesque mirificas secum ferunt ab imaginationibus mentibusque

to material forms with which they bear a special affinity: the "baits" or "lures" that have been implanted in nature via the seminal reasons of the world-soul attract such celestial occult virtues. In turn, then, these occult virtues enhance the inborn powers of such material forms. A function of the magus is to harness and direct the flow of astral influences via celestial spirits in order to draw down such occult virtues into their appropriate material vehicles:

> Agriculture prepares the field and the seed for celestial gifts and by grafting prolongs the life of the shoot and refashions it into another and better species. The doctor, the natural philosopher, and the surgeon achieve similar effects in our bodies in order both to strengthen our own nature and to obtain more productively the nature of the universe. The philosopher who knows about natural objects and stars, whom we rightly are accustomed to call a Magus, does the very same things: he seasonably introduces the celestial into the earthly by particular lures just as the farmer interested in grafting brings the fresh graft into the old stock.[28]

By acting as magus, man can actively transform material substances all about him; these, in turn, may be used—for instance, in the form of medicines—to strengthen his own *spiritus*, thus making it susceptible to the virtues of celestial spirits. Or man may act, through "magic," to draw down occult virtues via the celestial spirit to transform and invigorate his own spirit directly.

In either case, Ficino's magus acts with astrological knowledge of potential sources of life-enhancing occult virtues to draw down such virtues, such "celestial life," through the medium of an all-pervasive cosmological spirit. Such magical knowledge becomes practice when the magus learns either to manipulate those "lures" placed in nature by the world-soul, or to replicate such lures through art—the "artifice" of the magician. Such artificial lures can act, presumably, by reproducing some similitude of an appropriate "species"—that is, by forming in some sense a replica of the very "idea" or "seminal reason" that finds material expression in a species. Such a replica becomes then the "bait" for sympathetically attracting the occult virtue of the corresponding celestial force via the agency of the cosmic spirit.

What might such a replica or lure be? One powerful source of magical attraction during the Renaissance clearly was images: pictures, statues,

---

coelestium, vim quoque vehementissimam ex affectu illorum valido motuque corporum rapidissimo; ac proprie maximeque in spiritum agunt coelestibus radiis simillimum."

28. Ficino (1989: 387). Note Pico's use of this metaphor in his *Oration* (1969: 1.328).

talismans, and the like. Ficino also relied upon the suggestive power of music to draw down beneficent influences.[29] Still another "lure," at least in theory, might well be words—and especially names—particularly given a Neoplatonic theory of language in which "names," properly understood and used, were indeed "images" or representations of things. Stronger still, such words might well constitute the essence, power, and nature of things themselves.

As we have seen, Ficino's ideas about language certainly point in this direction. The practice of drawing upon the magical power of words was, however, one which Ficino approached with a great deal of caution. As Ficino reports, words were often combined with images in order to enhance and intensify the latter's effects according to occult authors.[30] The problem with this use of words—a problem shared to a large extent with the magical use of images according to authorities like Saint Thomas Aquinas[31]—is that the spoken or written word, such as those used on talismans or in incantations, bears the highly unorthodox implication of appealing to intelligent forces. Hence, the magical use of words threatened to brand the magus as someone engaged in the illicit practice of demonic magic.[32]

For that reason, Ficino tends to downplay the actual use of words in his magic. Careful to distance himself from illicit use of words in incantations and talismans, Ficino nonetheless suggests the possibility of a powerful, and licit, magical use of words. His emphasis upon uses of song, indeed, quietly recuperates the role of words in magical practice.

Thus, after raising the objection that words implicate the magus in demonic practices, Ficino deliberately reiterates a conception of words resonating with the view of language found in his earlier Platonic commentaries, namely, "that a specific and great power exists in specific words." This conception of words Ficino attributes to the likes of Origen, Synesius, Al-Kindi, Zoroaster, Iamblichus, the Pythagoreans, and the "Hebrew doctors of old" (1989: 355). Clearly, what Ficino has in mind here is not a "power" dependent upon demonic intervention, but, rather, one flowing from the "specific" nature of the word itself.

---

29. Walker (1958) provides the classic account of Ficino's use of music and images in his magic. Also see Yates (1964).

30. See Ficino (1989: 354–63 [book 3, chapter 21: "On the Power of Words and Song for Capturing Celestial Benefits and on the Seven Steps That Lead to Celestial Things"]).

31. Ficino (1989: 340–43). On this problem see the previously cited works of Walker, Yates, and Copenhaver.

32. Ficino (1989: book 3, chap. 21, e.g., 354). Also Walker (1958: esp. 80–81); and, more generally, Thomas (1971).

Such a power, I would suggest, might act much like the "particular lures" Ficino states that the magus uses when "introduc[ing] the celestial into the earthly" (1989: 387). In other words, one could grant to words the possibility of a licit magical use, if words inherently contained specific powers enabling them to operate by entering the chain of created "species" as yet another instantiation of a divinely originating power. As part of such a chain, individual words would then constitute another specific link among all those related species, or "material forms," that share their origin in one and the same specific seminal reason and, ultimately, divine idea. Such a word—precisely because of its "Neoplatonic" status as reflection of a divine idea and, indeed, of the "Word of God"—can then function as a "specific lure" drawing down the "natural" forces of the celestial region into mundane things.

While laying the groundwork for the possibility of such magical uses of words, Ficino remains exceptionally cautious by avoiding explicit endorsement of particular magical practices relying solely upon words. Instead, Ficino assimilates the licit use of words to his better known— thanks to D. P. Walker—use of music as a practice in his spiritual magic. By contrast to incantations, the effects of singing, Ficino asserts, may be attributed to nature (1989: 354–57). By "nature" Ficino here means, of course, the totality of celestial and sublunar material forms together with the pervasive *spiritus* that joins them together into an inter-communicating network. Singing, then, acts upon the spirit of individual humans, making it receptive to specific astrological influences attracted and transmitted through celestial spirit (Walker 1958). It is *not* a form of demonic magic since its effects do not call into play demonic intelligences.

Because of its connections with harmony, microcosmically in the harmony of man's soul, macrocosmically in the astrological harmony of celestial bodies, Ficino's emphasis upon song and music in his *De vita coelitus comparanda* might more easily assimilate itself to the benign, if somewhat heterodox, operation of natural magic than might magical use of words. Perhaps that is why Ficino reintroduces the power of words into his astrological medicine and spiritual magic clothed in the form of song.

Indeed, "words, songs and sounds" occupy a central position in Ficino's magic. Of the "seven steps through which something from on high can be attracted to the lower things," the fourth and "mean step" is occupied by words, songs, and sounds (1989: 355, 357). Unlike the first three steps, the fourth does not consist of "natural" objects or material forms—that is, words, songs, and sounds are not preexistent instantiations of seminal reasons or divine ideas in specific material entities. Rather, they are vocal expressions produced by man that somehow mimic the specific natures of celestial things and hence acquire the power to draw down the "life," or beneficial occult influences, of the stars. How do words, songs, and sounds accomplish this magical effect?

In the case of sounds and of that dimension of songs dependent upon sounds, the artifice of man, aided perhaps by the divine (1989: 357), creates and imitates the very tones characteristic of the stars themselves. For Ficino the harmony or music of the spheres is more than a pleasantly ornamental metaphor. The very "form" of individual celestial bodies, and hence specific "celestial powers," are captured in sets of characteristic tones. Hence, human sounds and the musical chords of songs produce magical effects not by "worshipping the stars," but, on the contrary, by "imitating" the stars themselves. Such imitating enables the magus to drawn down beneficial celestial influences (1989: 357).

But songs also contain words, and these words are themselves intrinsic to the beneficial, magical effects produced by songs. Such words function in Ficinian songs not through their tonal affinity with the music of the spheres, but, rather, through the "meanings" they embody and represent (1989: 357ff.). Here Ficino does not spell out a theory of the magical uses of words, perhaps for fear that his use of words will be mistaken as a form of demonic magic.[33] Nonetheless, he insinuates into his astrological and magical use of songs the magical power of words to represent, imitate, and hence draw down the nature and occult virtues of the stars. In this manner, Ficino's magic cautiously reflects his philosophical views on the power of words, his Neoplatonic theory of language in which words and things may be said to reflect one another in an intrinsic and essential manner.

Neoplatonic Language Theory, Magic, and Power

Ficino's example illustrates the affinities between a Neoplatonic theory of language and magic in the Renaissance. Yet his work also suggests deeply troubling and problematic dimensions to the marriage of language theory and magic in that period. At the core of such difficulties

---

33. For Ficino and "angelic" magic, see Walker (1958); Yates (1964); and Gandillac (1960). While conceptually distinct, "natural," "spiritual," magic and angelic magic are not easily separated in practice. Ficino seems to presume some divine/natural power in words that can then be used to advantage in natural magic.

are tensions engendered by the very relationship among language, knowledge, and power that the existence of a tradition of magic grounded in Neoplatonic theory served to highlight. Simply put, Renaissance magic highlighted tensions between views of language as the key to knowledge (*scientia/sapientia*) and as a vector of magical power.

Implicit in a Neoplatonic theory of language is the conviction that a proper understanding of language, of the words used to name things, will yield true and essential knowledge about nature. Such knowledge may for the present be hidden from human understanding, either because humans do not have access to the true names of things, or because they lack the requisite spiritual qualities or skills to interpret language correctly.[34] Nonetheless, such a theory supposes the existence of a pure, uncorrupted language as the key to all knowledge, which humans may unlock if only they can uncover its existence and recognize it for what it truly is. Such a theory of language, hence, commonly led to attempts to uncover an originary, uncorrupted language and to unlock the mysteries contained within it. During the Renaissance, as a result, theorists of language and students of nature placed much emphasis upon both exegesis and illumination. Through exegesis—through, that is, both ordinary and esoteric methods of interpreting received language—learned men, mystical theologians, and magicians attempted to strip away the vulgar meanings of words and texts to uncover their primitive sense. Such exegetical techniques were usually directed to texts regarded as, in some ideological sense, especially laden with meaning, and hence privileged. Among the candidates for exegetical analysis were the texts of the ancient philosophers, the works of supposed sages of great authority such as Hermes Trismegistus, and, of course, the Word of God itself, the Bible. Supplementing exegetical practice, however, was illumination— itself a "divine" gift—which allowed humans to unlock the mysteries contained in the true names of things supposed to have been discovered through exegesis.

As the key to knowledge, a Neoplatonic theory of language incorporating the convictions described above need not lead to magic at all. Indeed, it is quite possible that such a theory could conclude that the proper understanding of language necessary to true knowledge of the

34. See Chapter 3, below, for discussion of the range of possible Renaissance options regarding human access to originative, or Adamic, language and of the interpretive strategies available to fallen man. Such possibilities, I argue, are defined and legitimated by culturally framed and reworked narratives.

universe is beyond the grasp of human understanding, or that the conditions for its fulfillment are not yet present. And even if times and conditions were ripe for the full grasping of such knowledge, as it indeed appeared to be for certain sixteenth- and seventeenth-century thinkers, such knowledge might well be used for the understanding and nonmagical control of natural processes rather than for magical manipulation. Language might well lead to knowledge, and knowledge to power, but without canonizing words as, in themselves, vectors of magical power.

Yet, through the marriage of Neoplatonic language theory and Renaissance magic, words were always threatening to become just that: vectors of magical power. Stretching back to the Middle Ages and beyond to antiquity, magic constituted a variety of practices and traditions. Ficino's attempt to define a benign, spiritual magic which might serve to ennoble man by fulfilling his dual nature as divine-like contemplator and agent of beneficent change nonetheless drew upon questionable practices and suspect traditions. Although he filled his *De vita* with disclaimers intended to distance his own beliefs from unorthodox ones (1989: 238–41, 280–81, 320–21, 340–43, 398–99), Ficino's text resurrects the possibility that words may be used in talismans, incantations, magical rituals, and other suspect practices. In the wake of Ficino's Neoplatonic magic, Renaissance authors like Agrippa did in fact pursue highly unorthodox dimensions of magic.[35] Words, understood Neoplatonically as real representations of individual things, were transformed from sources of hidden esoteric knowledge of the universe to actual agents of magical practice through which the magus could harness the innate, often latent, powers of things to transform himself, nature, or others.

By wedding a Neoplatonic theory of language to magic, then, Renaissance thinkers helped to create a tension between a view of language as constituting the basis for a hermeneutics of nature and thus as the key to knowledge, and an alternative view in which words became the direct means by which humans harnessed occult powers. The former view empowered humans to understand nature through exegesis and illumination. Such understanding of, for instance, the true nature of specific plants or mineral substances and of their affinities with the organic constitution of human beings might, of course, lead to power over nature through the exercise of such knowledge to create effective

---

35. On Agrippa, see Nauert (1965); Keefer (1988); Zambelli (1966, 1976). Also see Chapter 5, below.

medicines. But, while based upon a Neoplatonic theory of language, such a harnessing of natural, if occult, forces would not depend upon words to *produce* the intended effects themselves. By contrast, the latter view relies on words as vectors of magical power insofar as words are considered to effect desired manipulations of cosmic forces through their very utterance or use in some specifically crafted magical object or ritual.

Complicating this neat differentiation of views, however, is the fact that the *learned* magic of Renaissance thinkers like Ficino, since it incorporated a Neoplatonic theory of language, wittingly or not encouraged *both* views of how language might be useful to the student of nature. As a result, the Renaissance magus is often enormously difficult to distinguish from the natural philosopher. Both may rely on a framework in which language becomes the critical element in a hermeneutics of nature; likewise, both may slide unwittingly toward the practice of using words as vectors of magical power. Such fluidity and slippage are most likely to occur where students of nature employ certain practices (like the Cabala) or hermeneutic schemes (such as the doctrine of signatures) in which exegesis and illumination purport to play a large role. In such instances exegesis and interpretation may well lead not just to esoteric knowledge (which may then be used in a benign, nonmagical fashion), but to identification of words, signs, and/or images that, in turn, are used *performatively* by the magus or natural philosopher.

What I am suggesting, then, is that the boundaries between magical (in a manipulative, nonpneumatic, and nonpsychological sense) and nonmagical uses of Neoplatonic language theory are not sharply drawn. Indeed, I want to suggest that, by its very nature, the scope and import of any Neoplatonic theory of language in the Renaissance cannot be grasped or even implied merely from some abstract formulation of the theory as a set of statements about how words and things are mutually related. Rather, I want to argue that by its very nature any Renaissance theory of language—Neoplatonic or otherwise—is incomplete without exposition of the narrative framework within which its account of words and things is embedded. I stress the importance of the narrative framing of language theory precisely because, for the Renaissance, what remained critical to any *use* of language was understanding of the nature and extent of human access to language. Such considerations, given assumptions about man's status as God's creature, inevitably entailed some narrative legitimation of any theory which purported to define how humans used, or might use, language. By embedding language theory in specially constructed narratives, Renaissance and early modern thinkers exposed and explored relationships between the "Word of God" and the "languages of man." They also situated language with respect to multiple cultural and social discourses. This cultural location of language narratives consequently heightened the stakes and deepened the "localized" implications which seemingly "abstract" theories of language might hold for contemporaries. Such stakes and implications are very much part of the story of the interconnections between early modern science, medicine, and language.

For Ficino, the narrative frame of Neoplatonic language theory is very much that of the *prisca theologia*, the "ancient theology."[36] In its most general sense, the notion of a *prisca theologia*—or of a tradition of theological wisdom circulating in ancient times among pagan peoples—served to some significant extent to repair the rupture between pagan and Judeo-Christian traditions that was a fundamental feature of the Christian understanding of history. If, as the sharp division of sacred from profane history suggests, Christian wisdom stemming from God's Revelation and Christ's presence and teaching among humans constitutes the exclusive originary source of human knowledge of the divine plan for the universe and God's creatures, then the so-called wisdom of pagan authorities and traditions is reduced in stature and severely restricted in its significance for humans. At best, the most worthy and pious learning of the ancients may reflect the highest knowledge to which unaided human reason may aspire. But even that attainment falls far short of divine wisdom, of theology in its noblest sense. In effect, the wisdom, knowledge, and language of the ancients is irreparably deficient, cut off from any substantive intercourse with divine wisdom and sacred tradition.

The notion of an ancient theology, however, to which Ficino and others in the Renaissance subscribed, held, by contrast to this dichotomized vision of history, that the greatest, most wise of the ancient pagan thinkers were themselves privy to some form of the divine revelation. While the exact details of the transmission of this ancient revelation varied significantly among Renaissance authors,[37] the general contours of such narratives of the transmission of ancient wisdom supposed an esoteric tradition of divine learning passed down from some genuine source within the strict line of biblically sanctioned

36. As noted earlier, the classic study is that of Walker (1954). Yates, Copenhaver, and many others draw upon this notion of a *prisca theologia*. In addition, see Schmitt (1970).

37. Again, see Walker (1954) for examples of how the lineage, etc., of the ancient theologians was variously interpreted.

patriarchs—such as Moses—too some individual pagan authority of unusual wisdom and piety. Thus, the legendary Hermes Trismegistus, for instance, became a real Egyptian luminary who, in some accounts, gained privileged access, either directly or through intermediaries, to the Mosaic Revelation. The "divine" Plato of less remote antiquity was believed by many to represent not merely the highest attainment of human reason but the embodiment of Mosaic wisdom, stemming, in all probability, from his sojourns in Egypt where he drank from the font of divine mysteries itself.[38]

Such narratives cast in a new light the learning, texts, and, ultimately, the language of human traditions. What one might have regarded as irreparably defective, one might to some degree now see as a reflection of the divine. Thus, not only might the teachings of ancient Egyptians like Hermes, or ancient Greeks like Plato, contain a fundamental, if veiled, core of divine wisdom, the very languages of such ancient traditions might themselves bear an obscure affinity to the divine Word, the *verbum Dei*. Such understandings of the "ancient theology" consequently open the door to deeper understanding of language, and of Neoplatonic theories of language. Indeed, the very idea of a *prisca theologia* becomes a hermeneutic starting point for constructing narratives—explicit or implicit—which may account for the *degree* of man's access to the pure, uncorrupted, originary language of creation itself.[39]

38. See Walker (1954, 1972). On the significance of Egypt for the Renaissance, see Dannenfeldt (1959); Grafton (1979); Iversen (1961); Valeriano (1556). The role of Egypt in the ancient world as understood by the Greeks and the fortunes of this ancient view in European history have been highlighted by Bernal (1987).

39. Ficino's compatriot within the Platonic circles of Renaissance Florence, Giovanni Pico della Mirandola, deepens and extends this hermeneutic quest for the divine, eternal, and originative in the thought and language of human traditions. Pico seeks such wisdom, including the kind of secret knowledge of nature afforded by natural magic and "higher" forms of occult wisdom, through the Neoplatonic tradition, the hermetic tradition, and the tradition of esoteric Hebraic wisdom in the study of the Cabala. If space permitted, it would be of considerable interest to discuss Pico further in the context of this study. Pico's influence centrally affects figures, such as Reuchlin, Agrippa, and Paracelsus, pertinent to our history in the late Renaissance. On Pico and esoteric wisdom, see, in addition to previously cited studies, Wirszubski (1989); Idel (1989, 1988); "Reuchlin: Pythagoras Reborn," in Spitz (1963); Scholem (1946).

3

The "Word of God" and the "Languages of Man"

Renaissance Cultural Narratives, the Reformation, and Theories of Languag

Pico and Ficino optimistically proclaimed the status of the languages of man late in the fifteenth century. During the sixteenth and seventeenth centuries, the relationship of human languages to the Word of God generated both cautious scrutiny and elaborate speculation. As humanists turned their attention to the Bible—the *verbum Dei*—and reformers raised questions regarding the possibility of man's salvation, the problem of language took on new interest and urgency. In the wake of the Protestant Reformation and humanist conceptions of the past, individuals, communities, and emerging nation-states forcibly experienced widespread questioning of authority and legitimacy. New conceptions of human history and its relationship to biblical history, and of God's power in relation to man, forced reconsideration of the very bases of religious, cultural, and sociopolitical authority. Within this context of cultural self-reflection, the problem of language—of its status and power as a source of meaning and cultural legitimation—became foundational to an entire epoch of Western history and its efforts to engender a new order. At the center of contesting inclinations were redefinitions of the relationship between the Word of God and the languages of man. To understand clearly what was at stake, however, we must first consider the sense in which theories of language constituted a problem for the Renaissance.

The Problem of Renaissance Theories of Language

Language itself became an object of increasingly sophisticated study during the sixteenth century. Scholarly, philological study of Latin and Greek, critical attention to European vernacular languages, examination of Hebrew and "oriental" languages, and the attempt to develop a comparative linguistics are among the phenomena that mark the sixteenth century as an epoch driven by a fascination with language.[1]

In addition, a number of critical developments influenced the relationship between language and science, shaping subsequent strategies for interpreting nature and practicing science during the sixteenth and seventeenth centuries. The development of a historical and comparative approach to culture and language prompted in part by humanist philology, textual criticism, and historiography; the reinterpretation of biblical narratives, ancient chronology, and the origins and diffusion of culture prompted by religious reform and the Reformation; and the reevaluation and critique of occult conceptions of language and nature all contributed to the construction of theories of language whose impact upon the study of nature has yet to be critically assessed.[2]

What theories of language did the Renaissance generate, and how did they affect the study of nature? Though seemingly straightforward, this question is fraught with difficulties for the historian, who, inevitably, approaches this subject with a set of assumptions about science and culture that predispose one to look for certain patterns in the relationship between language and science. For example, Brian Vickers has recently examined occult conceptions of language and their critique in the late sixteenth and seventeenth centuries. Vickers begins with a general assessment of occult and scientific modes of thought:

> In the scientific tradition, I hold, a clear distinction is made between words and things and between literal and metaphorical language. The occult tradition does not recognize this distinction: Words are treated as if they are equivalent to things and can be substituted for them. Manipulate the one and you manipulate the other. Analogies, instead of being, as they are in the scientific tradition, explanatory devices subordinate to argument and proof, or heuristic tools to make models that can be tested, corrected, and abandoned if necessary, are, instead, modes of conceiving

1. For overviews of some of these developments, see Apel (1963) and Dubois (1970). Gravelle (1988) provides a useful, brief survey and bibliography for earlier backgrounds.

2. In addition to studies cited in the preceding note, see Aarsleff (1964) (I have used the MS on deposit at the Warburg Institute, University of London); Céard (1988); Dubois (1977); Gombrich (1978); Grafton (1983); Simoncelli (1984); Vickers (1984); Walker (1958, 1972); Waswo (1987); and Wilcox (1987).

relationships in the universe that reify, rigidify, and ultimately come to
dominate thought. One no longer uses analogies: One is used by them.
They become the only way in which one can think or experience the
world. (Vickers 1984a: 95)

Vickers' analysis rests, I would argue, upon two questionable foundations: first, a theory of metaphor that permits one to deny any constitutive role to metaphorical language in scientific thought; and second,
a view of language that is far too restrictive to account for the complexity of scientific discourses found in the history of science and that implicitly denies the interpretive dimensions of scientific thought and
practice. The first element, the theory of metaphor in science, is one
that I have addressed elsewhere (Bono 1990a). The second element,
however, lies at the heart of my concern in the present chapter with the
role of language and interpretation in the Scientific Revolution.

What view of language is implicit in Vickers' analysis of occult and
scientific modes of thought? He himself makes reference to Robin Horton's analysis of scientific and traditional African thought and to
Saussure's analysis of "the relationship between word and referent"
(1984a: 97). Horton's views, as expressed in a well-known essay,[3] are
highly problematic; I shall not attempt to rehearse them here. Vickers,
however, sees Horton's distinction between "the traditional thinker,
who . . . sees 'a unique and intimate link between words and things' "
(Vickers 1984a: 95), and "modern science," which believes "that 'while
ideas and words change, there must be some anchor, some constant
reality. . . . words and reality [are] independent variables' " (1984a:
96), as analogous to his distinction between "science" and the "occult."
That is too say, for Vickers Renaissance occultism regards language as
"natural," as opposed to an emergent modern science which regards
language as "conventional," and this distinction between the two traditions is analogous to Horton's distinction between "traditional" and
"western scientific" thinking. In Saussurean terms, Vickers sees two
possibilities: either one implicitly embraces a view of language in which
signified and signifier are sharply distinguished (as he thinks practitioners of modern science do), or one lapses into the mode of thinking
(namely, the occult) in which the two are confused, indeed, identified.
As Vickers asserts, "In the occult and magical traditions the line [be-
tween the linguistic sign and its referent] is removed—or, rather, it is
never inserted: word and thing are not discriminated" (1984a:97).

Vickers' use of Horton and Saussure is problematic. More specifi-

3. Horton (1967). See Horton (1982) for responses to critics.

cally, Vickers' analogy between the Renaissance occult tradition and Horton's traditional African thought is strained and misleading. Horton's African traditional thinking, which (he claims) operates within a universe in which words in *ordinary* discourse are univocal and intimately attached to, indeed expressions of, the essential nature of things, has no exact parallel in the Renaissance occult tradition. Quite to the contrary, the Renaissance occult tradition—as a learned tradition based upon the study of texts as well as of nature itself—recognizes the equivocal and conventional nature of ordinary human discourse. The power that the occult tradition attributes to words—the power, for some thinkers, to manipulate things and natural/spiritual forces, or the power, for others, simply to reveal the essential nature and architecture of the universe—does not spring from an originary and uncontested "form of life." Magic for the Renaissance occultist is, in other words, a *discovered* or *constructed* mode of thought and action. The possibility of magic for the Renaissance occultist rests not in ordinary human languages but, rather, in the extra-ordinary language of magic and its extra-ordinary experiences of nature.

Let me be clearer about what I am suggesting. Vickers' analysis seems to rest upon the too-restrictive assumption that one must view language as either exclusively "natural" or "conventional" and that such mutually exclusive views dominate different modes of thought. Taken in its crudest, most abstract formulation, natural theories of language regard words and things as intimately, and necessarily, conjoined; words somehow capture and reflect the very essence, or nature, of the things they signify. By contrast, conventionalist theories regard words as mere arbitrary signs of things, as historically contingent and socially defined; in themselves, words bear no essential meanings. Horton's traditional African thought is an example of an exclusively "natural" conception of language in which "words" and "things" are identified. My contention is that Renaissance occultism does not embrace an exclusively "natural" conception of language.

To be sure, the Renaissance occult tradition does aspire, as an ideal, to *uncover* a language of magic in which words and things are, in some sense, identical. Ficino, as we saw in the preceding chapter, is an example. Its ideal, in short, is the discovery or construction (these may be the same) of *the* language of nature. But this ideal remains in tension with the recognition that the variety of human languages are, to some large degree, conventional. In Saussurean terms, Renaissance occultism does not confuse signified and signifier, does not fail to distinguish the linguistic sign and its referent. Rather, the Renaissance occult tradition starts with the fact that signified and signifier, word and thing, are divorced from one another in the languages of man and seeks to erase that difference, to reconstitute the union of signified and signifier in a perfect marriage of word and thing.

What Vickers has missed, or ignored, by merely opposing (following Horton) "natural" and "conventional" views of language is the fundamentally historical and hermeneutical character of the Renaissance occult tradition. That tradition does not spring from blind acceptance of a "natural" view of language, no more than modern science springs in any simple way from an open and conscious espousal of the view of language as "conventional." Rather, both occult and scientific traditions owe their origins to hermeneutic strategies implicit in the complex conceptions of language that inform each tradition. Those conceptions of language, moreover, arose from various Renaissance resolutions of the perceived *tension between* "natural" and "conventional" views of language.

To be precise, Renaissance occult and scientific traditions share a common hermeneutical problem converging, ultimately, upon the specific theories of language embedded in each tradition. The problem is a historical one with enormous theological and philosophical consequences: what is the relationship between the "Word of God" and the "languages of man?" The answers to this question not only determined specific theories of language within which magicians, scientists, physicians, and others operated, but invested each theory with a hermeneutical framework that shaped interpretations of earlier texts and of nature itself.

As a consequence, I would argue that it is impossible to distinguish between occult and scientific traditions in the manner Vickers claims. It is not a simple matter of science controlling language while language controls magical thought. Nor is language merely transparent to science. Rather, science and Renaissance occultism are both interpretive activities, and what distinguishes them, if anything, is the manner in which the theories of language by which each is informed determine its hermeneutical practices and thus defines it as a *discourse*.

The critique of occult uses of metaphor or analogy in the Renaissance is but one instance where we may glimpse the mutual construction of language theory and science as inscribed in scientific discourse and practice. If my contention that Vickers' analysis of occult and scientific mentalities falls into the trap of projecting unexamined assumptions onto these earlier discourses is correct, we should counter his analysis by reconstituting, where possible, dimensions of sixteenth-

and seventeenth-century language theory that his argument obscures or ignores. One problem is that Renaissance language theories rarely, if ever, find austere expression as abstractly "natural" or "conventional" definitions of language. Rather, the very meaning of Renaissance language theories is bound up with narratives that lend them legitimacy by inserting them into the larger framework of "master narratives" of Western, especially Judeo-Christian, culture itself.[4] Thus, narratives of origin, of creation, of salvation and redemption become the backdrop against which the story, and meaning, of language itself can be told. The organizing and legitimating categories for considering the significance of language become, for the Renaissance, not the "natural" and the "conventional," but rather the "Word of God" and the "languages of man." If we consider, then, narratives concerning language constructed in the wake of late Renaissance historiography, mythography, and the Reformation, we may find ourselves in a better position to suggest the significance of changing relationships between language and science during the Scientific Revolution.

Theories of Language and Renaissance Cultural Narratives

For much of the Renaissance, language acquires, or loses, its authority by virtue of its origins, antiquity, perfection, purity, and unity. The more humble its roots, recent its birth, corrupt its expressions, and diverse its forms, the less capacious and privileged its meanings. The key, then, to the authority, legitimacy, and significance of language lies in the relationship between specific "languages of man" and the pristine "Word of God." As a consequence, theories of language tend to embed themselves, explicitly or implicitly, in stories—narratives—that delineate the origins of languages and their descent from some pristine and lost past. Their links to that lost past determine their legitimacy: language acquires meaning, then, by virtue of its very history.

Not just any history will do. For the sixteenth and seventeenth centuries, it is the story of creation, the history of man as creature of God, that dominates discussion of language. The relationship between the "Word of God" and the "languages of man," subject to numerous narrative reconstructions in early modern Europe, provided the necessary framework for theories of language. By investing language with a cer-

4. For the notion of a "master narrative," I have drawn upon Lyotard's discussion of narrative and legitimation, as well as Fredric Jameson's introduction, in Lyotard (1984). Galison (1988) uses Lyotard's notion of "master narrative" in his discussion of "modernist" science and philosophy of science.

tain cultural legitimation and power, such reconstructions enabled scientists and physicians to construct interpretive practices and strategies for studying nature.

Certain assumptions about the *verbum Dei* and the "languages of man" inform the various narrative reconstructions of their relationship in the Renaissance. The "Word of God" is, of course, the Bible. But it was *also* the *Logos*, God's creative Word as manifest in His creation of the universe. This link between the Bible, biblical language, the creative Word of God, and the created universe fostered the idea of a divine language that, in turn, reflected the divine ideas "contained" in the divine mind (*mens*). These ideas were the essences or principles (*logoi*) that constituted the divine archetypes for created things. God's "creative Word" instantiated these divine ideas, bringing them into existence in material form. As we have seen, Ficino's cosmology and magic depend upon just such ideas. The "Word of God" could, then, be likened to a divine language of creation and of the created universe.[5]

This trope of a divine word or language clearly draws attention to the status of human languages, the "languages of man." The status of human languages was not, of course, a question first raised by Christian thinkers. In the Western tradition, the question finds its classical expression in the works of Aristotle and especially, and noted in the previous chapter, Plato, where the distinction between "natural" and "conventional" views has its roots.[6] Yet, the circulation of the biblical and Christian-theological trope introduces, I would argue, added complexities and a distinctly hermeneutical element to the question of the status of human languages.[7] For now we have no simple opposition of "natural" versus "conventional" languages. Instead, we have a new notion of a "super-natural" language, a divine language whose status or relation to human languages transforms debate concerning the "naturalness" or "conventionality" of human language or discourse.

When compared with the "Word of God" all human languages must appear defective and, to a large degree, conventional. On the

5. Dubois (1970:39) notes the significance of the link between the *verbum Dei* and Creation: "La Parole de Dieu, dans le récit mosaïque, est l'acte même de la création, laquelle se confond avec une suite de paroles. Ce caractère créateur attribué au verbe sera un des rêves constants de l'homme."

6. Vickers (1984a) has surveyed this tradition. I argue that Vickers' binary opposition of "natural" and "conventional" views of language is inadequate and misleading for an understanding of Renaissance and early modern science.

7. Contrast this view to Vickers (1984a). For a general discussion of religious discourse and the emergence of early modern forms of scientific knowing see Funkenstein (1986).

other hand, belief in the creative divine word, in a "supernatural" language which gave shape and substance to all created things, sustained belief that the "Book of Nature" could be deciphered if only man could (re-)discover or (re-)construct the language of nature. That there was a language of nature was a corollary to belief in the creative "Word of God." The only question that remained was whether, and in what manner, the "languages of man" were linked and had access to the "Word of God." In other words, the important question concerning human language was *not* whether it was natural or conventional (as Vickers assumes) but, rather, what its status was with respect to the "Word of God."

During the Renaissance and Scientific Revolution various theories of language emerged that, at least implicitly, addressed and answered this question. To the extent that magicians, natural philosophers, physicians, and scientists "accepted" a particular theory of language, each as a consequence engaged in a highly specific hermeneutical activity that structured readings of the "Book of Nature" and of the received texts of earlier traditions. Instead of allowing, as Vickers' analysis does, for just *two* options in engaging the natural world, the way of metaphorical language and its entangling web of meanings or the way of observation and its liberating access to things, the problematic of the relationship between the "Word of God" and the "languages of man" gave rise to a variety of approaches to the study and interpretation of nature that reflects, I believe, the richness and complexity of scientific discourses in this epoch of the history of science.

Let me then suggest how speculation regarding the relationship between the "Word of God" and the "languages of man" led to theories of language which ultimately helped to shape the hermeneutical practice of various students of nature. One starting point for consideration of this relationship must be the question of how sixteenth- and seventeenth-century students of nature understood the language spoken by Adam before the Fall. Typically, Renaissance authors contend that the Adamic language mirrored the nature of things themselves, that the "names" Adam gave to creatures in the Garden of Eden according to Genesis 2.19 captured their true essences. In other words, the language of Adam seemed to suggest to sixteenth- and seventeenth-century theorists something much like a natural language (in Vickers' sense of the term). But is this reading sufficiently nuanced to capture the meanings that language held for such theorists?

Here we must remember that the very idea of an Adamic language presents itself to Renaissance theorists in the guise of a narrative—

perhaps *the* "master narrative" of Western culture—the biblical narrative of creation, innocence and temptation, the Fall, spiritual and social
degeneration, and, finally, both the prospect and promise of redemption and salvation. Even if most commentators regarded the language
of Adam as a pristine language that mirrored nature perfectly and empowered Adam's dominion over all creatures, its significance could be
grasped only in relation to the larger narrative of which it is a part.[8] It is
therefore essential to remember that, beyond consideration of Adam's
naming of the creatures and the question of his access to divine words,
knowledge, and power, Renaissance cultural narratives focused upon
a number of problematic themes. What happened to the Adamic language after the Fall? What access did postlapsarian man have to the
pristine knowledge contained in the original, Adamic language? How
did the "confusion of tongues" resulting from the biblical episode of
the Tower of Babel affect the languages spoken by man? What were the
consequences of the confusion of tongues for man's search for the pristine, original language of creation? Does man have access to the originative divine Word, or is he hopelessly cut off from his prelapsarian
state of earthly perfection? Renaissance commentators, language theorists, scientists, and occultists, drawing upon such themes and questions, subjected their key cultural narrative—the biblical stories alluded to above—to a multitude of interpretations.

At the crux of such interpretations was the question of man's fall; its
consequences assumed enormous religious, intellectual, institutional,
and even political importance in the wake of the upheavals surrounding the Protestant Reformation. At stake was the very nature of authority, legitimation, and power themselves. The status of language before
and after the Fall, and cultural narratives of origins, therefore took on
an importance that could be immediate and far-reaching.

Consider, for a moment, the range of viewpoints available in the
divide separating such figures as Marsilio Ficino and Giovanni Pico
della Mirandola on the one hand, and Martin Luther on the other.
While refraining from denying man's fall, Ficino and Pico both stress
the continuing possibility of man's access to divine knowledge through
a hidden, esoteric, though nonetheless unbroken chain of divine revelation handed down from Moses through the so-called *prisci theologi*. This
tradition of a *prisca theologia* ensured man access to the very secrets of
nature and the cosmos through *magia* and Cabala, which would unlock
the nature of things themselves by revealing their true names. In effect,

8. For commentaries on Genesis see A. Williams (1948).

man's state of ignorance—a consequence of his fall—could be repaired and the divine language of nature restored to the benefit of mankind.[9]

By contrast, the Fall, for Luther, irreparably cuts man off from his prelapsarian state of knowledge and wisdom. While the external form of the divine, Adamic language may have survived, true understanding of it was lost forever; moreover, man's weak, mortal mind is incapable of fathoming, intellectually, God's Word. Yet the *verbum Dei* is salvific: man has only to cling to God's Word and he will be saved through, and by, his faith in it.[10] The absolute power of God dwarfs, for Luther, man's ability to know the divine mind or to save himself through volitional acts. God's absolute power is manifest in his Word, which man can apprehend spiritually, but not intellectually. While the Bible and human languages can give no special access to the essential nature of things, and while man's mind is too weak to penetrate the divine ideas, man can, through embracing the ethical and spiritual force of His Word, let God into his heart (Waswo 1987; Arndt 1983; Meinhold 1958).

Renaissance cultural narratives negotiate between the poles of Piconian exaltation of man's abilities and Lutheran insistence upon man's limitations. Such narratives inculcate views of language that, to varying degrees, regard humans as alternately enabled or disabled by words, texts, discourses. How such narratives and theories of language operate in the sixteenth and seventeenth centuries to promote scientific change has yet to be studied critically.

Before the Fall

As I suggested earlier, most Renaissance thinkers believed that the language Adam spoke in Eden mirrored the nature of things themselves. But this near unanimity did not preclude important differences in stories constructed about Adam and the origin of language. These differences, moreover, would have consequences for efforts to describe postlapsarian man's access to language and nature in the sixteenth and seventeenth centuries.

9. See Chapter 2 for fuller discussion and bibliography.

10. See Dubois (1970:52–53). For Luther, the perfect knowledge Adam enjoyed in paradise was a gift of God, implanted in his very nature. This knowledge enabled Adam to name the creatures in paradise. However, with his fall—as a result of *sin*—man lost this ability (Luther 1958:119–20). While Luther grants, in his commentary on Babel and the confusion of tongues, that the Adamic language survived within the family of Eber (Heber), corruption of man's understanding overshadows its potential significance for postlapsarian man (Luther 1960:215).

Let me start with the example of the early-sixteenth-century French philosopher and mathematician Charles de Bovelles, a disciple of the eminent French humanist Jacques LeFèvre d'Etaples (Victor 1978). Richard Waswo's recent argument concerning Bovelles's recasting of the story of Adam's naming of the creatures transforms it into an extreme example of Renaissance language theory. Bovelles stresses Adam's naming of the creatures in the Garden of Eden as an exercise of his divinely endowed free will.[11] In Waswo's argument, this freely chosen imposition of names upon things becomes an instance of the Renaissance's discovery of the conventionality of language, of the arbitrary nature of signs, and, more radically still, of the thoroughly relational nature of language and meaning. For Waswo, language becomes a human creation in Bovelles's reading of Genesis, and its status as "natural" in Vickers' sense is thereby called into question (Waswo 1987: 284–290).

Since Bovelles also dabbled in the Cabala and espoused a Neoplatonic philosophy and mystical theology that shared fundamental assumptions with the tradition of Renaissance magic, Waswo's argument, if correct, would provide a direct counterinstance to Vickers' strong identification of "natural" theories of language with the occult. While I hesitate to accept the full import of his argument regarding Bovelles, Waswo has at the very least cast doubt upon any universal equation of occult thought with a simple understanding of the Adamic language as "natural."

Such doubt remains even if we offer an alternative account to Waswo's by suggesting that Bovelles need not have linked Adam's free will with the radical conventionality, and arbitrary nature, of language. Adam, for Bovelles, even while acting as an autonomous creature, might nonetheless have access to the very language of nature, the *verbum Dei*, in his unforced choice of words to name things. This alternative reading of Bovelles's account of the biblical episode would then suggest some unspoken affinity between the mind of man, as *imago Dei*, and God himself. Far from being arbitrary and conventional, language for Bovelles—or, at least, the Adamic language—would mirror the perfection and power of the divine Word. To the extent that Adam's pure, unfallen nature enabled him to choose freely and spontaneously words capturing the true nature of things implanted in his mind as *imago Dei*, Adam's language was *the* language of nature.

11. For a modern reprint, see Bovelles (1973: esp. 46–47, 124–25). On Bovelles's text and ideas about language, see also Colette Dumont-Demaizière's introduction and Margolin (1985).

But Adam's access to the language of nature does not mean that his language was a *natural* language; nor was it, by default, merely conventional. Indeed, the very problem that Bovelles's account raises, and that Waswo's argument perhaps unintentionally uncovers, is the question of how humans have access to the language of nature, of how the Adamic language and the Word of God intersect. Thus, one of the narrative problems that faced those who, during the sixteenth and seventeenth centuries, would retell the biblical story of Adam naming the creatures in Eden was to give an account of the origins of Adam's language, of the very names he imposed upon creatures. As it turns out, there were a number of available alternatives.

Rather than cite numerous examples, let me turn to but one little-known text, Johannes Buxtorf's *Dissertatio de linguae hebraeae origine et antiquitate* of 1644.[12] Buxtorf acknowledges the two most obvious possibilities—that Adam's language was natural, or that it was conventional, that is, established by Adam's own prescription and imposition—but only to contest them as adequate explanations of the "efficient cause" of the "first man's . . . words or language" (fol. A4r). Rejecting the notion that the first language arose from nature, Buxtorf argues that Adam's language was "preternatural," the creation and gift of God to man.[13] Rather than arising from man's own free will, Adam's ability to impose names upon all creatures—names that captured their true natures—derived from his close relationship to God. How then did Adam receive this divine, creative word, the language by means of which God created the entire universe? The opinions that Buxtorf surveys alternately suggest that Adam learned the divine language through conversing with God, that Adam had this language stamped or inscribed upon him by God, or that the Adamic language arose from God's infusing man with his Word or spirit.[14] Hence, in addition to the possibilities that the Adamic lan-

---

12. Buxtorf (1644). This work is actually part of Buxtorf's *Dissertationes philologico-theologicae* which includes works dated from 1642 to 1645, many treating topics relevant to our discussion, for example: I. *De linguae hebraeae, antiquitate & sanctitate* (1644), and II. *De linguae hebraeae confusione, & plurimum linguarum origine* (1644).

13. Buxtorf (1644: fol. Bv): "Nec quicquam patrocinatur huic sententiae, quod primi nostri parentes Deum, certam linguam loquentem, statim a creatione intellexerunt, omnesque res suis propriis vocabulis appellarunt. Nam illa cognitio in illis non fuit naturalis, h.e. ex naturae principiis orta, sed praeternaturalis, hoc est, singulari Dei gratiam illis concessa, ut mox videbimus."

14. See Buxtorf (1644: fols. Bv–C4r). Note also, "Cum ergo utrique in eo conveniant, linguam primam neque fuisse Homini naturalem, neque ex nuda voluntate, arbitrio ac placito Adami principaliter institutam; sed a Deo creatam, &, vel generaliter secundum universalia linguae principia, vel specialiter, Adamo impressam & quasi concreatam;

guage arose from nature or from man's free and arbitrary creation (which Buxtorf rejects), we see in Buxtorf's account a number of distinct avenues through which the first man, Adam, may have gained access to the *verbum Dei*. What significance might such prelapsarian modes of access have for postlapsarian man and his efforts to study and conquer nature?

After the Fall

The question of postlapsarian man's access to the Word of God assumed pressing cultural importance in the wake of the Protestant Reformation of the sixteenth century. The question itself tended to engender either a nostalgic quest for the unity of a lost, primitive, Adamic language and its commerce with nature—God's works—or acceptance of the diversity of such works as necessary to fallen man's efforts to trace the inscription of God's Word in His Book of Nature. The possibilities associated with either path—the "search for unity" or the "way of diversity"—were numerous. Each assumed certain kinds of language theory and consequently adopted a specific hermeneutic stance toward God's Word and works.

Such positions and possibilities crystallized around diverse recastings of the biblical stories of man's fall and the confusion of tongues resulting from divine displeasure with the construction of the Tower of Babel.[15] Each biblical story raised significant questions regarding the survival and status of the original, primitive language of Adam—questions that drove sixteenth- and seventeenth-century theorists and historical actors to rework their master narrative differently.

The case of the fall of Adam and Eve was markedly simpler than that of Babel. The Fall clearly precipitated Adam's loss of dominion over nature, over the creatures inhabiting Eden. That loss entailed a clouding of Adam's intellect and the withdrawal of the perfect knowledge and perfect understanding of the true names of things that he enjoyed in his state of prelapsarian grace and innocence. But was the divine language of creation, humankind's primitive or mother tongue, itself lost?

Accounts vary, and are not always explicit about the full extent of the consequences of man's fall for the pristine, Adamic language. For the

---

dispiciendum porro, Quomodo cum hac sententia conveniat locus ille Genes. cap. 2. vers. 19. 20. de Nominum impositione ab Adamo facta?" (fol. C2r).

15. On interpretation of the Babel story and its cultural significance, see Borst (1957–63). For brief discussions, see also Allen (1963); De Grazia (1978); A. Williams (1948). For the biblical story of the confusion of tongues, Genesis, chapter 11.

sake of simplicity, however, let us assume three basic types of narratives. In the most optimistic accounts, the language spoken by Adam in Eden survives the Fall intact. But, while the words and grammar used by Adam and Eve and their progeny survive the Fall, the full meaning associated with that language in its prelapsarian use has been lost to the fallen human intellect. Without understanding the divine meaning embedded in the very names of things, "man's" dominion over creatures, his knowledge of and power over nature, are profoundly limited. With the relationship between words and things now obscured, human access to the divine, creative language of nature—the *verbum Dei*—is impaired. A second type of account shares with the first the belief that the Adamic language survives the Fall, though in this case the very form of that language has been corrupted by the Fall. Hence, in addition to the impact of the Fall upon a now weakened intellect and its ability to understand divine things, humans now speak an imperfect form of the primitive language. Nonetheless, as in the first account, that language still bears traces of the Adamic language and is closely linked to it. Finally, a third type of account sees humankind's loss as catastrophic and final. The language of Adam does not survive the Fall, and humans are left to descend into the contingency, arbitrariness, and contentiousness of corrupt, fallen languages of man.[16]

That descent was not necessarily immediate. Rather, the kinds of stories associated with the Fall, in turn, produce a variety of different readings of the Tower of Babel episode in Genesis. A narrative that stresses the complete loss of the Adamic language at the Fall would obviously have a more difficult task in attempting to give meaning to the confusion of tongues. One strategy for coping with this sort of difficulty, however, would entail claiming that postlapsarian man, before Babel, still enjoyed easy intercourse allowed by a common, if corrupt and merely human, tongue (indeed, the biblical story itself points to a single, common language). The confusion of tongues then destroyed that merely human, social harmony.

16. An important, synthetic text for surveying late Renaissance theories of language is Duret (1613). Duret surely summarizes the most common view, that the Adamic language survives the Fall intact. But he also alludes to the belief that this Adamic tongue underwent change, even corruption, in the postlapsarian period before the confusion of tongues at Babel. Such corruption, among those thinkers, who subscribed to it, frequently became associated with Noah's progeny after the Flood. See Duret (1613:26 [sic, 42]). For a brief discussion of alternatives to the idea that only one language existed before Babel, see Pererius (1592:390). I have not found strong evidence for descent into arbitrary "languages of man" after the Fall. Such results are typically seen as a consequence of the confusion of tongues that led to divisive, contentious human behavior.

Rather than elaborate on the many permutations of the Tower of Babel episode, let me again simplify the task by assuming a narrative that starts with the implicit or explicit belief that the descendents of Adam and Eve spoke something like the Adamic language up until the construction of the Tower of Babel. The consequences of the confusion of tongues at Babel as depicted in sixteenth- and seventeenth-century narrative reconstructions can once again be roughly classed into three categories.

One major class of narratives paradoxically describes how the originary Adamic language survives the confusion intact. Two narrative strategies lead to this result. Either the story assumes that the primitive language was present at the confusion at Babel, but was untouched by it, or it assumes that some of the descendants of Noah were not present at Babel and continued to speak the Adamic language. In the first case the salient feature of the confusion was not a universal corruption of the unitary language spoken by mankind, but rather the divine punishment whereby distinct groups of mankind were, in effect, alienated from another by their sudden inability to understand the discourse of others. Such a scheme allowed for the possibility that the Adamic language survived among at least one such group. The second case assumes, by contrast, that the confusion at Babel was profound and pervasive: language was corrupted by divine fiat. But the punishment imposed by the confusion extended, in this account, only to those who participated in the events at Babel. Only those assembled on the plains of Shinar were cursed by corruption of their tongues; one branch of the descendants of Noah, not present at Babel, escaped this curse and continued to speak the Adamic language, preserving it for posterity.[17] In both of these cases, of course, the problem presented to the early modern period is that of identifying which among the apparent "languages of man" is in reality the language of Adam.[18]

17. For example, see Webb (1678). Webb argues that "China was after the flood first planted either by Noah himself, or some of the sons of Sem, before the remove to Shinaar" (1678:31–32). The language of Adam survived the Fall without "any whatever alteration either in the Form or Dialect and pronunciation thereof, before the Confusion of Tongues at Babel" (1678:16). Since these descendants of Sem were not present at Babel, they did not share the guilt of the other descendants of Noah, and their Adamic tongue survived without confusion. On connections between European sinology and language theory, see Mungello (1985).

18. For many later commentators, Saint Augustine provided authority for the belief that the original Adamic language survived the confusion of tongues at Babel, having been entrusted to the family of Heber, the progeny of one of Noah's sons, Shem. See

A second group of narratives takes a slightly different, if not less paradoxical, tack. Assuming that the biblical curse of the confusion of tongues extended to all of mankind, such narratives assert that the emerging babble of languages preserved, to a lesser or greater extent, some vestiges of the primitive Adamic language. The confusion of tongues, in this kind of account, might be thought of as having introduced into language processes of variation and change normal to a human linguistic system, albeit in a rather sudden and accelerated fashion. That is to say, humankind experienced at Babel, through divine fiat, the effects of linguistic change subsequently produced by patterns of geographical migration and isolation, by processes of sociolinguistic variation, and by different sociocultural uses of language, among other pressures engendering change. Consequently, while nearly all post-Babylonic languages were viewed as human, and therefore contingent, historical, and to a large extent arbitrary, such languages were often presumed to be rooted in a pristine, originary language. In other words, human languages, given such narrative sanction, could be theorized as corruptions of an Adamic language retaining within them more or less significant traces of the *verbum Dei*. Properly understood, human languages could provide access to the language of nature, to a divine understanding of things.[19]

Finally, one class of narrative rereading of the confusion of tongues simply regards its effects as total and pervasive, resulting in the utter loss of the Adamic language. Thus the confusion establishes itself as absolute, extending to all groups of humankind despite geographical dispersion and to every aspect of language. Linguistic transformation was complete, leaving no traces of the divine word, of the originary

---

Augustine, *De civitate dei*, book 16, chapter 11. In addition to Webb cited above, some champions (among many) of the survival of the Adamic language include Buxtorf (1644); Casaubon (1650); Duret (1613); Pererius (1592) (Pererius provides a full account on 382–88; on 388–91 he discusses various opinions concerning the mechanism of the confusion); Stillingfleet (1662) (on 579ff. he discusses the confusion of tongues at Babel and the variety of ancient languages and traditions). For a comprehensive contemporary treatment of Babel, see Kirchner (1679).

19. For example, Leigh (1656:50): "Of languages, the Hebrew as it is the first and most ancient of all, so it alone seems to be pure and sincere, all the rest almost are mixt: for there is none of them which hath not certain words derived and corrupted from the Hebrew." Citing Bibliander (see below), Leigh again stresses the connections between Hebrew and other languages: "The Hebrew language was the first and most ancient, and the onely language before the building of the Tower of Babel. . . . The Hebrew tongue kept its purity, and remained uncorrupted, though other tongues were added to it, and derived from it" (1656:55).

language, available for future discovery or reconstruction. From the chaos of Babel arose an array of merely conventional, arbitrary human languages. Subsequent to Babel, the process of change initiated by the divinely decreed confusion proceeded of its own accord by contingencies of historical change and the distribution, isolation, interaction, and use of languages. Given such profound linguistic transformations, such narratives suggest that the past is not recoverable from the present state of languages.[20]

Humanism, Luther, and the Reformation

If the events at Babel figured a descent into history, with humans now divided into separate linguistic communities whose very modes of communication were subject to corruption, growth, variation, and the specific influences of geography, climate, politics, and social differentiation, then the study of language might acquire a historical and contextually specific coloration. Without raising the question of their relationship—"historically" or spiritually—to the Word of God, the languages of man could become the object of intense critical, textual, contextual, and, in short, historical scrutiny.

One of the major achievements and epochal innovations of the Renaissance was just such a critical-historical development within the movement known as humanism. The recovery and study of ancient texts entailed many things. For some, such recovery meant the rebirth

20. Dubois (1970:68) asserts, "rares sont ceux qui, comme Joachim Périon, considèrent que la langue originelle est définitivement perdue et que, dans la confusion linguistique de Babel, sont nées des langues absolument nouvelles." See Perion (1554). The great classical scholar Joseph Scaliger articulated a model for comparative European linguistics that asserted the existence of eleven (four major) matrices for European languages. While not discussing the confusion of tongues or the Adamic language, Scaliger seems to stress the separate, historically contingent nature of the major language families (e.g., "matricum vero inter se nulla cognatio est, neque in verbis, neque in analogia"). Scaliger's comparative linguistics does not assume that a primitive, or Adamic, language, common to all the "languages of man," survived the biblical confusion. See Scaliger (1610:119–22, "Diatriba de europaeorum linguis"); on Scaliger, Grafton (1983). While belief in the survival of the Adamic language was widespread, by the late seventeenth century no less a figure than Wilkins (1668) assumed that linguistic change and corruption were a fact and, at the very least, cast strong doubt upon the survival of any immediate, post-Babylonic language. To the question "whether and how any of the Mother-tongues have been quite lost since the Confusion" (1668:6), Wilkins answers "that if in some few hundreds of years a Language may be so *changed* as to be scarce intelligible; then, in a much longer tract of time it [mother-tongue] may be quite *abolished*, none of the most radical and substantial parts remaining: For every *change* is a *gradual corruption*" (1668:8).

of a lost, or at least an obscured, understanding of ancient wisdom, the return *ad fontes*, to the pure, unadulterated origins of a golden age of learning, eloquence, and deep human, or even divine, understanding of things political, ethical, poetic, philosophical, even theological. For others, recovery of ancient texts meant contact with cognate, but now past, if not dead,[21] civilizations whose differences and affinities with our own time and place could be traced in the very vocabulary, grammar, expressive modes, orthography, and historically specific meanings layered in their textual artifacts like so many fossilized deposits in distinct geological strata. Humanism embraced both these possibilities as well as a full spectrum of views and practices arrayed between these two poles.

Humanist exploitation of the possibilities opened up by the figuring of Babel as representing a historical and psychological divide among different groups and epochs of mankind coincided largely with the second strain within humanism noted above. Figures such as Lorenzo Valla turned to ancient texts with the conviction that their myriad elements were capable of bespeaking the specificity of another time, place, and habit of life and expression (Waswo 1987; Camporeale 1972; Jardine 1977; Kelley 1970). Moreover, Valla, and those like Angelo Poliziano who followed in his footsteps, developed critical tools and practices for weighing just such specific differences. For Valla and his followers, then, the languages of man emerged as product of human history, that is, of specific social conditions, experiences, values, and practices of an identifiable group of people in the context of a shared discursive community. Language, in short, was the product of

21. Terence Cave provides a subtle and illuminating discussion of the "powerful image of the death of antique cultures" (1979:69) and its link to the complex reception of the Tower of Babel legacy during the sixteenth century. In particular, he dwells upon the example of Du Bellay (1948), itself dependent upon Speroni (1978), originally published in 1542 (see Cave 1979: chap. 2, esp. 60–77). Cave's analysis nicely captures how the image of Babel continually reasserted itself and (even in the face of admitting irreparable change and loss—and, hence, variety—within human cultures) led to resurgent dreams of forging a unified culture. "The vernacular thus appears in this perspective as a 'natural' language, guaranteeing the authenticity of an 'art de bien dire' constructed with its native materials. Moreover, this exercise of conceiving of the ancients as speaking a *vernacular* tongue releases a crucial insight for the French poet, who must be made to conceive of culture as residing not in the past but in an imminent future; as a discourse which he is about to utter in his own 'domestic' language" (Cave 1979:70). This emphasis upon *making culture* as a prospective goal bears important affinities to the search for a post-Pentecostal unity and universality of discourse among such figures as Paré, Rondelet, and Bacon. See my discussion of tensions between diversity and unity among such naturalists and philosophers in Chapters 6 (following Jean Céard) and 7 below.

social use, stamped forever (in or upon texts) with the specificity of its provenance.[22]

Erasmus likewise approached human languages not as the repository of ancient, veiled wisdom, but, rather, as testimonials to social and cultural contexts that gave rise to specific uses of language by human agents. Erasmus also applied his philological techniques to the elucidation of the meaning of biblical texts in service of a search for a purer text than the Latin Vulgate provided. In this search, Erasmus was well aware that the textual critic of the Bible must attend to the original languages of that text—Hebrew and Greek—in order to correct the traditional Latin version and establish, as nearly as possible, the force of the biblical text as an expressive mode.

Erasmus' famous emendation of the beginning of the Gospel of Saint John, in which he rendered the Greek *logos* not as *verbum*, but as *sermo*, illustrates both his desire to capture the contextual sense of ancient texts and his own larger appreciation of linguistic phenomena. Rather than isolating a single term, *logos*, and rendering it by its seemingly literal Latin equivalent, Erasmus probes what he takes to be the sense, in context, of the Greek in offering his alternative, *sermo*. Thus, as Marjorie O'Rourke Boyle notes, Erasmus' translation carries a force that the Vulgate does not, that is of the *Logos*—Christ—"as the revealing discourse of the Father."[23] This philologically sensitive translation also suggests Erasmus' own view that language serves an expressive function. As such, it is revelatory of intent, meaning, and affect as associated with the context of use within which a community produces and expresses meanings in language. Erasmus does not take language as discrete units, or words, which constitute signs of things, ideas, or essences.[24] Erasmus' view can be regarded as antithetical to the traditional rendering of *logos* as *verbum*, in which the "Word" could be read in its association to *ratio* (Kelley 1970:28). In

22. For humanistic textual scholarship, its practices, premises, and implications, see Grafton (1991) and Kelley (1970).

23. Boyle (1977:25). Boyle provides an excellent account of the *logos/sermo* controversy and Erasmus' role in it in chapter 1, "Sermo" (1977:3–31). My own discussion in this paragraph is indebted to her account. See also Bentley (1983); Chomarat (1981); Harth (1970); Margolin (1971); and Payne (1969).

24. See Waswo (1987:220) for this view of Erasmus' orientation away from meaning as captured "in single, discrete words" toward "larger units that presuppose the semantic importance of usage and context." See also Waswo's extended discussion of Erasmus (1987:213–35) and his bibliographical references for issues that I can not begin to touch upon here.

such a reading, the expressive function of language—a function intimately tied to local meanings and specific contexts—gives way to an
essentialist view. Rather than expressing the affect of the early Christian community's understanding of Christ as son of God, the Greek
term, *logos*, comes to signify the Word as essence—as the "mirror" or
"reflection"—of the divine, particularly of the divine *mens*. *Logos*, the
divine "Word," becomes "reason."[25] With this rendering of *logos* comes
an entire metaphysics and cosmology unrooted in the specific time
and place that produced, and first read, the Gospel of Saint John, and
that therefore appeared anachronistic to Erasmus.

Instead, as Boyle notes, "Erasmus restores the verse Jn I:I *ad
fontes*" (1977:25) by unearthing a more ancient understanding of *logos*
than that articulated by Origen and subsequent Church Fathers and
enshrined in the translation of the Latin Vulgate. Erasmus' understanding of this key New Testament passage draws instead upon the
authority of Irenaeus. Renowned as "the first Christian theologian"
(1977:25), Irenaeus discloses meanings closer to the original intent
understood by early Christian communities which initially received
Saint John's Gospel. Thus, Erasmus employs humanist philological
methods to rework his translation of Scriptures and to capture a "pristine" contextual meaning associated with a specific historical discursive community that gave currency to the very language of that sacred
text.

Such a critical and historical philological method exhibits humanist
assumptions that regarded language as thoroughly social and contingent, while nonetheless enabling Erasmus to fathom the "Word of
God" through the lenses of the "languages of man." Indeed, even as
Erasmus approached the language of the Bible with a lively humanist
sense of social use and context, he nonetheless allowed for a kind of
reading of the Bible that was hybrid in reaching beyond a literal,
critical, historical sense. As Richard Waswo has epitomized Erasmus'
approach:

25. Boyle (1977:24). "By the year 1505 [Erasmus] claimed to have read most of Origen,
whom he judged a quarry of original ideas and a master of the principles of theological
science. Yet Origen's celebrated theory subordinates the interpretation of *logos* as speech
and designates *logos* as reason. It is "that which removes in us every irrational part and
constitutes us truly capable of reason." Modern scholarship asserts that Origen "rarely
conceives the relation of the Father and the Son on the model of the relationship between
the understanding and its verbal expression, "exactly the model which Erasmus thought
so vital. The recent opinion of some Erasmus scholars that he is influenced by the Alexan-
drian *logos*-theology, especially Origen's, ought to be abandoned then."

Allegory is still necessary in general to preserve the absolute privilege of
the divine Word, its qualitative otherness from our speech, and in particular to relate the Old Testament to the New in terms of typology. Erasmus,
however, also insists on beginning interpretation by a careful weighing of
the precise grammatical and historical context. The establishment of this
"literal" sense (depending on the correct establishment of the text itself) is
the prior basis for "figurative" interpretation.[26]

Thus, we see the continued centrality of the precise relationship of historically situated, human languages to the "Word of God" in Erasmian
humanism. While insisting upon the variety and historical specificity
of the different languages of man, Erasmus nonetheless seeks within
the "fallen" language of the biblical text transmitted to man a prior,
divine, and therefore authentic, unity. Utilizing the tropes necessary to
man's postlapsarian state, the language of the Bible reverberates with
alternately clear and "obscure," or veiled, passages. Through juxtaposition, through paraphrase, through attention to context and use, Erasmus seeks not the historic meanings of human languages, but the
transhistorical meanings to be found in the *sensus germanus*[27] of Scriptures, the *verbum Dei*. Acutely aware of the fallen, limited nature of the
languages of man, Erasmus the Christian humanist must nevertheless
embrace them in negotiating his quest for the Word of God (Cave
1979:101–2).

To Erasmus' sociocultural understanding of language, the reformers
added yet another critical perspective: an intense appreciation of the
impact of man's fall upon human discourse. While the "Word of
God"—the divinely authored Bible—offered for some a font of righ-
teous wisdom (*sapientia*) in matters of faith and religious doctrine, the
Book of Genesis revealed to reformers the wide gulf separating fallen
man from God and from the divine *scientia* known to Adam in paradise.
If God's Word were salvific, a healing oration that could affect men's
souls, it was also His originative, creative discourse which called all

26. Waswo (1987:221). For an excellent analysis of the conception and use of language, human discourse, interpretation, and related issues in Erasmus' texts, see Cave
(1979: esp. 78–124).

27. Cave (1979:90). Says Cave, "*Germanus* denotes close relationship, consanguinity,
and a consequent affective reciprocity. It replaces the publicly authorized systems of allegorical commentary with a private, inwardly experienced bond between reader and text:
an androgynous union, as it were, revivifying the dead fragments of a Scripture turned
inside out." Yet, Erasmus' quest for such authentic meaning acknowledges its elusivity:
"Erasmus' logocentric optimism—his conviction that there *is* a *sensus germanus*—is always balanced against an awareness of the problematic nature of the pursuit of sense"
(1979:101).

things, all creatures, into existence. This creative Word, mirrored for some in the language with which Adam named the creatures of paradise and gained dominion over them, was lost to human understanding, to his fallen reason. Where Pico and Ficino optimistically sought to repair man's lost knowledge by unlocking, through *magia*, the secret wisdom contained in the languages of man, the reformers judged man's mind too corrupt for such a task.

Indeed, man's fall called into question the very premises of that task. Not only was human reason unsuited to comprehending the divine nature, the very languages of man had seemingly been cut off from the source of all true knowledge and wisdom, the divine Word. Human discourse, like human nature, was corrupt, a mere babble of confused tongues. While salvation was still possible through faith, the prospect of a repaired knowledge of God and nature remained in doubt. How could fallen man, with his corrupt reason, hope to recapture lost knowledge—the pure, unfallen language of Adam? Must fallen man abandon such a task itself?

For Luther, as we have noted already, such a view of man as utterly divorced from his prelapsarian perfection was foundational. He therefore placed no stock in the ability to reconstitute Adam's understanding of the original language. In the quest for the Adamic language he even saw, as Claude-Gilbert Dubois (1970:52–53) notes, a sign of man's vanity, and associated the linguistic fall and confusion with man's sorry lot in life—his political and social corruption and divisiveness. In vain, Luther declares, does man try to seize hold of the *verbum Dei*; rather, man must allow himself to be seized by the saving message of the Word. The lost, Adamic language and the knowledge and power it held for the first man are beyond our grasp, a futile dream of the old Adam. But Christ's new dispensation, and the Pentecostal gift of tongues granted to his evangelizing disciples, bring promise to humans of the salvific effects of Christ, the *verbum Dei*, whose spirit engulfs those whose simple faith makes a home for the Word in their hearts.

Thus, Luther denies any special status to a sacred, originary language, but at the same time invests all language with a special, even sacred, role. For it is "by the means of languages that" the Gospel "comes [to us] and has been spread; by them must it therefore also be maintained and secured."[28] For Luther, Hebrew, Greek, and the vernacular—through translation of the Gospels—carry to humans

28. Luther (1524:37–38). Quotation and reference from Aarsleff (1964).

God's saving message. While language itself is neither privileged nor divine, but merely human, it is, if you will, the vehicle through which individuals hear, and are seized by, the Word. Without any intrinsic power or hidden source of *scientia*, languages convey the spirit of the Lord to humans, a spirit that can transform a human being into a vessel of the Lord. In short, the effect of the Word—and the language that gives humans access to the Word—is moral and salvific, rather than cognitive (Waswo 1987:235–49). It brings spiritual insight and knowledge of the Lord to transform man's will, *not scientia*, not knowledge of the inner nature or essences of things themselves. Language does not become a privileged repository of knowledge to be mined for understanding of nature.

Humans consequently do have access to the Word of God, but for Luther that access is affective: humans may come to feel and experience God's healing oration. And it is to that Word that humans, through faith, must turn: *not* to language or to mere human books, however ancient they may be.

What options do readings of the Fall and of the Babylonic confusion of tongues informing humanist sociocultural understandings of language and Reformation interpretations of the Word of God leave postlapsarian man in his search for divine knowledge and wisdom? Humanist views like those of Valla and Erasmus would seem to restrict humans to a mere workaday knowledge of ordinary human things as captured in the socially and historically contingent languages of man. But even such humanistic views could recuperate the notion of a special, originative language—a lost Adamic language, if you will—if those producing and receiving such views interpret them to apply *only* to human languages, or to those aspects of the languages of man that represent post-Babylonic corruptions introduced by sociocultural change. The French theorist of linguistic variety Charles de Bovelles, who was also a proponent of mystical theology and various occult ideas, may well be representative of such a recuperation. On the other hand, those who take this humanist stance as applying universally to all linguistic phenomena to which humans have even potential access would by that very narrative logic be forced to accept the thoroughly historical and contingent nature of all language. As an example of this sort of response, we shall discuss William Harvey in Chapter 4.

Luther's view of man as forever cut off from the Adamic language leaves virtually no room for recuperating language itself as a source of divine wisdom. But his views—and those similar to them—did leave sixteenth- and seventeenth-century thinkers with at least two options. First, one could conclude that humankind ought to turn away from language and mere human texts as false repositories of hidden divine knowledge. Preoccupation with them was but testimony to a weak, prideful human nature. Instead, humankind must turn to the very text of God's creation—to His Book of Nature—in which one can begin to read, in the symbolic language God has stamped upon things, the true essences of things that mirrored God's nature and creative *mens*. Alternatively, a second and related option would have human beings turn to the Book of Nature and read in it a contingent language of things inscribed by God in that text. Whereas the first option links God and nature in an essential manner, the latter supposes but a contingent and contractual, or covenantal, link between an all-powerful and inscrutable God and his creation, capitally "man" himself.[29] We shall discuss examples of the first option in the figures of Paracelsus and (in a more complicated fashion) Oswald Croll in Chapter 5. Sir Francis Bacon provides an example of the second in Chapter 7.

Significantly, Luther opens the door (without decisively entering through it) to attempts to recuperate such a search for "divine" knowledge in the language of the Book of Nature, when he declares:

> We are now living in the dawn of the future life. For we are now again beginning to have the knowledge of the creatures which we lost in Adam's fall. Now we observe the creatures rightly, more than popery does. But Erasmus does not consider them, and is not concerned how the fruit is formed in the mother's womb, how it is shaped and made. . . . We, however, by God's grace are beginning to see his glorious work and wonder even in the flower, how almighty and good God is: therefore we shall praise and thank him. We see the power of his word in the creatures. As he spoke, then it was, even in a peach-stone; for though its shell is hard, still it must in time give way to the tender kernel that is in it. But Erasmus does not note it and pays no attention to it, for he looks at the creatures as a cow at a new barn-door.[30]

The narrative problem of human access to the "Word of God"—of its relationship to "languages of man"—thus converges upon the very metaphor of the "Book of Nature" in the late Renaissance, informing

29. For a brief introduction to the general impact of Reformation theology upon scientific attitudes and resulting views of the contingent and covenantal nature of human knowledge of the natural order, see Deason (1986). Such attitudes toward God and nature were pervasive in seventeenth-century England and will be discussed in Volume 2. For now and in general, see Hunter (1981); Kroll (1991); and Webster (1975).

30. Martin Luther, *Tischreden*, vol. 1, item 1160, as quoted by Aarsleff (1964:23–24).

attempts to develop strategies for learning how to read that divine text properly, as we shall soon see.

Postlapsarian Readings of the Book of Nature

Narrative authority—the presumption, or assertion, of privileged interpretive access to the master narratives of Western culture—lay at the center of the clash of religious and sociocultural ideologies during the sixteenth and seventeenth centuries. Rereadings of man's fall and, especially, of the biblical story of the confusion of tongues were of crucial strategic importance in establishing the narrative authority of medical, occult, philosophical, and scientific discourses and texts in their claims to read the "Book of Nature." Given the variety of narrative recastings of the Tower of Babel story schematized above, we can begin to suggest how claims concerning access to the "Word of God," to the divine language of nature, and, thus, to the very "characters" composing the Book of Nature were invested with narrative authority in early modern Europe.

To the extent that science, in order to construct itself as a distinctive discourse, requires a hermeneutics of nature or, in other words, a set of interpretive practices or strategies for reading the Book of Nature, we may find it economical to divide our consideration of the different modes of access to the "Word of God" available to sixteenth- and seventeenth-century students of nature according to certain affinities that will soon become apparent. Thus, narratives that stress the survival—partial or whole—of the Adamic language after the confusion of tongues become the basis for an array of interpretive strategies that are, by and large, fundamentally divergent from those stemming from narratives that insist on the irrecoverable loss of the primitive, originary tongue and/or of the human ability to decipher it. Let me then start by discussing the modes of access to the "Word of God" available to the former.

With their conviction that the language of Adam has left traces of itself among historically existing languages of man, narratives proclaiming the survival of the primitive language beyond biblical times valorize the search for origins and for the pristine relationship between words and things known to Adam in the Garden of Eden. The direction of that search, and the chances for success that specific hermeneutic strategies enjoy, depend upon the relationship between the languages of man and the "Word of God" and upon the kind of access to the latter that relationship permits. Using Adam's access to the divine Word as a model (following Buxtorf's discussion), how might postlapsarian, post-Babylonic man, following the logic of such narratives, gain access to the *verbum Dei* and the language of nature?

Where Adam's access was immediate, postlapsarian man gains access to the true wisdom of the *verbum Dei* only through the mediations of language, nature, Scriptures, and, perhaps for some pious few, through God's illumination. Where Adam, for example, spoke with God and learned the divine language of creation, postlapsarian man, with few exceptions, was cut off from direct discourse with God. For Ficino, Pico, and some of their Neoplatonic followers, however, the source of true *scientia* derives ultimately from the oral, Mosaic revelation. Hence, the languages of man, insofar as they incorporate traces of the Mosaic language and revelation, do provide access to the divine Word. Much of the impulse behind the sixteenth-century fascination with the Hebrew Cabala stems from the conviction that the Hebrew language is essentially connected to the lost Adamic language, or, at least, to the divine language of creation revealed to Moses, and that the gulf between ordinary human understanding of Hebrew and the true Adamic (or Mosaic) understanding can be bridged by cabalistic methods of exegesis orally transmitted to and from Moses.[31]

For the most part, however, cultural narratives proclaiming the post-Babylonic survival of (elements of) the Adamic language suggested to early modern readers that human access to the "Word of God" springs from sources other than direct conversation with God. Humans must take an active role in searching for divine wisdom. They must look, instead, to the "traces" that God has stamped upon language and things in their search for the perfection of Adam's prelapsarian knowledge and dominion over nature. If, as these narratives suggest, the

31. On Moses, Hebrew, and the Cabala, see Pico (1969); Reuchlin (1964); and Duret (1613). The divine transmission of Hebrew and cabalistic lore via Moses was widely noted in the sixteenth century, not just by cabalist authors; see, for example, Le Roy (1577:fol. 17v). Thomas Tymme (1963:11–15), ties together nicely the Adamic language, man's fall, Hebrew, Moses, and the Cabala and even asserts their connection with Egyptian wisdom and heiroglyphs: "The greatest worthy among mortal men, Moses, was brought up in the Schooles of the Aegiptians at the cost & expences of Pharaos Daughter, to learne these Scyences, & the learned & excellent prophet Daniel in the doctrin & wisdome of the Chaldeans, become a perfect Cabalist, the wisdome of Gods spirit dwelling in him, whereby he expounded these misticall words *Mene Mene Tekel Upharzin*. The tradition of this Cabalisticall Art, was much in use among the ancient Sages, whereby they learned the true & right knowledge of God, and walked the more firmly in his lawes & comandements" (1963:13–14). For scholarly studies, in addition to the works of Walker and Yates cited earlier, see Clulee (1988); Copenhaver (1977, 1978, 1990); Dan (1977); Elsky (1984); Idel (1989); and Zika (1976).

Adamic language survives in some shape or form, then humans ought first to search for its traces among the languages of man.

Where might such traces be found? Early modern narratives about the confusion of tongues prompted students of language and of nature to look in a number of places. If the originary Adamic tongue survived Babel, one needed to identify its closest descendant among the existing languages of man. Numerous people took up this challenge and gave birth to an entire genre of studies that attempted to compare ancient and contemporary languages in order to retrace the links between specific languages and an ideal, if still hypothetical, originary language.[32] The "myth of a lost language" proved seductive to many, including some of the best-known scholars and scientists of the sixteenth and seventeenth centuries, who produced elaborate arguments for the primitive status of particular ancient languages and often for the ancient roots and privileged status of a single contemporary language of man.[33]

Most generally scholars accorded ancient Hebrew special status as the original language of the Bible and the very tongue by means of which God called all creatures into existence. It was therefore also most frequently identified as the original Adamic language.[34] But other ancient languages received the attention and support of scholars, including other Near Eastern and Semitic languages and, closer to "home" for

32. Two early sixteenth-century linguists helped to establish a kind of comparative linguistics. The French polymath and occultist Postel devoted a number of works to the search for an originary language; for example, Postel (1538). At Zürich, Zwingli's successor Theodore Bibliander (1548) attempted to study the linguistic legacy of Babel systematically. Gesner, who avoids discussion of Babel, cites these authors as his chief inspiration in undertaking a comparative study of linguistics: Gesner (1974:245 [fol. 78r]). See also Bouwsma (1957); Egli (1901); Eros (1972); Metcalf (1974, 1980); and Peters (1984).

33. For example, Gesner, Leibniz, and Simon Stevin. I am indebted to Dear (1988:178) for the reference to Stevin. Duret represents a student of language theories who contributes to science; see the final chapter, "Des sons, voix, bruits, langues dex Animaux, & Oyseaux" (1613:1017–30). For the language of animals in the Renaissance, Dubois (1970); Gray (1990).

34. This trend follows the example of Saint Augustine. For some early modern examples, see Casaubon (1650); Chéradame (1532); Duret (1613:39ff.); Pererius (1589:372–73); Postel (1538); Tyard (1587, 1603). Bibliander (1548:38) succinctly suggests what was at stake in the notion of an originary language when he links together man's origin and end: "Quemadmodum enim ante linguarum divisionem in turri Babel sanctam universa terra unius erat labii: ita in resurrectione generali omnes homines una loquentur lingua, Hebraica scilicet, qua primi parentes in paradiso loquebantur, & qua electi omnes in statu innocentiae locuti fuissent, si ipsi primi parentes in gratia dei stetissent."

Europeans, ancient Greek.[35] More atypical was John Webb's (1678) argument for Chinese as the descendant of the Adamic tongue. Indeed, one all too frequently finds arguments for particular European languages—for example, German, Dutch, Danish, Swedish—that barely, if at all, conceal narrow, nationalistic intentions.[36]

Apart from such exercises in self-serving boosterism, the search for the lost, primitive language had a serious intent. If one could identify the Adamic language, one could begin to recover Adam's lost wisdom and harmony with nature by attempting to revive his understanding of the essential link between words and things. Man's access to the "Word of God," in this scheme, depended upon the survival of a divinely established link between the *verbum Dei* and at least one language of man. In practice, however, few claimed to have found the uncorrupted Adamic language. Even if it had survived the confusion of tongues, most theorists assumed that it had been subject subsequently to all kinds of intervening processes leading to linguistic change. At best, then, one might hope to find a single descendent of the language of Adam that might become the object of intense study, reconstruction, and hope. Alternatively, given the probability of historical borrowings, corruptions, and transformations, it was far more likely that traces of the originary language might be found among any number of human languages, with perhaps greater frequency and less distortion among the most ancient languages.[37]

For all intents and purposes, then, this view of the relationship between the "Word of God" and the "languages of man" merged with the implication of those narrative recastings of the biblical episode of the confusion of tongues that asserted that the common tongue of the descendants of Noah had survived Babel only in a corrupted form. This corruption of the Adamic language might well be found among a number of languages of man and, again, particularly among the ancient

35. See Gesner's statement concerning Postel: "Samaritana lingua, prisca Hebraica est, Postellus. Idem Hebraicam & Phoenicum linguam eandem facit, omnium antiquissimam" (1974:184 [fol. 47v]). Dubois (1970:83) notes "Syriaque" as one candidate for the mother tongue in this period. See also Metcalf (1963, 1963a).

36. Jan van Gorp (1518–72), Simon Stevin (1548–1620), and Leibniz champion Germanic languages; see Goropius (1569); Stevin (1955:58–93); Courtine (1980); and Walker (1972a) reprinted in Walker (1985). For Danish, see Ole Worm (1588–1654) (Wormius 1636). As Elert (1978) argues, Kempe's case is complex: he champions Swedish, but comically has God, Adam, and the serpent in paradise respectively speak Swedish, Danish, and French!

37. See Leigh (1656:50, 55), quoted earlier in notes; and Gesner (1974:94, 184 [fols. 2v and 47v]).

languages. Hence, for the sixteenth and seventeenth centuries, one could assume that human access to the Word of God was made possible by God's stamping the imprint of his Word as traces upon the languages of man.

The implications of this view for reading the divinely created Book of Nature were specific. If God's Word found expression in his authorship of the Book of Nature, and if that same creative Word—known to Adam immediately through his grasping of the divine meaning implanted in his own language—survived in corrupted or veiled form in language(s) accessible to postlapsarian man, then humans could potentially read the secrets contained in the Book of Nature through exegesis of discourse about nature, particularly discourse written in ancient tongues. Hence, the pursuit of *scientia*, scientific knowledge, as a process of exegesis that sought to rediscover the true, pristine meaning of words and thus their intimate relation to the essential nature of things was validated by the narrative authority claimed for such theories of language as veiled traces of the divine Word.

If the basis for such a hermeneutics of nature as exegesis lay in God's stamping of His Word upon human languages, the possibility that individual readers of texts could reconstitute lost meanings through exegesis was sometimes thought to rest upon other grounds as well. Even though God's Word, the meanings that allowed one access to the Book of Nature through language, was available to all postlapsarian exegetes, successful exegesis was often linked to the status of the individual reader. Human discourse remained opaque, and any hidden divine dimension to language might be regarded as largely inaccessible to clouded human intelligence; the ability to lift the veils hiding traces of the divine Word in such fallen human discourse was by some theorists conceived to depend upon God's illumination of individual humans. Such views claimed to privilege only certain people as exegetes who could read the divine message of the Book of Nature within the languages of man.[38]

38. "Exegetical" hermeneutics, applied either to language or to the Book of Nature, frequently traffics in metaphors of "light" and "illumination." The register of such metaphors varies considerably, however. Jean Fernel (see Chapter 4, below) attempts to read the texts of ancient, pagan medicine and philosophy from the vantage point of the Christian who looks upon ancient monuments and sees them whole—even their most remote recesses—thanks to the bright light that Christ shines upon human history. For Fernel, all Christians enjoy access to such illumination; scholarly, especially philological and philosophical, skills distinguish those who are privileged to read texts correctly. See Fernel (1550: fol. 3r): "Nunc vero quando Dei Opt. Max beneficio, nobis per Christum lux ipsa veritatis affulsit, multa simul nobis sunt divinitus allata, a veteribus animo non

Indeed, the limits of postlapsarian man as exegete and interpreter of mere human discourse gave rise to another spectrum of hermeneutic strategies closely tied to the more restricted, textual strategy of exegesis legitimated by the most optimistic of narratives regarding man's fall and the confusion of tongues. This alternative set of hermeneutic strategies envisioned man as reader not simply of texts, in the restricted sense of actual written discourse, but additionally of the "text" of nature itself. By investing the trope of the "Book of Nature" with special meaning, language theorists and students of nature transformed nature into a text that could be read exegetically, much as some would subject ordinary human discourse to exegesis.

Thus human access to the "Word of God" also derived from the act of God's "stamping" his Word upon *things* in nature. Since the archetype for both the Adamic language (words) and entities created by God in the world (things) was the divine ideas—tropologically figured as a divine language and rendered narratively as His creative Word—the notion of "things" as material elements or instantiations of the "language of nature" made it possible to think of nature itself as embodying the Adamic language. As a result, the search for origins, for the lost originary language, might look to nature itself, not just to the languages of man, for its traces.

For students of nature in the sixteenth and seventeenth centuries, the disruptive, disintegrating effects of the Tower of Babel could be repaired by careful exegesis of both the languages of man and the Book of Nature. Both bear the imprint of the originary divine language of creation known to Adam. The degree of importance attached to one or the other during this period reflects differing valuations of language itself and, in turn, of the spiritual and intellectual effects of the Fall. The most optimistic reading, described above, asserts the existence of a close relationship between the "Word of God" and the "languages of man": exegesis of language can, in itself, lead to restoration of the Adamic dominion over nature through grasping the true names and essences of things. Alternatively, various degrees of pessimism regarding human

---

integre percepta." By contrast, those who turn increasingly from exegesis of "human" texts to symbolic exegesis of the divine "Book of Nature" tend to invoke not a "general" Christian light, but the special "illumination" God provides to only the chosen few—the "elect" or perhaps the "pious." The different registers struck by such metaphors reveal different discursive and ideological formations. The ideological excesses associated with some more radical forms of symbolic exegesis contributed to growing disdain for exegetical hermeneutics of science later in the seventeenth century. I shall discuss these implications in Volume 2, focusing on England from 1640 to 1670.

ability to repair the effects of Babel were mirrored in other theories of language and in the hermeneutics of nature they respectively pre-scribed. Such theories ranged from regarding the languages of man as closely, but imperfectly and complexly, related to the Word of God to theories in which the relationship was, at best, distant and marginally accessible to human apprehension through exegesis. Paralleling such theories, then, exegesis of the traces of the divine language in the Book of Nature itself ranged from being considered optional (though desir-able) to being considered absolutely necessary and, indeed, superior to the exegesis of texts.

What significance did such theories of language have for science in the sixteenth and seventeenth centuries? How did they lead to herme-neutic strategies that in turn constructed various kinds of approaches to the study of nature—indeed, various kinds of science itself? Before suggesting how these questions might be answered by pointing to a typology of early modern hermeneutic strategies and to a range of ex-amples from among contemporary natural philosophers, physicians, and scientists, we need to bring back into the picture one last category of narratives. More specifically, we need to examine how narratives that insist that the Adamic language was lost forever after the Fall and the confusion of tongues at Babel construed the possibility of man's access to the "Word of God" and, in concert with this construal, framed theories of language and hermeneutic practices.

Narratives that stress mankind's loss of the originary, Adamic lan-guage in effect deny the legitimacy of the Renaissance search for ori-gins (Quint 1983; Rothstein 1990). God's creative Word is not stamped upon the languages of man, awaiting rediscovery and full recovery through exegesis of texts and discourse. The resources available to man are limited. The languages of man bear no relationship to the "Word of God" except for a negative one: that of absence and utter difference. In this sorry state, the languages of fallen man hold out no prospect of penetrating the recesses of nature to reveal her innermost secrets. Lan-guage is merely conventional, an arbitrary social construct, a historical artifact (Waswo 1987).

We have here, then, at the threshold of the Scientific Revolution, a thoroughly secularized view of language, a triumph of the modern: or do we? Retrospectively, language is much as we have described it—cut off from its Adamic, divine origins. Yet, despite this chasm, the ideal of a pure, Adamic coincidence of words and things remains, if only as an eschatological, and hence distant and prospective, goal. In other words, the languages of man are cut off from the Word of God and, as such, devalued as any significant element of a hermeneutic strategy for interpreting nature. Conversely, if one privileges nature—"things," not "words"—the possibility of constructing a new language, one based upon things, suddenly looms large. Hence, narratives of loss, of estrangement from a prelapsarian Adamic paradise of language and power, paradoxically empower humans to forge a new relationship with the Word of God. For while humankind cannot speak with God, cannot uncover and read symbols stamped upon nature as revelations of His Word, humans can turn to the Book of Nature and find there God's hand manifest in His works. Rather than describe this prospective presence of the Word in nature through the metaphor of stamping, the trope of "inscribing" may work better. In short, one consequence of the radical narrative emplotment of man as estranged from his prelapsarian past is to read the Book of Nature anew as representing God's plan, His inscription. Rather than bearing "signatures" of the divine (reflecting God's nature and ideas) that man must interpret symbolically, nature—the Book of Nature—has been "written" by God in the contingent language of things. The inscription of things, the plotting of their relations to each other, constitute the "text" that science must uncover, reveal, and represent in language. By mirroring things, language at some future state of maturity—perhaps at the climax of a millenarian age—can approach the perfection of the Word of God as inscribed in Nature's Book. First, as we shall see, man must labor; thus attending not to God's Word, but to His works.

Language and Interpretation: Hermeneutic Strategies and the Scientific Revolution

Before moving on to the next chapter, let me extract from the foregoing a somewhat crude, but I hope useful, typology of hermeneutic strategies made available to students of nature during the sixteenth and seventeenth centuries by the emergence of revisionist cultural narratives. In subsequent chapters I shall then try to suggest the import of such a typology by reference to examples illustrating how the narrative authority lent to language theory consequently validated hermeneutic strategies and practices that contribute to scientific change in early modern Europe.

The types of hermeneutic practices I shall highlight sometimes appear distinct from, and sometimes overlap with, a spectrum of other principles and practices. At their extremes, however, the differences between hermeneutic practices tend to outweigh the commonalities, and were frequently portrayed as polar opposites in the disciplinary rhetoric employed by the highly motivated historical agents of scientific change in the period. Thus narratives, language theory, and hermeneutic practices play a decisive role in both the rhetoric and the "reality" of scientific changes that mark the "Scientific Revolution."

What I have termed "exegesis" represents a kind of interpretive strategy in which language—"words" and texts—plays a central role in the practices that, it is presumed, will yield true knowledge of nature. In the sense in which I use this term, exegesis should perhaps be distinguished from the more familiar, if not hackneyed and caricatured, fashion in which "words" have been regarded as central to Aristotelian Scholasticism of the late Middle Ages and Renaissance. While much scholastic debate does indeed revolve around the proper definition of key terms, such terms themselves merely "stand for," or represent, major concepts reflecting essences that constitute the very being and necessity of a substance.[39] Such essences, and their corresponding concepts, are abstracted from lived, sensual experience of actual substances. Hence, the "reality" of such substances or essences and our knowledge of them are in no necessary way, generally speaking, identified with or derived from the "words" that name and represent them.[40] In this sense, Scholastic Aristotelianism could well regard language, including "names," as thoroughly arbitrary and conventional, while installing the proper definition of words and names as central to its dialectic investigation of nature.[41]

Of course, the extent to which both Scholastic science and the "new"

39. For these points, which are not of direct and immediate concern to this chapter, consult Schmitt et al. (1988); Kretzmann et al. (1982); and Colish (1968 [1983]) for an overview of medieval and Renaissance Scholastic semantics and epistemology. For further inquiry, see Arens (1980); E. J. Ashworth (1980); Gehl (1984); Koerner (1980); Padley (1976); and Stock (1983).

40. For a neo-Aristotelian treatment of language and grammatical theory that rejects the "natural" view of words, see Scaliger (1540). On Scaliger, see Jensen (1985, 1990).

41. I am aware that this brief discussion does violence to the complexity and variety of Renaissance Aristotelianism. I have consciously chosen not to attempt to situate Aristotelianism in the present chapter, as it would introduce considerations that lie only partially within, and chiefly outside of, the hermeneutic strategies I dwell upon here. Indeed, as the late Charles Schmitt never tired of saying, there are many Aristotelianisms in the Renaissance. For an introduction to Schmitt's vast knowledge of this tradition, see the reprinted collections of his essays (Schmitt 1984, 1981), as well as Schmitt (1983, 1983a). See also Henry and Hutton (1990) and Kessler et al. (1988). For the problem of words and things in Aristotelianism, see the discussion by the late-sixteenth-century skeptic Sanches (1988), who trained in and taught natural philosophy and medicine at universities.

Renaissance exegetical science revolve around "words" and texts in their different hermeneutic practices allowed for a certain ease among scholars in transferring from one set of practices to the other. Thus, the social, institutional, and educational contexts of practitioners of Scholastic and of exegetical science in many instances overlapped. The textual, scholarly nature of exegetical approaches to nature perhaps attracted a more elite, university-trained, if not university-based, set of practitioners than some of the other hermeneutical strategies highlighted below. The sixteenth-century physician Jean Fernel, who will serve as example of this exegetical hermeneutics of science in Chapter 4, also exemplifies this social and institutional tendency. Fernel, trained in Scholastic logic and science at Paris, juxtaposed Scholastic elements of analysis with an exegetical approach to texts and nature in his medical and philosophical works written while a professor in the faculty of medicine at the University of Paris.

Exegesis, to sum up, prescribed a hermeneutics of scientific practice that focused upon the interpretation of language and texts as bearers of a lost, but recoverable, Adamic and divine understanding of nature and things. But by being transferred to "things" as the "pages" upon which God stamped His mysterious, veiled symbols (or "signatures") in creating His Book of Nature, it also led to a related exegetical hermeneutics applied directly to that Book. Within this important Renaissance hermeneutics of nature, the analogical, symbolic universe of the Middle Ages was concretized. That is to say, the "abstract symbolism" of the Middle Ages was transformed into a concrete "symbolic literalism" in which material things themselves came to be seen as symbolic and as constituting an elaborate tropological network linked together by intersecting, metaphorically conceived relationships in which the distinction between the "metaphorical" and the "literal" was continually effaced.[42] Such a hermeneutics of nature bears strong resemblances to what Foucault (1973) characterized as "similitude" and "resemblance" and, more recently, William Ashworth, Jr. (1990), has called the "emblematic world view."[43]

42. I have briefly discussed "abstract symbolism" and "symbolic literalism" in Bono (1984:100–101). This distinction is then taken up and applied to the metaphorical discourse of Renaissance medicine and its tendency toward a kind of slippage from the metaphorical to the literal in Bono (1990), which forms the basis for Chapter 4 below. The impulse toward such a "symbolic literalism" may be partially glimpsed in Fludd (1659:161) (see Chapter 5 for the pertinent quotation from Fludd).

43. For a criticism of Foucault's view of the Renaissance, see Huppert (1974). While Foucault's typology of how language—words—map onto things is still too crude and ahistorical, he nonetheless points toward an important variant in the hermeneutics of

The "text" of nature was therefore poetic. Nature, like poetry, was written in a highly symbolic language and required the interpreter to read beneath its surface for meanings that linked together its different objects and strata. Nature's text was poetic in another, root etymological sense: nature was creative. It was productive of diversity and of a sort of playfulness in fashioning objects that often wryly "reflected" other strata of nature and that Paula Findlen (1990) has recently called attention to as *lusus naturae*, or jokes of nature.

Hence, the exegetical hermeneutics of nature as symbolic text led to a pronounced affinity for the particular and for the diversity of "things" in the Book of Nature. Such concerns manifested themselves variously: as the pious, empirical quest for knowledge of the hidden powers of individual minerals, plants, or animals in Paracelsian medicine (Pagel 1935, 1982, 1982a, 1984; Webster 1982); or as the naturalist's wonder at capturing the plethora of organic forms and the plurality of meanings they embodied.[44] But, just as the exegesis of texts and language traced a path from the diversity of human languages back to the unity of the creative Word of God manifest in the originary, Adamic language, so, too, the studious attention to the diversity of things in the Book of Nature led back, ultimately and ideally, to the unity of the text of nature. Those who embraced this exegetical approach to nature, the "emblematic world view," found that unity in the poetic, tropological structure of nature's text: in the system of correspondences, similitudes, resemblances that underlay the semantic unity of God's visible Book and that underwrote this entire hermeneutics of nature.[45]

By contrast, narratives of loss, of estrangement from Adam's prelapsarian grasp of the *verbum Dei* and commerce with things, gave narrative authority to hermeneutic strategies that championed diver-

---

nature available during this period. W. B. Ashworth's excellent paper breaks important historiographical ground in insisting upon an "emblematic world view" grounding the practice of at least certain kinds of natural history during the Scientific Revolution. The dimension of narrative of language theory advanced in the present chapter is needed, I think, to situate that emblematic worldview among the variety of hermeneutic strategies then available, or coming into practice. This dimension is also important, I argue below, for attempts to understand changing "worldviews" or discourses during this period. See also Harms (1985, 1989).

44. Again, see works by W. B. Ashworth, Findlen, and Harms cited above. They point to important texts such as Gesner (1551–58); Aldrovandi (1599–1603, 1602, 1599–1668); and Topsell (1607).

45. I shall discuss more fully the interplay between unity and diversity, and the tensions inherent in that interplay, in succeeding chapters, and especially Chapter 6. See also Céard (1980) and Dubois (1970).

sity without underwriting a search for an immanent unity of meaning. Looking toward God's "works" rather than His "Word," some students of nature refused to read the Book of Nature as a text written in symbolic, veiled language mirroring the Word of God and thus embodying concretely the lost Adamic language. Rather, they began to read nature as a text in which God had inscribed, not His deepest mysteries, but merely His scheme for the order and functioning of created things. The language of this text was therefore a language of things, not one of symbols enclosing a vast metaphorical network of meanings. The very metaphoricity of the text of nature experienced a profound shift of meaning, one intimately dependent upon a shift in narrative formulations of God's role as author of the Book of Nature. Within this story of God as creative author of the universe, the virtual identity of author and text—in which "divine idea or archetype" and "thing" mirrored one another—had been challenged and eroded. Instead of a divine author who stamped some fragment of His very image upon individual things in the text of nature, God became an author who distances Himself from His text. As author, God, instead, was figured as free, all-powerful will, as unconstrained by the necessity of a single rational order of divine ideas to create a world that reflected only that one, ideal, rational order. Instead of stamping that divine "image" within the Book of Nature, God as autonomous author may imagine as He will a fictive order, and, through His free act of creation, transform that "fictive" text into the "real" Book of Nature. In short, God as author "inscribes" his text, using His own chosen instruments, in a highly arbitrary, but nonetheless ordered and coherent, language of things. Such a language of things then characterizes the natural order as humans come to experience, manipulate, and know it. But it is always a *contingent* language bespeaking a *contingent* natural order wholly dependent upon God's will. God's Book of Nature is thus an artifice of His writing, His *contingent inscription of things;* as such His works must be read by scrutinizing carefully their details for clues to their internal order. The very diversity and differences marking things must be grasped in order to understand that text, to grasp the language of things inscribed by God. The language of things might be particular and empirical or it might be relational, even mathematical.[46] In any case, this theory of a language of

46. The particularity of nature, and the need to grasp the order of nature through particulars, encouraged an empirical approach to the language of things. One finds such an approach, for example, in Baconianism. A mathematical, or relational, view of nature locates *scientia* not in the grasping of the inherent essences of individual things (e.g.,

things, of God's inscription of nature, valorized a hermeneutic practice of "de-in-scription," or *description*, of nature. Where exegesis championed unity and the turn to words and symbols, the hermeneutics of deinscription gloried in the particularity of things and processes, championing diversity and the turn to empiricism—to God's works.

"Aristotelianism"), nor in the correspondences and resemblances linking individual essences together into an overarching symbolic, but real, network, but, rather, in the system of relations, operations, and actions that constitute the phenomenal world of things. The latter are not, in themselves, knowable as essences—that is, apart from the web of relations in which they are enmeshed. By contrast to the exegetical/symbolic approaches to nature I have described, this latter view of the language of things seeks knowledge of the "grammatical" or "syntactic" dimensions of the language in which God has inscribed the Book of Nature. One example of this kind of approach to nature is Marin Mersenne (see Dear 1988 and Chapter 8, below). Galileo represents, I think, another example (Chapter 6). For a suggestive and important view of the relationship between mathematics and language, see Knoespel (1987).

4

The Priority of the Text

Bookish Culture and the Exegetical Search for Divine Truth (Fernel versus Harvey)

Near the end of his last major treatise, *De generatione animalium*, William Harvey confronts for one final time the vexing philosophical difficulties attending the physiological activity and generation of living organisms (Harvey 1651). Harvey's analysis results in an overt, if somewhat speculative and incomplete, theory of living matter that complements his noted insistence on the primacy and inherent vitality of the blood (Pagel 1967, 1969–70, 1976; Webster 1967a; White 1986). This theory of matter—more precisely, some extraordinarily suggestive musings about living matter—is framed by Harvey's rejection of alternative explanations of vital activity. With force and telling specificity, he informs us that

> Scaliger, Fernelius, & others, having not thoroughly considered the excellent endowments implanted in the Blood, have phansied other Aerial or Aetherial spirits, composed of an Aetherial and Elementary substance, to be a more excellent and diviner Innate heat then blood; which they conceited to be the most immediate instrument of the Soul, most proportionable to all its operations: grounding their opinion upon this opinion; namely, that the blood (as being a substance compounded out of the Elements only) cannot perform any action beyond the sphere or activity of the Elements, and such bodies as are framed out of them. Hereupon They feigned a distinct spirit, and innate heat, which is of a celestial extract; namely, a most simple, most subtle, most thin, swift, lucid, and aetherial substance, partaking of a fift essence. But they have no where demonstrated that there is any such substance, or that it doth act beyond the

power of the Elements, or execute greater things then the blood alone is able to accomplish.[1]

Harvey's conception of the blood thus stands in sharp opposition to Renaissance theories of medical spirits; as a consequence, he starkly contrasts his understanding of vital activity to the ignorance of those who lay stress upon such *spiritus*. Harvey's *Second Essay to Jean Riolan* (1649) leaves no room for doubt about the matter:

> With regard to . . . spirits, there are many and opposing views as to which these are, and what is their state in the body, and their consistence, and whether they are separate and distinct from blood and the solid parts, or mixed with these. So it is not surprizing that these spirits, with their nature thus left in doubt, serve as a common subterfuge of ignorance. For smatterers, not knowing what causes to assign to a happening, promptly say that the spirits are responsible (thus introducing them upon every occasion). And like bad poets, they call this *deus ex machina* on to their stage to explain their plot and catastrophe. (Harvey 1963:149, 1649:59–60)

In proposing his own view of blood and vital activity, Harvey rejects Renaissance medical pneumatology while locating the source of its errors in a flawed methodology. Faults of reasoning, failure to ground entities like *spiritus* upon empirical evidence for their actual existence, and unwillingness or inability to observe the properties of the elements and of such vital matter as blood are all criticisms that Harvey levels against Renaissance spirit theorists (Walker 1958a). As twentieth-century observers we find it frightfully easy to read Harvey's criticisms as *simply* methodological: when one employs, as Harvey did, the Ockham's razor of observation to cut through the obfuscations of metaphysical entities, such dubious medical spirits are naturally banished from physiology and generation. However, though Harvey surely used the *strategy* of observation to undermine the theory of medical spirits, his criticism of such spirits cuts much deeper. His commitment to a conception of the blood as immanently active, which is implicit in the above-quoted passage from the *De generatione*, surely disposed him to criticize Renaissance medical spirits; his quarrel with Fernel, as we shall see, concerns itself with the role of transcendent *versus* immanent

---

1. The translation is from Harvey (1653:448). In subsequent cases, where I have altered the translation I have placed changes within brackets. Since it is more easily available in a modern reprint, I shall also cite appropriate page numbers from Harvey (1847); see Harvey (1847:502–3) for passages cited. For the Latin text, see Harvey (1651:245). In general, see Bono (1990)—an earlier version of this chapter—for *full* citation of Latin texts of Harvey and Fernel. To save space, I have eliminated many Latin passages.

causes in physiology and generation. Yet, underlying both the methodological and the substantive criticisms of *spiritus* theory, I shall argue, are fundamental differences between Harvey and his opponents concerning the nature of theoretical medicine as an enterprise. Thus, when Harvey criticizes the medical spirits and methodology of Fernel he implicitly rejects wholesale Fernel's conception of medical, scientific, and philosophical *language* and with it his conception of the relationship of philosophical and medical truth to the classical tradition and to the interpretation of nature and texts.

Not only do Fernel and Harvey present us with examples of different "scientific" programs informed by radically different theories of language; they also illustrate how contrasting exegetical and deinscriptive hermeneutical practices authorized by such narratively embedded language theories produce distinctive discourses of theoretical medicine with their own technologies of reading and associated "methodological" practices. In this chapter, I propose to examine some of the these differences by focusing upon the contrasting views of Harvey and the sixteenth-century French physician Jean Fernel (1497–1558). Harvey, as we have seen, prominently mentions Fernel in connection with the theory of medical spirits; and, indeed, Fernel is perhaps the most renowned of the academic physicians of his time who wrote extensively on *spiritus*.[2] His theory of medical spirits, moreover, both displayed some unusual features and was of some influence and importance.[3] My

2. For the literature on Fernel, see works cited in Bono and Schmitt (1979) and Bono (1981). Fernel is quite frequently cited as an authority in early modern Europe. For example, the late Scholastic "Coimbra Commentaries" on the works of Aristotle (themselves rich resources of Renaissance medical thought not sufficiently explored by historians) published under the auspices of Iberian Jesuits use Fernel's works extensively. See, for instance, Coimbra (1616:247–57, 593–99). Fernel's works became the subject for a number of learned commentaries by Jean Riolan the elder of the Parisian faculty of medicine (see the "List of J. Riolan the Elder's Works on Fernel" in Sherrington 1946:179) and by Jacques Aubert. At Paris, numerous medical theses of the sixteenth century bear the stamp of Fernel's influence, to judge from the titles of disputed questions preserved in manuscript. Beyond Paris, Felix Platter testifies to the influence of Fernel in his exams for the M.D. at Basel (September 20, 1557): Platter (1976: 308) (I am indebted to Dr. Karen Reeds for this reference). Finally, Fernel's stature is perhaps best indicated by the frequent printing and widespread diffusion of his works. See "List of editions of the writings of Jean Fernel," in Sherrington (1946:187–207). Françoise Lehoux's (1976) detailed study of the quotidian world of sixteenth-century Parisian physicians notes the preeminent centrality and popularity of Fernel's works among contemporary physicians; excepting ancient authorities, Fernel was the best-represented author, save for Girolamo Cardano, in the libraries of the physicians she surveyed.

3. See below, and my own work cited above. Pagel (1984:132–34) stressed the importance of Fernel's notion of *spiritus*.

plan is to expose, briefly, Fernel's varying attitudes toward observation, in order to underline my contention that it is his conception of theoretical medicine as an enterprise, *not* simply his methodology, that is at issue in the contrast between his theory of spirits and Harvey's views. Fernel's methodology differs from that of Harvey. More fundamentally, however, that difference, as well as their disagreements over the theory of spirits, springs from the radically divergent ways in which Fernel and Harvey emplot the foundational problems of theoretical medicine and contextualize historically and culturally the relationship between scientific inquiry and earlier traditions and texts.

I shall then characterize Fernel's understanding of his enterprise by discussing the goals, motives, and philosophical/cultural assumptions informing it. With this background in mind, I shall next summarize those aspects of Fernel's notion of *spiritus* that are pertinent to my argument and turn to consideration of Harvey's criticism of them. Harvey's rejection of the Fernelian spirits springs from a deeply ingrained monistic and Aristotelian perspective; his work first and foremost tackles problems of natural philosophy and theoretical medicine from that perspective.[4] Yet, strengthening and, to a subtle but important extent, transforming that perspective are significant convictions about the nature and purpose of his enterprise. Harvey narratively construes ancient sources and the language of natural philosophy and medicine in a manner totally opposed to Fernel. In this opposition I believe we can emplot Harvey as a reformer of Renaissance medicine and an antagonist to the philosophical and cultural wellsprings of reform imbibed by Fernel. I hope to suggest that Harvey's enunciation of a narrative of reform—a reform that prescribed the adoption of observation as a strategy rather than being itself legitimated by observation—pervades his critique of Renaissance spirits and articulation of an idea of active "biological" matter.

Fernel's Attitudes toward Observation

Approximately a century separates Fernel's major medicophilosophical works from Harvey's *De generatione animalium*. Institutionally, philosophically, and in certain significant professional ways, the differences between Harvey and Fernel are great (Sherrington 1946; Keynes 1966). Yet each, in his own way, attempted to introduce certain unify-

4. On Harvey as an Aristotelian and a monist, see the literature cited in the first paragraph of this chapter and also Schmitt (1984a).

ing and central conceptions into physiology. Each also claimed that observation had a role to play in the investigation of nature, although they construed its significance in markedly different ways not unconnected to their opposed evaluations of the role of spirits and blood.

Fernel, as an eminent professor of medicine at the University of Paris,[5] was fully capable of displaying a clearheaded and perfectly comfortable advocacy of tasks rooted, first and foremost, in observation. Thus, he could advise learned physicians, in his *De methodo medendi*, that "The knowledge, collection, choice, culling, preservation, preparation, correction, and task of mixing simples all pertain to the pharmacists; yet it is especially necessary for the physician to be expert and skilled in these things. If . . . he wishes to maintain and safeguard his dignity and authority among the servant of the art, *he* should teach *them* these things."[6]

Nor is this advocacy of observation merely casual or anomalous. As Guillaume Plancy, his protégé and biographer, tells us in the *Life of Fernel* (Plancy 1946, 1656), Fernel was an avid and insistent supporter of observation, and even active experimentation, in the pursuit of more effective medicaments appropriate to French medicine. Fernel, says Plancy, was of the opinion that "a number of remedies which Nature produces within our own countryside . . . have still escaped our knowledge" (1946: 162; 1656: fol. **4r). Fernel the physician could not rest content while so rich a lode of useful substances ready to be extracted from the earth itself remained unexploited. Hence, for the improvement of the medical art our Renaissance physician proposed that "these home-grown remedies are what call for *active search:* they should be brought into use and their names and virtues placed on record and kept in constant remembrance." Fernel—cut short, Plancy tells us, "by untimely death" (1946:162; 1656: fol. **4r–**4v)—was not to complete this program of collecting, describing, and observing the effects of native botanicals. Yet even while studying traditional medicaments preparatory to writing his *Therapeutics*, he engaged in painstaking observations. As Plancy again reports: "He was scrupulous in settling the right doses of the purgatives, and in examining their action on a number of individuals. He would not leave any state-

5. Sherrington (1946). On Fernel and the University of Paris, see Quicherat (1860) and Concasty (1964).

6. I am indebted to Dr. Karen Reeds for pointing out this passage, which she has translated in Reeds (1975:40). The Latin text (from Bauhin 1622) appears in Reeds (1975:233 n. 77).

ment of their action which had not been ascertained by repeated trial"
(1946:162; 1656: fol. **4v).

Despite the centrality of observation to pharmaceutical knowledge
and to practical aspects of medical therapeutics, when we turn to
Fernel's works of theoretical medicine and natural philosophy we find
that its role appears quite restricted. Indeed, Fernel frequently cites
received observations. There are, of course, a number of conceivable
ways to account for these very different emphases on observation. One
might, for example, appeal to the sharply distinct nature of the enter-
prises involved in each area with which Fernel concerned himself. The
knowledge of botanicals, of drugs made from them, and of therapeutics are all forms of *practical knowledge*—of *praxis*—in which the
very content of what is known is inseparable from the operations used
to acquire that knowledge. By contrast, such works as Fernel's
Physiologia and *De abditis rerum causis* aim for knowledge that is *theoretical* in character.[7] Consequently, observation need not be nearly so
important—provided that an acceptable groundwork of fundamental
data already exists for raising such a theoretical structure. Fernel *does*
come at the end of a long tradition of Galenic theoretical medicine and
Western natural philosophy. His theoretical work, accordingly, has
been viewed as synthetic and conservative, rather than innovative and
tradition breaking: in short, the kind of theoretical enterprise one
would expect to minimize the role of observation.

However strong the temptation may be to accept the slender role
Fernel gives to observation in theoretical medicine as the predictable
artifact of a conservative, premodern mind, that temptation ought to
be resisted vigorously. For one thing, Fernel's orientation, though tra-
ditionally Galenic, demands more nuanced a reading than the anachronistic label of "conservative" provides. Then again, however correct
the above characterization of Fernel's understanding of the role of
observation given above may be (and I do think it largely accurate),
the truly significant historical question remains to be asked: What atti-

7. In the "Praefatio" to book 2 of his *Physiologia* (the text of this preface differs considerably between the original 1542 edition of the *De naturali parte medicinae* and the later 1554
text; see the following note for bibliographical details), Fernel explicitly defends the appropriateness of relying upon reason when seeking such theoretical knowledge, or
knowledge of causes, in medicine (see also Fernel's *Pathologia*). This, of course, is a commonplace notion in medieval and Renaissance medicine and philosophy. What is of
added interest, I think, is the degree to which Fernel's analytic practice is grounded in
conceptions of language, reading, and interpretation. This chapter focuses on the latter
rather than on Fernel's explicit remarks about reason. Of interest too is Harvey's rejection
of the hermeneutical assumptions of Fernel's practice.

tudes, motives, and commitments find expression in this characterization of observation's role? Put this way, Fernel's use of observation in theoretical medicine becomes problematic, for it raises the possibility that his enterprise as a whole is not simply "conservative" but rather rooted in problems specific to his various professional and cultural contexts.

Indeed, one should take the sharply contrasting views of Fernel and Harvey regarding observation as symptomatic of their respective attempts to rework and reconstruct the quite different contexts, assumptions, and underlying problems within which each worked. One can no more explain the richly empirical content of Harvey's work by appealing to an enlightened, "modern" scientific mentality pervading his thought than one can account for Fernel's relative disregard of observation by appeal to his traditional and "conservative" cast of mind. Neither attitude toward observation is more "natural" than the other; both are historically conditioned and local responses to specific contexts and problems. Historians need to uncover and explore these contexts and problems. In approaching that task, I wish to stress that what has changed between Fernel and Harvey is not simply a scientist's evaluation of the adequacy of traditional observational evidence, but, more fundamentally, the very configuration of hermeneutic strategies and practices, and with them conceptions of *what* theoretical work in the medical sciences ought to be, what sorts of analyses it entails, and what species of fruit it may, eventually, bear.

Fernel, of course, specifically disavows the centrality of observation to theoretical aspects of medicine, which seek the rational causes of things. Nonetheless, his *Physiologia* and *De abditis rerum causis* do provide observational accounts.[8] These accounts include a wealth of traditional, even formulaic, appeals to experiential data, a limited reliance on novel observations, and a not infrequent use of casual, common-sense data. One finds, for example, extended use of a modified, analogically amplified version of the Galenic lamp metaphor in Fernel's discussion of innate heat and spirits in relation to the fleshy parts of the body.[9] To appreciate the significance of Fernel's methods we must, how-

8. Fernel (1542), namely *De naturali parte medicinae*, was later renamed *Physiologia* (Fernel 1554) when published as part of his monumental *Medicina* (Fernel 1554a). His *De abditis rerum causis* (Fernel 1548) is of philosophical, cosmological, and medical significance. I have used the Venice edition (1550).

9. For a discussion of this metaphor, see the chapter entitled "Spirits and Physiology" in Bono (1981). See also Hall (1971); McVaugh (1974); and Niebyl (1971). I have discussed this topic at greater length in Bono (forthcoming).

ever, look beyond such uses of "experience" to the very aspirations underlying the program and content of Fernel's theoretical medicine and natural philosophy.

Fernel and the Idea of Reform

These aspirations may be summed up in one word: reform. The overriding concern, the single feature dominating the cultural landscape sketched by Fernel in the course of his medical career, was the idea—and ideal—of a reform of medicine.[10] It is easy to misconstrue the nature of this reform. Sir Charles Sherrington's book on Fernel captures only a partial aspect of his reforming spirit. For Sherrington, Fernel embodies the very spirit of the Renaissance in revolt against the Middle Ages and, more to the point, against unswerving fidelity to authorities. The image of this sixteenth-century medical reformer that he draws for us is that of a modern believer in "progress": Fernel's endeavor, in this interpretation, was to free medicine from the taint of irrationality and superstition associated with astrology, and to lay a solid groundwork for the growth of an ideal of medical progress through reason and experience. Throughout his career Fernel increasingly approached that ideal, leading Sherrington to conclude that, "[i]n effect, Fernel was among the relatively few, who, as they enter advanced years, grow more modern" (1946:52–53).

Sherrington has rightly noted Fernel's abhorrence of the *mere* imitation of, and subservience to, authorities. Yet the kind of "progress" hoped for in Fernel's *De abditis rerum causis* of 1548—from the prefatory epistle of which Sherrington draws support for his interpretation—does not entail the outright rejection of so-called authorities in favor of a brand of "modern" scientific individualism and empiricism.[11] Indeed, for Fernel the progress witnessed in his age can be extended to medicine and the other arts and sciences only through a careful and critical *cultivation* of authorities—particularly antique authorities. This cultivation of authorities demands attention to the texts of ancient authors, to their language and contexts.

10. I have discussed Fernel's "reform of medicine" and its relation to broader currents of reform in a chapter entitled "Humanism, the University, and the Crown in Sixteenth Century France: The Roots of Fernel's Reform of Medicine and the Galenic Language of Life" in Bono (1981).

11. Sherrington (1946:16–17) places great weight on a passage translated from the preface to Fernel (1550: fols. 1r and 1v). Citing Fernel's words, Sherrington shows him to be a defender of the "moderns." He then paints a portrait of Fernel as a "modern" opponent of ancient authorities, and especially of astrology.

While Fernel aspires to reform medicine and natural philosophy through a renewed engagement with ancient traditions, the ideal of reform he has in mind is not simply one of restoration. To be sure, it includes restoring a purified understanding of the literal sense of the texts involved. But, in addition, Fernel seeks to uncover what lies hidden or is only obscurely revealed: he seeks from the texts, in other words, a *renewed* meaning possible to one possessing the light of truth through Christ (Fernel 1550: fol. 3r). Not possessing such light, the ancients could not grasp wholly the meaning inherent in the truths they did articulate. Hence, Fernel seeks to liberate the "true" hidden meaning embedded within their texts from the hardened understandings of received traditions. What Sherrington saw as the liberation of the human spirit for the great adventure of science, was in reality Fernel's aspiration to disengage medical and natural philosophical language from the confines of pagan frameworks in order to discover Christian evidences of transcendent, divine activity (Fernel 1550: fol. 2v ff.).

At this point Fernel's aspirations—his ideal of reform—and the methods underpinning his theoretical medicine and natural philosophy converge, linked by his attitudes toward language, particularly the language of medicine, itself embodying the language of natural philosophy. For Fernel, the received "languages" of natural philosophy are all *deficient*, and hence the scorn that Sherrington notes in Fernel for mere imitators and slavish followers of authorities, antique and modern. Yet these received "languages" are deficient only as *incomplete* revelations of the nature of things. The adequacy of previous medical and natural philosophical traditions rests in the degree to which they properly embody the signification of key terms. (One such key term was *spiritus*, a term whose Fernelian signification Harvey adamantly opposed in the following century.) Fernel regarded many key terms as notoriously, even *intentionally*, obscure. In effect, then, the received languages of natural philosophy function not as a system of univocal signs, but rather as a system of "symbols" concealing the true, but only partially revealed, nature of the things signified.[12]

Given this attitude to "scientific" languages, one can readily specify both the direction and methods of Fernel's reform. Two goals define the task of medical reform: the restoration of the knowledge of ancient

12. See "Praefatio" for Fernel's notion of "veiled" philosophical language, and the need to study systematically and reinterpret the major traditions of ancient medical wisdom.

medical and philosophical texts to Fernel's contemporaries, and, beyond that, the interpretation of those texts' literal statements in the
light of Christian understanding in order to illumine the true sense of
the pristine wisdom spoken through their languages (1550: fol. 3).
When, for instance, Fernel's concern turns to the relations between
soul and body, or the influence of the celestial order upon mortal, elemental things, he finds the ancient Academic, or Platonic, philosophers most helpful. He specifically cites such followers of Plato as Nu-
menius, Philo, Plotinus, Iamblichus, and Proclus as having touched
upon the magnificence of "divine things."[13] Yet, even Platonic philosophers remain obscure, requiring Christian interpreters such as John,
Paul, Dionysius, and Hierotheus (the teacher referred to by Pseudo-
Dionysius' writings) to determine the true sense of their doctrines
(1550: fol. 3r).

The problem consequently articulated for theoretical medicine by the
task of reform highlights the quest for understanding, for *meaning*. Observation, in this context, can offer little support as an effective
archeological tool for chipping away the encrustations of several distinct strata of understanding. Instead, language itself becomes the
"tool" of "scientific analysis," a tool best utilized when subjected to
interpretation. Fernel, indeed, subjects received medical and natural
philosophical languages to close historical/philosophical *"exegesis,"*
and he grounds such methods, or hermeneutical practices, in a familiar
legitimizing story. Thus, he seeks to recover the true signification of
terms as they once appear in a lost, originative language—a *prisca
scientia*. Such narrative reconfiguring of his practices fuels Fernel's
hopes to *revive* medicine, to parallel the "rebirth" of the other arts and
sciences in his own day, and to allow mankind a glimpse of those "hidden" causes of things that are the foundations of true *scientia* and true
piety.[14] Fernel thus emplots the reform of theoretical medicine as an
unfolding narrative in which a pristine, divine language of nature once
present to man and now lost may again be uncovered through the re-
finement of an exegetical hermeneutics and its application to ancient
texts.

---

13. Fernel (1550: "Praefatio," fol. 3r). Fernel's concerns are evident throughout the
"Dialogues" and in much of his *Physiologia*, where he notes the limitations of antique
traditions together with their potential value to an illumined Christian understanding,
and stresses the significance of Platonic philosophy.
14. Again, Fernel's "Praefatio" (1550) best presents his notion of reform and revival
through seeking hidden wisdom, though the text as a whole presents evidence for such
ideals.

Fernel's ideal of reform, his exegetical approach to obtaining knowledge, and his wedding of *scientia* to piety and the search for divine things in nature acquired special significance and urgency in the context of sixteenth-century France. The humanistic reform of learning closely linked itself in France with reform of intellectual traditions and institutionalized disciplines of research and teaching within the university. During the early part of the century, in particular, the efforts of the great humanist Jacques Lefèvre d'Etaples to reform the arts curriculum at the University of Paris dominated the cultural landscape. Drawn toward the syncretistic and irenic tendencies of philosophers like Giovanni Pico della Mirandola and Marsilio Ficino, Lefèvre sought to renovate the philosophical traditions of the university. He initially centered his program for reform upon the revitalization of Aristotelian studies, but his interest was increasingly drawn toward the pursuit of magic and the Cabala as adjuncts to the higher intellectual realms of knowledge (*scientia*) and wisdom (*sapientia*) in the "divine" philosophy of Plato and the Neoplatonists. Indeed, following the paths opened to him by his chief spiritual mentors—Pico, Nicholas of Cusa, and Ramon Lull—Lefèvre's career progressed from university reform and an irenic, eloquent Aristotle to the reform of Scriptures, patristic studies, and spirituality: that is to say, to a kind of evangelical, Christian humanism.[15]

The starting point of Lefèvre's journey nevertheless was, and remained, Aristotle—not the corrupted Aristotle of Scholasticism, but the purified, eloquent Aristotle of French humanism. Lefèvre maintained that the contentious sophistry and verbal wranglings of Scholastic commentators had obscured Aristotle's true wisdom and eloquence by fragmenting the texts into so many isolated bits that their essential harmony and message were lost. The purity of Aristotle's language, the clarity and unity of his texts, were essential ingredients in Lefèvre's conception of an eloquent Aristotle. However, Aristotle's true eloquence had to be grasped not in the externals of the texts but in their core, their meaning, the exalted ends and virtue toward which they exhort us. By exhorting man to virtue Aristotle also prepared the way for a truly Christian life: one filled with piety. For Lefèvre, then, a restored Aristotle bespoke a vision of the universe and of man that instilled a love of higher, superterrestrial and celestial things (E. Rice 1970).

This vision of an eloquent Aristotle had its roots in a certain narrative

15. For background and references, Bono (1990: esp. 353 n. 29).

ideal of reform—*renovatio*—that enjoyed widespread currency in France during the first half of the sixteenth century. Like Ficino in Italy and his compatriot Symphorien Champier in Lyon, Lefèvre believed that the "rude" philosophy of the ancient Greeks had been touched and at least dimly illumined by the sacred revelations of a divinely originating, esoteric, oral tradition. For Lefèvre, this tradition of *prisca theologia* left its traces upon the greatest and most wise of Greek thinkers, *the* philosopher, Aristotle.[16] Lefèvre believed that Aristotle's work led from human, sensible philosophy to theology—that is to say, to the contemplation of divine, immutable things, to the ground of all truth and being. More important, Lefèvre infused this hierarchical conception of Aristotle with Christian significance, viewing the philosopher's work as the starting point of a spiritual journey that led from the best, "most divine" of ancient thought through the theology of the Church Fathers to the theology of Christian Neoplatonic mystics, the esoteric exemplarist theology of Ramon Lull, and the negative theology of Nicholas of Cusa. Ultimately, the chain of divine wisdom whose first link was Aristotle ended in the very *verbum Dei* of sacred Scripture itself.[17] Hence the reform of Aristotle—the humanist interpretation of the "eloquent Aristotle"—was closely tied, in Lefèvre's circle, to the reform and vitalization of religion and to a renewed spirituality grounded ultimately in the Bible and its interpretation.[18]

16. In his own day Lefèvre was praised for having transformed the "rude" inherited philosophy: "primus enim apud Gallos (ut Cicero apud Romanos) philosophiam rudem adhuc et impolitam cum eloquentia iunxit" (Rice 1972: 288). For the tradition of a *prisca theologia*, see the fundamental study of Walker (1954). For Ficino, Pico, and the Renaissance notion of an ancient, occult revelation, see Chapter 2, above. On Champier, see Copenhaver (1978). Lefèvre himself was cautious in appropriating the *prisca theologia* and the esoteric, magical tradition often associated with it, as Walker has noted. Nonetheless, Lefèvre was attracted to the hermetic corpus and was responsible for publishing a number of volumes, including commentaries on the *Pimander* and *Asclepius* (for details Rice 1972:539). See also Rice (1975). On a divinely inspired Aristotle, see Rice (1970:139–42).

17. See Lefèvre to Leonard Pomar, reprinted in Rice (1972:94–97). Rice (1972:xv–xvi) quotes Lefèvre's commentary on Aristotle's *Politics*. See also Lefèvre's discussion (1972:354–58) of the three kinds of philosophy ("mathematicum, physicum, divinum"), where he stresses the importance of Aristotle to those who seek understanding of God and "divine things."

18. For the importance of Lefèvre's emphasis upon Aristotle as a fitting introduction to the quest for Christian illumination, the works of his disciple Charles de Bovelles (Carolus Bovillus) are noteworthy. Unlike Lefèvre who increasingly turned to a public, reform-minded, evangelical Christian humanism, Bovelles chose the quiet path of contemplation and mystical theology. The examples of Lefèvre and Bovillus illustrate how common assumptions about Aristotle, antiquity, and Christianity can lead to quite different, though not necessarily incompatible, cultural programs and practical activities. We

Unlike Lefèvre, Fernel centered his reformist efforts upon medicine. He was not led in the direction of evangelical Christianity, to a mystical reform of theology and ecclesiastical practices such as is associated with Lefevre's later career and with the activities of Bishop Guillaume Briçonnet at Meaux.[19] Still, the spirit of Lefèvre's reformist mentality, the vision of antique philosophy and learning that it encouraged, and the narratively grounded ideal of a Christianized, renovated understanding of ancient texts based upon an appreciation of the hidden, "divine things" contained in them appealed to Fernel. His own work, after all, was itself a search for the "divine" in medicine, in disease: such an epic quest would reach its denouement in a renewed Christian society, a renovated French cultural and political life in which wisdom and piety would be joined in a manner befitting the Most Christian King and his divinely ordained *imperium*.[20]

The Theory of Medical Spirits: Fernel

Fernel's reformulation of the traditional notion of medical spirits exemplifies, I believe, his epic aspirations and hermeneutic practices. Substantively, Fernel's reform of theoretical medicine entailed a reworking of the "natural part of medicine"—that is, that part of medicine dealing with the *nature* or constitution of the animate body, what he came to call "physiology."[21] His *Physiologia*, first published in 1542, gives a prominent place to the theory of medical spirits.[22]

The theory of medical spirits was an inheritance from ancient traditions transformed by medieval Christian natural philosophers, who read their Aristotle and Galen through the lenses of medieval Arabic

---

should not be surprised, then, to find Fernel adopting a similar hermeneutics with respect to Aristotle, Galen, and other ancient authorities while adapting it to somewhat different ends and avoiding the evangelical fervor of a Lefèvre and the potentially quietistic, mystical impulses of a Bovelles. On Bovelles, the definitive work is Victor (1978).

19. On Lefèvre and Briçonnet, see Bono (1981, 1990) for citations.

20. For discussion of the role played by cultural and historiographical myths in shaping ideals of intellectual, religious, and sociopolitical reform in France during the reigns of Francis I and Henry II, and for bibliography, see Bono (1981).

21. The division of medicine into natural, nonnatural, and contranatural parts was a commonplace in premodern Galenism, which we need not discuss here. For the Galenic background, see Bylebyl (1971); Jarcho (1970); Rather (1968).

22. See especially book 4 of the *Physiologia*, entitled *De spiritibus & innato calido*. Fernel's theory of medical spirits is extensively discussed in Bono (1981: chap. 1).

and Christian Platonic systems of thought.[23] *Pneuma,* according to Galen, derives chiefly from the air taken into the body through inspiration, and from the blood (though this source of Galen's *pneuma* tends to be ignored) (Temkin 1951, 1977; Wilson 1959). It is, consequently, of a corporeal—that is *material*—nature, though the matter in question is of an exceptionally fine and rarefied sort, rather like hot vapor. The composition of *spiritus* makes it apt for functions deemed unsuitable to ordinary matter, with its heaviness and consequent sluggishness and lack of penetrability—such as functions associated with the senses and the nerves. Galen gives a special name to the spirits found in the brain and nerves: these are called "psychic" or "animal" spirits (*pneuma psychikon*). He refers also to "vital" spirits (*pneuma zotikon*), chiefly emanating from the heart. Later Galenists attributed a third type, "natural" spirits (*pneuma phusikon*), to him as well. All three spirits serve quite specific functions in the body, but otherwise Galen seems to attach little significance to them—they are merely another material instrument of the body (not unlike other "organs"), peculiarly well adapted to perform some functions but not others (Bono 1984: 92 n. 4).

The philosophers and physicians of the twelfth and thirteenth centuries, however, began to regard spirits—and especially the vital spirits—in a new light, as the primary and the essential instrument of life. By "reading" the languages of ancient medicine and natural philosophy through the transforming lenses of other discourses—through the "language" and discursive practices of theology and biblical revelation—they transformed these inherited traditions. Powerful metaphors and analogies—such as the biblical tropes of the spirit and flesh, *caro* and *spiritus,* and of Jacob's ladder—worked to unsettle and reform the signification of medical terms like *spiritus.* As a result, *spiritus* now became the *medium* of life: the tool whereby the body's immortal soul produced and sustained life. Nevertheless, for the mainstream of biological and medical authors up to the time of Fernel such spirits were still material, a subtle form of air, breath, or vapor, as they had been for Galen. Fernel transformed the role of medical spirits further still.

Throughout his discussion of *spiritus* Fernel is at pains to elucidate its true nature and significance. Little effort is taken to prove its existence; rather, this is regarded as a necessary requirement for the ac-

23. For fuller discussion of this process of transformation, see Bono (1984). See the bibliography there and in Bono (1981: chap. 2). The following two paragraphs are drawn from the latter.

tion and diffusion of the life-giving innate heat. As Fernel states—based upon intricate arguments we shall not consider here—the innate heat is *not* present in the body's fleshy parts as something connatural; on the contrary, it is something adopted and extrinsic. This makes it necessary to explain *how* this nonelemental, innate heat can be present in each part of the body and also how it can be diffused throughout the body as a whole. Both problems require that some vehicle contain the vital heat. In addition, to satisfy the requirements of motility and diffusion, the vehicle itself must be fluid (Fernel 1554:104). For Fernel no bodily humor is suited to this latter task; none can traverse the whole body with the great speed required. Only the traditional *spiritus* can contain the innate heat, provide its requisite locus in the individual solid parts of the body, and act to distribute it to the entire body.[24]

Interesting as the relationships among *spiritus*, innate heat, and the parts of the body may be—and Fernel treats them extensively in his *Physiologia*—at this point it is more instructive for us to consider Fernel's methods and motives for articulating the *nature* of *spiritus*. Essentially, he resorts to historical precedent: through a critical examination of traditional language and its implications, he offers a "true," synthetic interpretation of *spiritus*. Hippocrates, Plato, and Aristotle all speak of a kind of aetherial, invisible substance in conjunction with their discussion of important, life-sustaining, organic powers. They may not always, or clearly, attribute such powers ultimately to that aetherial substance, but their language, however dimly, does place them in close conjunction. Indeed, the very terms themselves—*aer* or *spiritus*—hint at a substance like the one Fernel has in mind (1554:104). Following these paths of ancient learning, Fernel then presses beyond the linguistic coinages themselves to plumb the nature of the substance underlying and enlivening them. Again his procedure is to examine, then reinterpret, the language of earlier authorities. Following the lead of Plato and Aristotle, two chief schools employed the term *spiritus* in their philosophical language of nature. The Stoics gave central prominence to a brilliant, transparent, and invisible substance called *spiritus*, while claiming it as the substance of nature and the soul. By contrast, the Academic, or Platonic, philosophers did not identify nature and the soul; they regarded man's natural body as perishable and the soul as

24. Fernel (1554:104). Chap. 2 identifies *spiritus* as just this sort of aetherial substance. In the rest of book 4, Fernel discusses the relation of *spiritus* to innate heat, the solid parts of the body, etc.

immortal. Hence, for these philosophers, *spiritus* was a substance intermediary between soul and body—a *tertium quid*.

From Stoic to Platonic natural philosophies Fernel envisions a gradual clarification, a partial stripping away of veils of human ignorance, in search of the true, hidden meaning of *spiritus*. The Stoic *spiritus* crudely identifies it with the rest of nature, whereas the Platonic understanding introduces the notion of transcendence and acknowledges the ontological gulf separating body and soul. Following the latter, Fernel interprets *spiritus* to be precisely the substance that enables the soul to inhere in the parts of the body as they are formed and to remain in its corporeal prison throughout its tenure on earth.[25] *Spiritus* is not then simply a splendid, aetherial substance which happens to be the vehicle of life-giving heat in the body—its meaning is much broader, its significance only suggested by that formulation. Fernel's *spiritus* assumes great importance in theoretical medicine and natural philosophy as *the* unique substance ontologically wedding body and soul, and, thus, the primary instrument of *all* organic functions (1554:105). The *meaning* of *spiritus* lies in its role as the key to relations between the divine and the created, material order of *things*. In the microcosm, man, it is *spiritus* that allows something divine to enter into the workings of the organic body—that allows the body to act with an efficacy *beyond the power of the elements*.[26]

Fernel's theory of spirits consequently presses the medieval view of their importance to an extreme. They are for him more than an essential mediator of life: spirits act upon the body with a power that the latter's merely elemental composition is incapable of producing. In his words, *spiritus* is "an aetherial body, the seat and container of the innate heat and faculties of the soul, the primary instrument for the execution of organic functions" (1554:105). Spirits are *the* active agents and *direct* causes of *all* vital functions.

The central, all-encompassing nature of medical spirits prompts Fernel to place great importance upon his distinction between "wander-

25. Fernel (1554:104–5). For the full Latin text see Bono (1990). This passage was highlighted and translated in Walker's fundamental article (1958a:119). For further discussion of spirits and astral bodies see Walker, and Bono (1981).

26. As we have seen, Harvey in *De generatione animalium* later attacks Fernel for having recourse to a mysterious substance acting with an efficacy beyond the power of the elements. This charge certainly seem justified, as Fernel (1554) explicitly attributes the functional properties of different organic structures not to their material composition, but, rather, to distinct and stationary spirits dwelling within the various solid parts. See especially chap. 10 (1554:119–20). Fernel also explicitly regards *spiritus* as a divine substance throughout book IV.

ing" and "fixed" spirits. The former are the traditional animal, vital, and natural spirits associated with the three chief organs, the brain, heart, and liver. These traditional spirits are fluid, wandering throughout the body as a whole. The fixed spirits, by contrast, locate themselves in certain specific "sites" or "seats" within the organic body (1554:119). Fernel links this notion of fixed spirits to a detailed analysis of the various substances that, together, compose the organic body. A solid substance provides strength, support, and structure; it manifests itself as the fibers, or characteristic "lines" or "threads," that ramify throughout the body. A fleshy substance provides bulk to the body. Finally, Fernel locates a specific "spirituous substance" within the solid substance. Contained within those very fibers which give structural strength and support to the body, this spirituous substance—the "seat" of the fixed spirits—is the creator of all functions within the body (1554:109–10). Underlining the centrality of spirits still further, Fernel adds one more characteristic to his notion of fixed spirits: these fixed spirits are themselves variegated and *specific* in their nature and functions. That is to say, Fernel insists that each solid part of the body has its own characteristic spirit which, in turn, is the source and cause of the organic functions of that part only and no other. Not only, then, are Fernel's spirits the essential mediators of life, they are the very cause of life functions within the body (1554:119).

In his complementary medicophilosophical work, *De abditis rerum causis* (1548), Fernel draws microcosm and macrocosm—man and universe—together analogously into a similar dependency upon divinely originating power. Drawing upon Aristotle's dictum that the frothy matter of semen contains a kind of *spiritus* whose nature is analogous to that of the stars,[27] Fernel claims that terrestrial substances receive the "divine" forms of things *via* the agency of the stars and, specifically, through the mediation of celestial spirit. Hence, in addition to the elemental constitution of natural things, including man, there is also a divine substance, or *spiritus*, whose activity in nature and in the body is "hidden" or "occult."[28]

---

27. Aristotle, *De generatione animalium*, book 2, chap. 3 (736b35–737a1); Fernel (1554:105); Fernel (1550:187).

28. In book 1, Fernel faces the vexing problem of how passive matter was made active; the problem is resolved through demonstrations of the mediating role of celestial spirits. The role of *spiritus* in preparing matter to produce activity is stressed continually; see, for example, Fernel (1550:54 [55] ff.). The link between the stars, *spiritus*, and "divine" forms of things is also emphasized throughout Fernel's work; see, for example, the following sections from book 1, chap. 8, "Rerum omnium formas, primasque substantias de coelo duci, ex Aristotele" (1550:101–8); and chap. 9, "Aristoteles supra

Fernel's method—his hermeneutic practice of "exegesis"—
together with the relatively minor role accorded to observation,
reaches its full flowering in this work, as he appeals not only to Aristotle, Plato (especially the *Timaeus*),[29] and the Stoics, but to a host of
other classical authorities, such as Vergil, and Christian sources, such
as the Pseudo-Dionysius and sacred Scripture.[30] In his preface, Fernel
speaks of ancient philosophers as touching upon or approaching divine things. Their ideas, however, present themselves in some respects as mysteries—their truths obscured, as it were, by veils.
Hence, one must scrupulously interpret their *sententiae* to reveal the
truths they contain. Perhaps most telling of all, Fernel's *De abditis
rerum causis* is itself an interpretation of the Hippocratic saying that
there is something divine in disease, for his concern in these "Dialogues" is to fathom the "divine" meaning in Hippocrates.[31] Once
again, the word *spiritus* provides the key to reinterpreting veiled, ancient wisdom. Fernel believes there is indeed something "divine" in
disease. While most diseases are the result of humoral imbalances, a
number of "occult diseases" (plague, contagion, pestilence) have
quite different causes. The substances causing such occult diseases

---

naturam philosophatus statuit formarum divinam originem: Deum maximum coelos
astraque condidisse, hisque gignendarum rerum vires indidisse: quae etiam Platonica
sunt sacris literis consentanea" (1550:108–20). Finally, Fernel's treatise highlights "occult causes" and the "hidden" nature of *spiritus*, especially book 2; see, for example,
chaps. 7 and 8 (1550:186–203).

29. Fernel regards Aristotle and Plato as authoritative sources for cosmology, emphasizing the presence of a divine *spiritus* as source of transcendent activity; see, for example, Fernel (1550: book 1, chaps. 8 and 9). He utilizes Plato's *Timaeus* to evoke an image of
the universe created and controlled by God (1550:113). Fernel regards the ancient Aristotelian tradition as following this viewpoint (1550:121). Note, too, that the dictum "Sol et
homo generant hominem," from Aristotle's *Physics*, book 2, chap. 2 (194b14–15), is
quoted by Fernel (1550:100) and used as locus classicus for celestial influence upon the
sublunar world.

30. For example, see Fernel (1550:123). In addition, Vergil's *Aeneid* 6. 724–32 provides
a crucial text for Fernel. Finaly, Fernel is at pains to direct his discussion of pagan philosophy and poetry toward an understanding of Christian wisdom. Thus, toward the end of
book 1, he makes increasing use of Pseudo-Dionysius and Scriptures in chapters 9, 10,
and 11.

31. Fernel (1550: fols. 3r–3v). The dictum Fernel alludes to derives from the Hippocratic work *Prognostica*, which includes the line "ἅμα δὲ καὶ εἴ τι Theîon ἔνεστιν ἐν τῆσι
noúsoisi" in the introductory section. The Loeb edition eliminates this line from the
Greek text, including it only as a footnote (Hippocrates 1923:2.8); nonetheless, ancient
and medieval authors cite and comment upon this line. See bibliography in Bono (1990).
Unlike Galen, Fernel uses this saying as justification for divine and celestial influence in
medicine.

do not act by virtue of their manifest, qualitative properties; rather their very *essence* opposes itself to the body's "total substance" acting by means of hidden virtues conveyed by the "divine," celestial spirits.[32]

The Theory of Medical Spirits: Harvey

Harvey's views of spirits stand as quite opposed to Fernel's and illustrate well the change in hermeneutic practices and consequently in theoretical orientation—in assumptions, aspirations, and methods—that, I would claim, Harvey marks in Renaissance theoretical medicine. His most pointed attacks on Fernel and *spiritus* appear in two late works: *The Second Essay to Jean Riolan* and *On the Generation of Animals*. In the former, Harvey, as we have seen, criticizes the *use* to which medical authors have put the spirits: holding dubious, even conflicting, views of their nature, these authors—"like bad poets," according to Harvey—use spirits as their *"deus ex machina"* to explain all happenings in the theater of their little world, man (Harvey 1963:149). Immediately after this condemnation, Harvey refers to Fernel by name. His criticism in the *De generatione* is still more explicit and decisive: he berates Fernel for ignoring the blood's own "excellent endowments" in favor of an imagined *spiritus*. Because he regarded the blood as simply elemental in composition, Fernel erroneously concluded that it "cannot perform any action beyond the sphere or activity of the Elements," and he had recourse to a celestial *spiritus* whose nature could account for the animate, superelemental properties of living things. Yet, Harvey dryly comments, nowhere has Fernel proved the existence of *spiritus*, or its superiority to the elements and to blood.[33] The contrast between the theories of the two men is stark.

Where Fernel's viewpoint is "animistic" and dualistic, importing transcendent causes into physiology, Harvey's (as Pagel has emphasized) is vitalistic and monistic.[34] Moreover, while Fernel's views arise from the exegesis of the symbol-laden language of traditional medicine and natural philosophy, Harvey's, it appears, derive from the careful use of observation to cut through the mystifications foisted upon physi-

32. The culmination of Fernel's exegetical musings upon the "divine" in disease occurs in the middle chapters of book 2 of Fernel (1550); see especially chap. 10, "De morbis totius substantiae . . ." (1550:214–23), chap. 11, "Occultorum morborum differentiae" (224–29), and chap. 12, "De pestilentibus morbis quod illorum occulta sit causa" (230–46). Also, see Richardson (1985) and Nutton (1983).
33. See note 1, above.
34. See, among others, works by Walter Pagel.

ology by imagined entities like *spiritus*. Such methodological sophistica-
tion, such judicious use of observation, we usually assume to have
resulted from the influence of the Renaissance anatomical tradition
upon Harvey. However, I believe that this explanation credits the anatomical tradition with more power to change fundamental conceptual
assumptions than is warranted. Observation was not so much a cause
of Harvey's changed viewpoint as a tool useful in constructing supports for it.

Hence, we should note that Harvey's concern in these passages has
less to do with fostering an ideal of observation than with the task of
discovering an adequate language of natural science through reconstructing *words* as true *signs* for things. Like Francis Bacon, whose
Novum Organum preached the power of observation and experiment
as a new instrument for overcoming the abuse of human language in
the study of nature, Harvey considered observation to be a *means* dic-
tated by the perception of a fundamental problem in natural philosophy and medicine. That problem no longer has at its core the mere
exegesis of received, traditional languages of natural philosophy and
medicine; rather, the locus and source of truth and meaning has
shifted from the past to the present. Unlike Fernel, Harvey no longer
implicitly regards those received languages, however corrupt, as
linked historically to a distant, but recoverable, divine, Adamic language of nature. Since the past for postlapsarian and post-Babylonic
man is no longer a privileged guide, Harvey, again like Bacon, turns
to the present for instructions and to the task of recreating language
through *naming* things.[35]

Bacon's program is also, like Fernel's, one of *restoration:* but, in evoking an ideal and original state of human knowledge and existence,
Bacon—as we shall see in Chapter 7—finds himself drawn not to pagan antiquity, nor to any historical, literate period, nor to surviving
fragments of a "lost" original language, but, on the contrary, to the
prelapsarian state of an Adamic earthly paradise. This difference between Fernel and Bacon is a radical one.

Harvey, like Bacon, rejects the oracular nature and veiled wisdom of
ancient texts, instead bearing witness to an important cultural shift that
concerned the nature and adequacy of traditional languages. Such a
shift, while emphatically not the consequence of questioning the adequacy of traditional observational data, did, however, tend to legiti-

35. For Harvey, language, and things, see below. For Bacon, see Chapter 7, below,
and *The Great Instauration* and *Novum Organum* in F. Bacon (1857–74): 4.7–33, 39–248.

mate the strategic use of observation in the de-in-scription of nature. The shift exemplified by Harvey's work had its roots in a number of critical developments during the Renaissance and seventeenth century which we surveyed in the preceding chapter. Where precisely Harvey fits into these late Renaissance developments is not a question that I can answer in the present chapter, in part because I have chosen to focus only upon Harvey's late works where his most explicit statements on language occur and where one finds quite full discussion of Fernel and spirits. Full assessment of Harvey's place must take into consideration the entire range of his works, both early and late, and also build upon the kind of historical investigation of the interdependence of language theory and science during the sixteenth and earlier seventeenth centuries attempted in this book.[36]

However, even apart from such full contextualization of his views, I believe that Harvey regards the received languages of natural philosophy and theoretical medicine as not just deficient, but *defective* as well. They are not somehow rooted in a transcendent revelation, nor may pious exegetes purify them through the light of Christian wisdom alone; rather, they are first and foremost historically contingent human artifacts. Harvey, of course, uses the "language"—the terms and "names"—of the Galenic and Aristotelian traditions. But he does not invest that language with "veiled" and transhistorical meaning. The words used by ancient authorities are clearly helpful tools that they have used to name things, to communicate their discovery of things, and Harvey, too, uses these words as tools, always mindful that their legitimacy lies not in their privileged status as part of an originative, divine language, but rather in their role as accurate human constructs, or representations, for the things themselves studied by the ancients.

Words, for Harvey, do not then carry the transhistorical significance implicit in Fernel's endeavor: they are not symbols alternately concealing and revealing the true nature of things. Consequently, the *exegetical* practice of interpreting received natural philosophical languages is an empty enterprise, since behind their defective, obscure terms lies no privileged level of meaning. Without such meaning, traditional terms constitute a debased coinage circulated all too freely by exegetes like Fernel. Moreover, this poverty of the hermeneutics of exegesis reveals the poverty of the narratively constructed theoretical task that gave

---

36. Céard (1980, 1988) has written suggestively on the links between sixteenth-century language theory and science, especially natural history. See especially Chapters 3 above and 6 below.

birth to it. For Harvey, the task of physiology as part of theoretical medicine is not to unlock the true meaning of words, but, by contrast, to construct a contingent, human language to reflect as accurately as possible the visible, transparent nature of things inscribed in God's Book of Nature. The language of science and medicine must form a mirror image of the text of nature. To mirror nature's text, words must be *signs* pointing unequivocally to things signified and inscribed in nature. The problem theoretical medicine must then face is the *defective* nature of existing human languages in comparison to the multiplicity and complexity of a divinely created and contingent nature. As Harvey himself states in the *De generatione:* "Whoever entereth this new, and unfrequented path, and inquires for truth in the vast volume of Nature . . . he meets with such a crowd of observations . . . for many things occurr which have yet no name; such is the plenty of things, and the dearth of words" (1653: preface, n.p.; 1847:166; 1651: fol. C3).

Finding words is easy enough, of course. Defining words in a manner suited to reflect some aspect of nature properly is more difficult. Harvey the quiet, even conservative, "revolutionary" still chose to mine classical sources and their offspring for words; his vocabulary is still the vocabulary of Renaissance theoretical medicine—of Aristotle, Galen, Fernel.[37] Yet, if Fernel in his attitudes toward language and the interpretation of nature conforms most closely to the esoteric traditions of Florentine Neoplatonists Ficino and Pico della Mirandola, Harvey instead follows critical tendencies like those of the Italian humanism exemplified by Lorenzo Valla.[38] Harvey, in short, looks upon his Greco-Roman and medieval-Renaissance sources as rich lodes to be mined, but ones whose ores the scientist must assay to determine their authenticity. Hence, Harvey subjects ancient medical authors like Galen to two kinds of critical tests. The first examines the *internal* consistency of signification in an author's use of scientific words, as in Harvey's treatment of Galenic pulmonary "veins" and "arteries"; the

37. See Harvey (1653: preface, n.p.; 1847:166–67; 1651:C3). Note particularly the closing paragraph.

38. See assessments of Valla's importance by Jardine (1977) and Waswo (1979). Valla's conception of language, according to Waswo, would be the antithesis of esoteric Neoplatonic views of language as "divine," but veiled, revelations: language is a human construct, constitutive of "truth." I do not suggest a causal connection between Valla and Harvey, but, rather, stress that the "interpretive practice" of scientists and physicians during the late Renaissance is central to our understanding of their "science" and of scientific change in this period. Such interpretive practices are rooted in basic notions about history, cultural transmission, and language. See Chapters 1 and 3, above.

second seeks external corroboration of the adequacy of these terms as signs.[39]

This latter test includes existential verification—that is, whether or not a term signifies an existent *thing*, a real entity or substance, as its *name*. Harvey's criticism of the theory of medical spirits, and in particular of Fernel's definitions of *spiritus*, focuses upon the appropriateness of the term *spiritus* as signifying an independent medical entity with a discrete function within the economy of the animal organism. And yet, while he rejects traditional definitions, including Fernel's there yet remains a proper use of the term *spiritus*, for in addition to functioning as signs—or *names*—of things themselves, words may also be used, more subtly, as terms signifying some *modality* of a thing. *Spiritus*, as it turns out, is a good example. For Harvey, it is not a separate celestial or material entity, but it does carry valid signification:

> What, however, is especially relevant to my theme after all other meanings have been omitted from consideration as being tedious, is that the spirits escaping through the veins or arteries are no more separate from the blood than is a flame from its inflammable vapour. But in their different ways blood and spirit, like a generous wine and its bouquet, mean one and the same thing. For, as wine with all its bouquet gone is no longer wine but a flat vinegary fluid, so also is blood without spirit no longer blood but the equivocal gore. As a stone hand or a hand that is dead is no longer a hand, so blood without the spirit of life is no longer blood. (Harvey 1963:150–51; 1766:117)

Spirit, then, signifies a visible quality, or modality, of matter: the liveliness or vitality of the blood, which imparts these same qualities to the blood as a whole.[40]

Harvey's criticism of traditional, particularly Fernelian, uses of the term *spiritus* is symptomatic of his abandonment of Renaissance cultural narratives legitimizing an ideal of reform as *renovatio* that authorized as its foundational practice the exegesis of received medical and natural philosophical languages. Fernel's hermeneutic circle is closed; within the narrative context constructing its view of language and its

39. Harvey relied upon analysis of Galen's own testimony to adduce evidence for the possibility of a pulmonary transit of the blood. Bylebyl and Pagel (1971); Harvey (1928:37–40).

40. Harvey's discussion of blood, spirit, soul, and primigenial humor in the last two exercises of his *De generatione* supplements his statements here. By identifying blood with "spirits" and "soul," Harvey may seem to attribute a transcendent quality to biological phenomena; however, Harvey's text attempts to redefine traditional philosophical-medical language in order to stress the special, vital properties of living matter itself.

practice of exegesis there is no need to break, and move beyond, that circle. Observation, at best, plays a strictly ancillary role in theoretical medicine. By contrast, Harvey, the discoverer of the blood's circulation, shatters that circle of interpretation. For Harvey, the reconstruction of natural philosophical terms as true signs of things forms the solid, incorruptible metal of that coin to which his work gave currency. Its two faces—one depicting the "modern" critic of speculative ancient medical texts, the other describing a method, or perhaps a mythology, of observation—have won the attention of his century and ours. Without that metal, however, without his narrative reconfiguration of postlapsarian man's access to the Word of God and consequent rejection of the "symbolic" nature of language and the "exegetical" discursive practices of speculative physiology, and without the conviction it nurtured that nature, to be understood, must be "named" properly, that familiar coin would have been insubstantial: incapable of being minted, or, at best, quickly debased. As it was, Harvey's new narrative and use of scientific language prompted a *re-forming* of medical traditions, the creation of a new language of animate matter. By requiring a critical approach to experience, it defined the context within which observation was utilized.

Harvey's Theory of Biological Matter

Harvey's rejection of Fernel's medical spirits is indeed grounded in wholesale abandonment of Fernel's exegetical approach to medical reform and scientific truth. More significant still, the assumptions about language and about the relationship between "words" and "things" in science underlying this rejection also constitute a powerful stimulus for Harvey's positive articulation of ideas about active biological matter. Harvey's speculations about both language and living matter become explicit only near the end of his life in the *De generatione animalium* of 1651. Both sets of speculative attitudes are, I would argue, essentially linked to the role Harvey accords to observation in medicine and physiology. Hence, an interlocking examination of these concerns may help us better understand Harvey's assumptions, methods, philosophical commitments, and discursive practices.

Undoubtedly Harvey's view of the contingent, human nature of scientific language evolved over a long period of time before he articulated it in the *De generatione*. Once it was articulated, the methodological necessity of observation became explicit. This necessity, moreover, had its foundation in the manner in which theoretical medicine and natural philosophy, for Harvey, acquired legitimacy. Legitimacy did *not* reside in observation itself. It is absolutely fundamental to grasp that, for Harvey, observation was only a "tool" serving a model of scientific speculation and theoretical explanation that itself had acquired legitimacy independently of any specific use of empirical evidence; rather, that model was based upon, and legitimized by, philosophical and, one might almost say, theological convictions.

Philosophically, Harvey's work breathes the conviction that all causes and effects in nature are manifest ones: they are palpable, patent, and thus, in principle, observable. I do not mean to suggest that he considers all causes in nature to be *immediately* palpable and therefore directly observable; he fully recognizes that "causes" are often hidden from direct observation, not open to our gaze, but he regards such causes and their activities as *manifest in* the organic structures, states, and textures of living things. Put differently, vital properties are themselves aspects, or "powers," of matter—*not* extrinsic forces "superadded" to matter—and are *expressed in* the various configurations and textures of living things.

Here, perhaps, one finds the deepest significance and true implications of Harvey's adherence to the new, Renaissance tradition of anatomy: the notion that "structure," "texture," and palpable physical properties of organisms are physiologically significant and bear a direct correlation to function. Hence, the *inherent* differences between organisms, or between various organs, tissues, or bodily fluids, are expressed in their varying structures. When "occult causes" were located in a "faculty" of an autonomous soul separable from its material body, or in a "faculty" attributed to Galen's nature, or in a transcendent, "celestial" *spiritus*, as with Fernel, such emphases upon hidden, as opposed to manifest, causes allowed traditional "dietetic" physicians to downplay the theoretical significance of anatomical structures. By contrast, Harvey's new perspective as anatomist and physician led him to interpret organic activity in terms of the morphology, composition, and properties of matter. This perspective in turn *demanded* and legitimized observation.[41]

Of course, the Aristotelianism Harvey imbibed from his Renaissance sources buttressed such a perspective insofar as it regarded phenomena as the result of immanent causes discoverable in the things

41. For the important distinction between Harvey and Galenic "dietetic" physicians, see Bylebyl (1977). Andrew Wear (1983) has published an important article concerning the impact of the anatomical tradition upon Harvey's use of observation.

themselves that change and are subject to change. Moreover, such philosophical convictions were deepened and given added significance by narratively emplotted "theological" convictions regarding nature. For Harvey, God created different realms of nature—including animate nature—to operate wholly and independently, sui generis, as integral strata of a functioning universe. Within each stratum—within the stratum, for example, of living organisms—each entity had the inherent capacity to function according to its kind, its specific nature.

This "theological" conviction allowed Harvey to develop a more radical "Aristotelianism." In effect, it permitted him to be, as Walter Pagel has said, "more Aristotelian than Aristotle," chiefly through undertaking a "vitalistic criticism of ancient materialism."[42] For Harvey eschewed and criticized the dualistic tendencies of Aristotelian emphases upon an "ordinary matter" transformed by a substantial "form" into an animate being; instead, he interpreted the Aristotelian *entelechia* quite literally as the *activity* of an organic body: the activity displayed by biological matter. Indeed, I would argue that, to a large extend, the speculative and ontological significance of Harvey's anatomical approach was reinforced, and even became consciously apparent to him, as a result of his engagement with Aristotelianism and its monistic perspective.

A different monistic viewpoint, in some respects similar to Aristotelianism, was also available in certain contemporary Neoplatonic speculations, as Pagel has stressed,[43] and such strains undoubtedly influenced Harvey in addition to his Aristotelianism. Yet I believe it is significant that his Neoplatonic tendencies always stop short of embracing a full-blown monistic cosmology: the Neoplatonic identification of matter and activity, for Harvey, extends only to the realm of biological matter! Here, I believe, theological convictions are apparent: the ordinary matter of the universe, elemental matter, displays no exceptional, inherent active abilities, nor is the cosmos at large suffused by active, "divine" impulses, an all-embracing world soul. God is still transcendent author and controller of the universe, though in living things he

42. Pagel (1967:332). Pagel's chapter "Harvey's Vitalistic Criticism of Ancient Materialism" (251–82) is especially pertinent. Harvey's radical Aristotelianism purged Aristotelian biology of its dualistic tendencies—slight as they may be—by defining biological matter (which Aristotle saw, like all matter, as inert and requiring a form or *entelechia*) as inherently active and monistic. Neoplatonic elements in Harvey are operative precisely in this notion of biological matter.

43. See the works by Pagel cited in the first paragraph, and, in addition Pagel (1982a), and Bono (1984a).

has created a form of matter, a kind of being, that can *act*, can create, sui generis. The kind of activity Neoplatonism attributes to the universe as a whole—as an expression of the interpenetration of the divine and its emanations—Harvey localized in one particular stratum of created beings, namely, living things. By creating a special kind of biological matter in the *humidum primigenium*, God is the author of an immanently active, truly monistic, material substance, which, because it is unique, cannot compromise His essentially transcendent nature. In this way, Harvey ensures an autonomous, anatomically based physiology and medicine whose language, now created anew, can and must be independent of the transcendent language of theology.[44]

Harvey's monist and immanentist perspective as a natural philosopher—his belief, that is, that organic entities act by virtue of powers that are inherent in them and not mysteriously imposed from without—thus depends upon his firm theological conviction regarding the role of the divine Creator in nature. This Creator is the root cause of all generation, of all life. Yet His active power does not entail direct and continual intervention of the divine in the realm of nature; instead, divine power operates through the immanent agency of its created natural instruments. The Book of Nature, for Harvey, is a divinely ordained and lawful order in which both inorganic and organic beings act of their own inherent power in determinate and specifically appropriate ways.[45]

This specifically Harveian perspective is voiced throughout the *De generatione* and significantly informs the concluding sections, "Of the Innate Heat" and "Of the Primigenial Moisture" (1651: 244–51, 252–56). I would suggest that we also witness in these sections the intimate connections between Harvey's attitudes toward language, his theory of active living matter, his immanentist theological and philosophical views, and, of course, his deinscriptive hermeneutic practices for reading the Book of Nature with their consequent emphasis upon observation and his critique of Fernel's *spiritus*. Indeed, Harvey's explicit *purpose* in these concluding sections is to reform the meaning others have attributed to the terms "innate heat" and "primigenial humor." His

44. For Harvey's theory of biological matter, see Harvey (1651:252–56), and my discussion below.

45. Harvey (1651:146; 1653:267–68; 1847:369, 146–47; 1653:268–69; 1847:369–70). The point, for Harvey, is that the "powers" of natural/animate bodies are inherent; as immanently active entities they are instruments of the divine, acting to fulfill God's determinate and lawful order. It is emphatically *not* a question of divine or celestial agencies operating, from without, upon natural/animate bodies.

strategy involves demanding that the reader view these terms from Harvey's own immanentist perspective and simultaneously insisting that the reader abandon all inclination to invest these and other medical terms with a privileged esoteric and transcendent meaning. The result of this strategy is the intimation of a new theory of active biological matter.

Hence, Harvey's *De generatione animalium* concludes by articulating a theory that regards biological matter as *innately* active, sui generis, and utterly opposed to ordinary elemental matter. *Exercitatio 72*, entitled "De humido primigenio," appeals to the notion of a primigenial humor, not to introduce a discussion of "divine" and celestial forces, as with Fernel, but rather to argue for the existence of a special kind of vital matter whose properties and active abilities are themselves the manifest causes of vital, physiological activity.

This argument for a specal kind of vital matter is the direct consequence of Harvey's exposure, in *Exercitatio 71* ("De calido innato"), of the deeply embedded tensions within the traditional Renaissance language of life.[46] While purporting to account for the phenomena of life, that language subtly belies their reality. That is to say, the very terms used to describe the explain vital phenomena undercut their concrete materiality: manifest entities, through a kind of linguistic slippage, become drained of ontological efficacy and are, at best, relegated to mere ciphers of some hidden, as yet undecoded, realm of power.[47] Harvey's

46. I use the phrase "traditional Renaissance language of life" to refer to the "language" constructed during the twelfth and thirteenth centuries and utilized by subsequent physicians and natural philosophers in discourse about living things. Arising from the active reception of ancient medical and philosophical language by medieval Christian authors, this language reflects the shaping force of metaphors and values imbibed from biblical religious language as well as other "influences." For a much fuller treatment, see Bono (1984).

47. Here I would like to point to an aspect of this Renaissance "language of life" I have discussed more fully elsewhere, in Bono (1981) and (1984)—namely, its hybrid nature and internally divided, even divisive, character. Most simply, that "language" (more properly speaking, discourse) embodies elements that are in tension with one another: e.g., immanent versus transcendent impulses, or "materialist" versus "spiritual" tendencies. These tensions and oppositions are the result of its attempt to assimilate (to put a complex matter simply) the languages of Greek medicine and biology into an overarching system of transcendent meaning found in the medieval Christian discourses of theology and biblical religiosity. As a result of its dominantly transcendent use of metaphor and analogy, that hybrid language of life tends to oscillate between immanentist and transcendent understandings of biological terms (like "blood" and "spirits"), and consequently it exhibits a recurrent pattern of linguistic slippage. Terms that seemingly designate "concrete" entities become metaphorized and transposed into terms that designate transcendent, even occult, entities. One result is the intimation of hidden, often unfath-

attack in *Exercitatio 71* is an attack precisely against such linguistic slippage and its consequences.

Let me turn to Harvey's work in order to explain these remarks. "I perceive many men to please themselves much with these two Notions [*nominibus*], when as (according to my judgement) they do not understanding their meaning."[48] This complaint introduces Harvey's own consideration of the terms in question, the *calidum innatum* and the *humidum primigenium*, in the concluding "exercitations" of his *De generatione*. Here his focus fixes upon *meaning*—specifically, upon the *distorted* meaning acquired by key terms in the language of life inherited from his Renaissance forebears. These terms, he notes, are closely linked with others: with the Hippocratic *impetum faciens*, and capitally with "spirits" and the "blood" (Harvey 1653:448; 1847:502; 1651:244–45). But distorted meanings have distorted those links. The heat and spirituousness *of* blood—vital properties manifest in the palpable blood itself—and "raised" into entities in their own right. What were "properly" understood as essential "qualities" of the material blood have acquired a distorted meaning. The very metaphoricity of the terms "spirit" and "innate heat" have been "worn down" and "forgotten": what remains is an unintended literalness.[49] Such slippage from "metaphorical quality" to "literal entity" has resulted in the draining of all ontological efficacy from the blood itself. To combat this slippage, Harvey must remind his readers that "the truth is, there is no need at all to enquire after any kinde of spirit distinct from the Blood it self, or to introduce any forraign heat, or invoke the Deities to appear in the fa-

---

omable, realms of activity and power capable of pervading and controlling the visible world. On theory, metaphor, and the history of science, see Bono (1990a).

48. Harvey (1653:447; 1847:501–2; 1651:244): "multos videam nominibus istis plurimum deliciari; cum rem ipsam tamen (meo quidem judicio) minus intelligant."

49. See note 47 above. The process alluded to above is more complex than I have suggested. In the process of becoming "metaphorized" and "transposed," terms like "*spiritus*" are *not* just simply *compared* with progressively more transcendent entities, in whose "meaning" they participate analogically (such as the set of terms "medical spirit," "celestial" or "quintessential spirit," "angelic spirit," "the holy or divine spirit"). Rather, the very transposition leads to a kind of *identification* of one term with another—of one "meaning," and hence one entity, with another. In this process, the metaphoricity of the metaphorized term is effaced, erased, forgotten; in its place is a meaning now taken as *literally*, or "naturally," the case. I have discussed this process with respect to both the twelfth/thirteenth and the sixteenth/seventeenth centuries in Bono (1984). In a long footnote (1984:100–101) I argue for a distinction between an "abstract symbolism" and a "symbolic literalism," the latter increasingly common with the rise of Neoplatonism, magic, and pantheistic tendencies. That discussion is centrally relevant to problems uncovered in this chapter.

ble." Rather, "the Blood alone is the true *Calidum Innatum*"; indeed, since "the Blood (the primogenit and principal part of the Body) is furnished with all these respective qualities, and endowed with active power beyond all other parts of the body," it "doth therefore deserve the name of a spirit."[50]

To combat this linguistic slippage, then, to combat the raising up of fictitious entities in medical discourse that threaten to supplant the visible blood as source of life, Harvey turns to an examination of language itself: specifically, of that language used in discourse about life. In a few words, he wishes to show "what a spirit is, and what it is to act above the power of the Elements, and likewise what is meant by these words."[51] We have already noted one major consequence of Harvey's analysis: the rejection of separate, superadded celestial spirits (like Fernel's) and the reiteration of the immanent life-sustaining qualities of the material, living blood. Spirit, as we have noted, signifies for Harvey a visible quality or modality of blood itself.

But Harvey's examination of the phrase "to act above the power of the Elements" reveals the even more fundamental grounds for, and consequences of, his reappraisal of spirits. For the distorted meaning attributed to "spirits," "innate heat," and their link to the blood—the tendency to wear down the metaphoric value of these terms and raise up a false, literal meaning in its stead—rests upon the inability to understand the proper meaning of this phrase. Certainly Fernel's errors with respect to spirits and the blood were, as we have heard Harvey say earlier, based on a misunderstanding of this same phrase.[52] Harvey's own analysis of its meaning unleashes his emphatic opposition: "they therefore who conceive that in a body compounded of the Elements, it cannot act beyond the power of the Elements . . . do seem to have built their reasons upon a very shallow foundation" (1653:456; 1847:508; 1651:248–49).

Several keys are necessary to unlock the proper meaning of this phrase. To understand how the blood can "act above the power of the Elements," Harvey insists upon a reconsideration of what it means to

---

50. Harvey (1653:447–48; 1847:502; 1651:244–45): "Non est opus profecto spiritum aliquem a sanguine distinctum quaerere, aut calorem aliunde introducere, Deos-ve in scenam advocare . . ."; and "ideoque *kat' eksochen* spiritus nomen meretur."

51. Harvey (1653:453; 1847:506; 1651:247): "Ut autem haec liquido magis appareant, liceat nobis a proposito aliquantulum digredi, & paucis ostendere, quid sit *spiritus*, & quid, *supra vires elementorum agere*. . . ."

52. See Harvey's remarks in the passage quoted at the beginning of this article and in note 1, above.

act, to be an agent: "In short therefore, this distinction is necessary, namely, that no Primary of first Agent, doth act any thing beyond its own power: but every Instrumental Agent, doth exceed its own power; for it acts not onely by its own power, but also by the virtue of the superiour efficient."[53] Hence the blood, as a material entity, *can* act in a manner "above the power of the Elements" without presupposing that some separate entity, such as the "spirits" of Fernel, have been added to it. For the blood, at the very least, may be considered an "instrumental agent." But in what exact sense is blood the agent of life? Harvey's discussion of the blood, bristling with tensions though it may be, offers us other clues to understand the meaning of the assertion that blood is superior in action to "the power of the Elements."

Harvey insists, for one thing, upon yet another distinction: that between extravasated blood and blood contained within its proper vessels. "The blood," he says,

> being considered absolutely in it self out of [i.e., outside] the Veins, as it is an elementary substance, and composed of several parts (namely, of serous, thin, crass, and concrete parts) is called Cruor, Gore, and doth possess a few onely, and those obscure abilities. But being in the veins, as it is a part of the body, and that also an Animate, and Genital part, and the Immediate Instrument, and primary seat of the soul: and also as it seemeth to partake of another more divine body, and is inspired with a divine animal heat; it is then endowed with excellent abilities, and answerable in proportion to the element of the stars.[54]

Living blood, Harvey seems to suggest, is fundamentally different from *dead* blood—from cruor and mere elementary fluid. But does Harvey's distinction offer us a real difference when compared to, say, Fernel? Here we witness a remarkable transmutation. For Harvey does use language remarkably similar to Fernel's and those of his ilk. That language—which speaks of the "soul," "divine animal heat," and the "element of the stars" (celestial *spiritus?*)—may easily suggest that, for

53. Harvey (1653:457; 1847:509; 1651:249): "Nimirum distinctione ista opus est: Nullum agens *primarium* primum-ve efficiens, supra vires suas operatur: omne vero agens *instrumentale*, in agendo proprias vires superat; agit enim virtute, non solum sua, sed etiam superioris efficientis."

54. Harvey (1653:458; 1847:510; 1651:249–50): "Sanguis nempe extra venas absolute et per se consideratus, quatenus est elementaris, atque ex diversis partibus (tenuibus scil. serosis, crassis, & concretis) componitur, *cruor* dicitur, paucasque admodum & obscuras virtutes possidet. In venis autem existens, quatenus est pars corporis, eademque animata & genitalis, atque immediatum animae instrumentum, sedesque ejus primaria; quatenus etiam corpus aliud divinius participare videtur, & calore animali divino perfunditur; eximias sane vires obtinet, estque analogus elemento stellarum."

Harvey, living blood can act as an agent of life only insofar as soul, innate heat, and spirits are *added* to it! Harvey's language here is every bit as slippery as Fernel's; the difference, I think, is that Harvey's hold on it is somewhat firmer, due to his theological convictions—which anchor that language, and prevent the slippage noted in Fernel from "metaphorical quality" to "literal entity."

Harvey, indeed, explicitly evokes that theological context when speaking about blood in *Exercitatio 71:*

> So likewise the heat of the blood is an animal heat, inasmuch as it is guided in its operations by the soul; and also a Celestial heat, as being subservient to the Heavens; and lastly, a Divine heat, in that it is the Instrument of Almighty God: as we have formerly said, where we also did demonstrate, that the Male and Female, are the Instruments of the Sun, the Heavens, and of God [*satoris*] himself, as being subservient to the generation of Animals.[55]

For anyone who has followed Harvey's arguments about generation closely, the force of this statement is not easily missed.[56] Blood, like male and female in generation, acts with powers immanent to it. These powers may, in fact, appear celestial, even divine, and beyond the power of the ordinary elements. They result, in short, in vital activity, just as male and female generate new life. The blood's power, its vital effects, signify to us that it is an instrument of the Creator and *Author,* of God. Like God in the universe, [57] "by a free perpetuall motion" the blood "doth conserve, nourish, and encrease it self."[58] But, again, as with God's action in the universe, the action of God—of the "divine"— in the body is remote. God does *not* act directly upon animate bodies; nor does He act indirectly, imposing foreign powers upon animate bodies through celestial or other intermediaries. Rather, God—author of the Book of Nature, and creator of animate bodies—has established a lawful order in which things themselves are capable of acting in accordance with His plan. For Harvey, blood *is* an instrument of the divine,

55. Harvey (1653:455; 1847:508; 1651:248): "Similiter & sanguinis calor est *animalis,* quatenus scil. in operationibus suis ab anima gubernatur: est etiam *coelestis,* utpote coelo subserviens; & *divinus,* quod Dei Optimi Maximi instrumentum sit: quemadmodum supra a nobis dictum est; ubi etiam marem & foeminam, Solis, coeli, vel Satoris summi instrumenta esse, perfectorum animalium generationi inservientia, demonstravimus."

56. For discussion of notions of instrumentality in Renaissance medicine, natural philosophy, and Harvey, see Pagel (1969–70).

57. See note 45 above.

58. Harvey (1653:458; 1847:510; 1651:250): "sed quod vago ac perpetuo motu se ipsum conservet, nutriat, & augeat."

but one whose powers do not come from outside it, are not superadded to it: those powers are, rather, created with it and are immanent characteristics of *living* blood itself!

Where, then, does this understanding of matters leave us? How is Harvey different from Fernel? Consider again Harvey's distinction between extravasated blood and blood contained within its proper vessels. The latter, for Harvey as (presumably) for Fernel, displays all the qualities—all the powers—associated with the terms "soul," "innate heat," and "spirit" in the Renaissance language of life. How are those terms to be read? What is their meaning? In the language of life adopted by Fernel, the terms assume a transcendent meaning, signifying independent, "occult" entities added to the blood. Harvey sees the very same powers signified by these terms as inhering, properly, in living blood itself. Beneath Harvey's rejection of Fernel's *spiritus* lies, then, a more fundamental rejection of Renaissance narratives on the relationship of the divine to nature and of language to meaning. For Harvey, God is the "Creator" and "Author" of the Book of Nature. Nature, in turn, is not continuously linked to God by a hierarchy of intermediary, "occult" entities of progressively more "spiritual" nature; rather, the created universe—the "material" universe—operates on its own, without occult intervention, without superadded "middle substances." As a result, the student of nature—the physician—must look to created matter itself, to the qualities, powers, and operations it manifestly displays as instrument of the divine. The "thing" itself—whether it be blood or some other visible entity—must be attended to. To look beyond "things," to search for "occult" entities, is to diminish the power of God to create an ordered nature—distinct from Himself—whose palpable entities, acting as His instruments, are endowed with immanent properties sufficient to their ordained roles in His creation.

Ultimately then, for Harvey, the blood *can* act with a force superior to the forces of the elements because it is "the Instrument of the Omnipotent Agent." Living blood, blood contained in its proper vessels in the living body, exhibits its own "admirable," even "divine faculties."[59] Indeed, "the blood," in Harvey's view, "seems to differ nothing from the soul; or ought at least to be counted that substance, whose act the soul is. . . . And therefore it comes all to the same reckoning, whether we say, that the soul and the blood, or the blood with the soul, or the soul

59. Harvey (1653:459; 1847:510; 1651:250): "Quoniam itaque sanguis supra vires elementorum agit, dictisque istis virtutibus pollet, atque summi Opificis instrumentum est; nemo facultates ejus admirabiles & divinas satis unquam depraedicaverit."

with the blood, doth performe all the effects in the Animal."[60] The words we use to speak of that entity whose activity is life may be multiple and, to some, charged with different semantic force. These words can mislead; their meanings distort the relations among things, and between things and the divine. But such traps, Harvey suggests, occur only if we fail to heed the nature of language, meaning, and the act of divine creation itself. For "[w]e are accustomed, neglecting things, to worship specious names. Blood, which is before our very hands and eyes, sounds neither lofty nor powerful; but we are struck with wonder at such noble words as spirits and innate heat."[61]

Herein lies Fernel's errror. To seek something "divine" in mere words, or "names"—in the act of "exegesis"—is to mistake the significance both of language and of God's relationship to nature. There is no hidden realm of occult entities in the universe; there are no noble, divine-like intermediary substances linking God to ordinary things; there are no "lofty and powerful" names to be admired above the rough and ready words used to describe the ordinary things before our eyes. There are only things themselves, and the language we humans create to name and describe them properly. To reject Fernel's spirits, one must reject *both* the narratively framed theological cosmology and the symbolic, exegetical view of language nourishing it. Such rejection— prompted by Harvey's own narratively emplotted theological convictions and attitudes toward language—leads ultimately to "things themselves" and to observation. "But when once the [mask] is removed from before" such noble words as spirits and innate heat, "as our error [vanishes], so [does] our wonder."[62]

Harvey reserves his wonder, not for some invisible fiction, but for God's visible creation: the living blood itself. Thus he ends the penultimate section of *De generatione* with a stinging criticism of Fernel's misguided exegetical habits:

> In like manner, if I should describe the Blood under the veil and covering of a Fable, calling it the Philosophers stone, and displaying all its endow-

60. Harvey (1653:459–60; 1847:511; 1651:250): "sanguis ab anima nihil discrepare videatur; vel saltem substantia, cujus actus sit anima, aestimari debeat. . . . Eodem ergo res redit, si quis dicat, Animam & sanguinem, aut sanguinem cum anima, vel animam cum sanguine, omnia in animali perficere."

61. The translation is my own. Harvey (1653:460; 1847:511; 1651:250–51): "Solemus, rerum negligentes, speciosa nomina venerari. Sanguis, qui nobis prae manibus atque oculis est, nil grande sonat: ad spirituum, & calidi innati, magna nomina obstupescimus."

62. Harvey (1653:460; 1847:511; 1651:251): "Detracta autem larva, evanescit, ut error, sic etiam admiratio."

ments, operations, and faculties in an aenigmatical manner; doubtless every body would set a greater price upon it, believing it to act beyond the Activity of the elements, would ascribe another and more divine body unto it.[63]

In the end, then, Harvey returns us to a word: the living *blood*. But his "blood" is not *just* a word—not just a disembodied meaning, a spiritual sense. Through that word shines transparently the brilliance of a *thing:* the living blood. In the beginning, there was the Word; in the end, here on earth, there are things, things that require naming and, hence, close scrutiny and observation. Theology and language are linked, for Harvey as for Fernel—but the links are different and the consequences, for each, are divergent. Language, for Harvey, is man-made, not divine. Words are traps: they ensnare man, mislead, distort. But in them things can also be caught: they are tools for dissecting, measuring, describing (de-in-scribing) *things*—*if* we attend to things, not words. Harvey's resonant cultural narrative, the primacy placed on God as Creator and on the independence and self-sufficiency of created things, breathes life into things and gives to words—to language—a definite if modest role.

The independence, self-sufficiency, and re-creative power of that most important of all things within the body, Harvey hopes—rather, strives—to capture in a word, "blood." His striving meets with obstacles, some of which we have already seen him struggle against. He can only point to the thing itself, to blood, to its observable characteristics and manifest effects. This blood, contained within the living body, "conserve[s], nourish[es], and encrease[s] it self," by virtue of its "free perpetuall motion"; this blood is the innate heat and "farther deserves the name of spirit, inasmuch as it is the radical moisture. . . ."[64] Additionally, this living blood is *different* from extravasated blood—which is

63. Harvey (1653: 462; 1847:512; 1651:251): "Ad eundem pariter modum, si sub fabulae involucro sanguinem alicui depingerem, lapidisque Philosophici titulo insignirem, atque omnes ejus singulares dotes, operationes, ac facultates aenigmatice proponerem; illum procul dubio pluris aestimaret; supra vires elementorum agere facile crederet, corpusque illi aliud ac divinius non illibenter attribueret." The whole context of this concluding paragraph suggests that it is meant to apply most directly to Fernel, whom Harvey mentions in the immediately preceding sentence. Moreover, in the preceding paragraph he cites a story recounted in somewhat different language by Fernel, though the quotation given in Harvey does not clearly indicate Fernel as source; instead, when Fernel's name is mentioned some lines below, both the Latin edition and Willis cite the wrong reference: it should read Fernel (1550:2.17, 284–85).

64. I have used Willis' translation in this one instance. Harvey (1653:458–59; 1847:510; 1651:250).

no more than cruor and elementary fluid—since in the latter one does not observe the characteristics and effects of blood.

To explain this difference, to capture the thing that is living blood in contradistinction to dead blood, mere cruor, Harvey needs other words: the radical humor, or primigenial moisture. For the living blood is "radical moisture"; and blood is the first part of the fetus to arise from the "crystalline colliquament" which Harvey calls "by the name of . . . the Radical and Primigenial moisture."[65] The critique of Fernel's *spiritus*, the Aristotelian monism of Harvey's natural philosophy, his narratively framed theological convictions and closely associated attitudes concerning language, all converge upon these words—the radical or primigenial moisture—and upon the implicit theory of living matter that his discussion of them suggests.

Harvey's discussion—to dwell only upon its conclusions here—establishes a profound disjunction between living and nonliving matter. Here perhaps lies the answer to an unspoken tension in Harvey's thought. For, as we have seen, the living blood's patent ability to act with a force beyond that of the elements results from the fact that it is an *instrument* of God, the divine Creator and Author. This instrumental capacity of living blood distinguishes it, functionally, from dead, extravasated blood. Yet, beyond this difference in function, Harvey allows of no superadded material or occult factors to distinguish the two sorts of "blood." How, then, can living blood achieve its difference in function? Since Harvey's theological convictions rule out any direct and continuing divine action upon the living blood as God's instrument, the answer does not seem apparent, for in all other respects the two kinds of blood ought to be identical. But here we ought to remember and carefully heed Harvey's own *practice* as anatomist and medical scientist. For, in his view, the functional differences between living and extravasated blood, and such palpable differences as their motility and heat, *do* signify differences of kind, of matter. Function and properties do matter: the difference is material.

Harvey, who refused to admit Fernelian spirits superadded to the material blood, admits instead a new *kind* of matter added to nature. While at once rejecting the interference of transcendent spiritual causes with matter and blood, he establishes an immanently active matter—

---

65. Harvey (1653:462; 1847:513; 1651:252): "Sanguinem jam *calidi innati* titulo ornavimus; aequumque pariter censemus, ut *colliquamentum crystallinum* a nobis dictum (ex quo foetus, ejusque partes prima immediate oriuntur) *humidi radicalis & primigenii* nomine insigniatur. Neque enim res alia aliqua in animalium generatione occurrit, cui potiore jure id nominis conveniat."

the primigenial moisture and the living blood—whose origin is the transcendent Creator Himself. In short, what Harvey proposes is no less than a separate category of *living, active matter* created originally by God and perpetuated, transmitted, and increased through the very process of generation.

The very "root" of life and vital activity is the radical and primigenial moisture. Through generation, it becomes the blood. It is a homogenous, nonelemental, active, generative, and creative matter. Even more, it is utterly distinguished from "ordinary" elemental matter:

> Since therefore I plainly see, that all the parts are fashioned and fed by this one moisture onely, (as the matter and first root of all) . . . I can scarce refrain my pen from rebuking those that follow Empedocles and Hippocrates also: (who will needs have all similar bodies to be generated by the congregation of the four contrary Elements: (as being mixt bodies) and dissolved or corrupted by their segregation).[66]

Rather than such a "congregation" of elements, Harvey the Renaissance Aristotelian insists that

the first rudiment of the body is onely a similar soft gluten, or stiff, substance, not unlike a [congealed spermatic fluid], or coagulated seed [*spermatico concremento*]: out of which (the degree of generation going on) being changed, cut in sunder, or distributed into several parcels, as by the divine Mandat as we have said . . . *out of an inorganical substance, was made an organical.*[67]

This is an instance, we must note, not of a living emerging from a nonliving matter, but rather of an organic emerging from an inorganic. Living, active matter generates a complex, differentiated organism, replete with appropriate and necessary *instruments*—the organs. Harvey places absolute temporal and ontological priority on living matter over and above elemental matter. Living matter, not elemental, is the sole and fundamental component of animate beings or living organisms:

66. Harvey (1653:467; 1847:516; 1651:254): "Cum itaque videam, ex uno hoc humido, (tanquam materia aut radice prima) partes omnes fieri, & augescere; . . . continere me sane vix possum, quin *Empedoclis,* & *Hippocratis* etiam sectatores (qui corpora omnia similaria, ex quatuor elementis contrariis (ceu mista) congregatione generari, & segregatione corrumpi volunt) vellicem, ac perstringam. . . ."

67. Harvey (1653:468 [emphasis added]; 1847:517; 1651:254–55): "primum corporis rudimentum, est similare duntaxat & molle gluten spermatico concremento non absimile; ex quo (procedente generationis lege) mutato, simulque secto, vel in plura divisio, ceu effato divino, ut diximus, . . . ex inorganico, fit organicum. . . ."

And this do I believe to be observed in every Generation; so that the similary mixt bodies have not their Elements existent in time before them, but are rather themselves in being before their Elements . . . as being in Nature more perfect than they. There are I say mixt and compounded bodies, even in respect of time before any Elements, as they call them, into which they are corrupted and determine; for they are dissolved into those Elements rather in order to our apprehension, then really and actually. And therefore those bodies called Elements, are not before those things which are made and generated; but rather after them, and their Reliques rather then their Principles.[68]

"Living blood," the "radical moisture," *is* such a principle, the divine instrument of life—indeed of all vital activity. Cruor, dead blood, is but the relic, the remainder of a more noble, and fundamentally distinct, active matter. All these things are there, in nature, to be observed. And it is the power of words, of human language, to speak of God's creation so.

68. Harvey (1653:468; 1847:517; 1651:255): "Idemque in omni generatione fieri crediderim; adeo, ut corpora similaria mista, elementa sua tempore priora non habeant, sed illa potius elementis suis prius existant . . . utpote natura quoque ipsis perfectiora. Sunt, inquam, mista & composita, etiam tempore priora elementis quibuslibet sic dictis, in quae illa corrumpuntur & desinunt; dissolvuntur scil. in ista, ratione potius, quam re ipsa & actu. Elementa itaque quae dicuntur, non sunt priora iis rebus, quae generantur aut oriuntur; sed posteriora potius; & reliquiae magis, quam principia."

5

Reading God's Signatures in the Book of Nature

Paracelsian Medicine and Occult Natural Philosophy (Reuchlin, Paracelsus, Croll)

> Vom namen, so die wassersucht hat, es sei zu latein, zu griechisch, arabisch, chaldeisch, lasz dich nit bekümmern in ir etymologia, dan do spilen miteinander die sprachen und scherzen, wie die kazen mit den meusen; es ist on nuz.
>
> Do not trouble yourself over the etymology of names, such as dropsy, in Latin, Greek, Arabic, Chaldean; for languages play with one another and joke, like a cat with a mouse. Such an exercise is of no use.
>
> Paracelsus

The use of language—of etymology and exegesis of the languages of man—were of capital significance to the sixteenth-century academic physician Jean Fernel. Despite man's fallen state, the light of Christian tradition gave access to the hidden secrets of nature transmitted by an unbroken chain of *prisci theologi* and by later written texts to readers living in this privileged age of reform and renovation. The Age of the Book, of the dawn of printing announced by Gutenberg's Bible, thus witnessed hopes, and claims, of access to the originary *verbum Dei* through language itself.

While Fernel exemplifies the exegetical turn toward texts and language in the interpretation of nature, other early modern investigators turned their exegetical skills toward nature itself to uncover the network of symbols linking the Book of Nature to its divine author. Some, including naturalists like Conrad Gesner (1516–65) and Ulisse Aldrovandi (1522–1605), found in natural objects complex echoings of texts as well—of fables, myths, and poetry.[1] Others, including occult physicians like Paracelsus (1493–1541), exalted the wonders produced by nature over the vain products of human pride.[2]

Johannes Reuchlin

Despite sharp differences, the learned French medical doctor Fernel and the iconoclastic Swiss physician Paracelsus share an interest in reading the hidden causes or occult properties that lie veiled beneath the text(ured) surface of the Book of Nature. Indeed, both may trace aspects of their hermeneutical enterprise to the learned German occult-ist Johannes Reuchlin (1455–1522), heir to the Neoplatonic magic and language theory of Ficino and Pico.[3]

Like Fernel, Reuchlin suggests that manifest properties of living things are not responsible for the operations that support life. Rather, essential physiological functions such as digestion owe their action to "occult properties" (*proprietate occulta*).[4] The visible effects of life have roots in the hidden recesses of natural things, in powers and virtues that are not directly accessible to our senses. In Fernel this dichotomy between the manifest and the occult, the visible and invisible, both produces and supports a reading of nature that turns to texts—to books—and to language for access to the "divine" in nature, for access to divinely originating occult properties. Access to the "Word of God" through human texts and linguistic traditions thus holds the key to unlocking the hidden mysteries of nature.

Reuchlin, as student of languages, especially Hebrew, and of the oc-cult, especially through the cabalistic art, opens possibilities for both Fernelian and Paracelsian exegetical practices in reading the Book of Nature. Thus Reuchlin undertakes to recover from the most ancient philosophies (*vetustissimae philosophiae*) those secret words (*secretorum*

1. See Chapter 6 below and now W. B. Ashworth (1990) and Findlen (1990). On Aldrovandi, see Olmi (1976) and Pattaro (1981).

2. On Paracelsus, Paracelsians, and van Helmont, see literature cited below. On "won-ders" in general during the sixteenth and seventeenth centuries, see Céard (1977); Park and Daston (1981).

3. On Reuchlin, see Zika (1976, 1976–77); Geiger (1871); Secret (1964); Spitz (1963), and consult their bibliographies. The discussion in Copenhaver (1977) touches upon issues raised below.

4. See *De verbo mirifico* in Reuchlin (1964:41 [sig. c5r]) (I shall cite either signature or folio numbers from original as well as pagination provided in facsimile edition). Reuchlin's example of gastric digestion (1964:41–42 [sigs. c5r and c5v]) underlines hid-den, occult properties in medicine. See Pagel (1982, 1984).

verborum) that reveal divine powers and possess miraculous operations.[5] The very possibility of this enterprise rests upon an assumption that humankind has access to ancient, divine wisdom through two, ultimately connected, sources. Following Ficino and Pico, Reuchlin envisions history as a continuous, if esoteric, chain linking ancient wisdom to the modern search for truth and power. In his own version of the *prisca theologia*, such ancient wisdom has been passed down from Moses to the ancient Jews; the Greek philosopher Pythagoras consequently imbibed such wisdom from the font of Jewish learning and became the source for the much touted philosophical systems of the ancient Greeks.[6] The second source is the Hebrew language itself, which Reuchlin regards as a sacred, divine language.

Although Fernel, in his own account of the transmission of ancient wisdom to the Christian culture of the sixteenth century, does not emphasize the centrality of Hebrew in the same fashion, Reuchlin's statements can nonetheless provide some support for an exegetical search for traces of divine wisdom in the texts of ancient philosophy and medicine. In this sense, Fernel's program can look to Reuchlin, Pico, and Ficino for its inspiration.

Yet the narrative informing Reuchlin's major works—the *De verbo mirifico* (*On the Wonder-Working Word*) of 1494 and the *De arte cabalistica* (*On the Cabalistic Art*) of 1517—differs in subtle but significant ways from that of Fernel. Although both Fernel and Reuchlin spring from the rich humanist culture of late-fifteenth- and early-sixteenth-century transalpine Europe, the intrinsic value humanists placed upon the language and texts of ancient Greece—a value shared by Fernel—gives way in Reuchlin to the greater antiquity and divine origin of the Hebrew language. Although connected by a genealogy of *prisci theologi*, among whom Pythagoras enjoys a pivotal place, the legacy of Greek thought is, at best, inferior to that of Hebraic wisdom. With perhaps the exception of an esoteric strain stemming from the secret survival of Hebraic wisdom, the dominant, exoteric tradition of Greek thought

5. Reuchlin (1964:7 [sig. a2r]): "Tantorum igitur uirorum motus ingenti amicitiae uinculo: tantas ausus sum tenebras & tam obfuscata sacratorum: immo secretorum uerborum latibula ingredi: & quasi de adytis oraculorum & uetustissimae philosophiae penetralibus: exponere nostro saeculo quantum nobis memoria suppetit. Vniuersa ferme nomina: quibus superiori aetate sapientes homines & miraculosis operationibus praediti utebantur in sacris: siue pythagorica fuerint & uetustiorum philosophorum sacramenta: siue hebraeorum chaldeorumque barbara memoracula: seu christianorum deuota supplicia. . . ."

6. Reuchlin (1964:156–57 [*De arte cabalistica*, fols. XXIIv and XXIIIr]). See also Zika (1976–77:238).

introduced extraneous elements of mere human origin into the received ancient learning. As a result, Reuchlin denigrates both the traditional learning of the Schools and the merely human eloquence of the ancient Greeks and Romans revived by the humanists of his own day.[7]

Indeed, where Fernel would see continuity between the *prisci theologi* and the rational philosophy and medicine of later Hellenic and Hellenistic authors and would attempt to restore the hidden divine wisdom they contained through exegesis, Reuchlin exhibits by contrast a more radically ambivalent view. That view pushes him further and further away from traditional texts toward full embrace of the esoteric possibilities of the sacred Hebrew language itself made accessible only through the arcane practices of the Cabala. In essence, then, Reuchlin's embrace of this occult cabalistic practice springs from his radical recasting of dominant Hellenophilic readings of Renaissance master cultural narratives. As Charles Zika provocatively states, Reuchlin reverses the Greek denigration of the "Other" as *barbari:*

> The Hebrews are the true *barbari* and Hebrew the *barbara lingua*. The *barbara*—the true, holy, primitive, divine language and the rough, simple unadorned truth, handed down to men before the pride and confusion of Babel led to the development of human language—is intrinsically antagonistic to humanly devised eloquence. Moreover, the ancient pre-Greek peoples—the Egyptians, Chaldeans, Syrians and Indians—through their contact with the Hebrews and by means of their careful conservation of tradition and religion which contrasts with the fickle-minded Greeks interested only in novelty, preserved these essential qualities and are therefore also termed *barbari*. Not only are the pre-classical easterners chronologically prior, asserts Reuchlin, they are also culturally superior. (1976–77:245)

Consequently, Reuchlin's discourse moves away from a Fernelian emphasis upon an exegesis of traditional texts open to all postlapsarian, Christian readers to an esoteric, exegetical practice closed to all but an elite privileged with access to the esoteric practices of the cabalistic art.

In its construction of a rift between Greek exoteric *scientia* and Hebraic esoteric *sapientia*, Reuchlin's system of thought precipitates a nascent division between mere human reason and inspired divine insight that threatens to denigrate humankind's unaided ability to uncover the divine, or "occult properties," within the Book of Nature. Although he

7. Reuchlin (1964:160–61 [*De arte cabalistica*, fols. XXIIIIv and XXVr]). Zika (1976–77:240). Zika (1976–77:243–46) argues that Reuchlin, as proponent of Cabala and occult philosophy, turns away from traditions of Renaissance humanism that nurtured his thought.

is optimistic about human access to the "wonder-working Word," with all that it entails, Reuchlin's optimism nonetheless rests upon premises that ultimately encourage a more pessimistic view of inherent human abilities. It is precisely this pessimistic view of human knowledge, reason, and exegetical skills coupled with a conviction that humans can transcend such limitations by turning to sources of divine inspiration that links Reuchlin's cabalistic practice with Paracelsian occult natural philosophy (Pagel 1982:290–94).

Let me briefly point to those views of Reuchlin leading implicitly in the direction of Paracelsus. Reuchlin dismisses the ability of mere syllogistic reasoning to penetrate the secrets of nature (Reuchlin 1964:160–61 [*De arte cabalistica*, fols. XXIIIIv and XXVr]; Zika 1976–77:240). While reason has its place in our attempt to fathom the world about us, it is nonetheless severely limited as a tool for generating true knowledge. Apprehending the truth of things, for Reuchlin, requires faith. For it is only through faith that the human intellect reaches toward and may perhaps attain the purity, lucidity, transparency, and radiance of the divine and supercelestial. Through faith man's mind can become conjoined with God and hence enjoy something like intimacy with things (Reuchlin 1964:24–25 [*De verbo mirifico*, sigs. b2v and b3r]).

By stressing the limitations of unaided human inquiry and the dependence of *scientia* upon faith and the divine, Reuchlin moves dramatically in the direction of an illuminationist epistemology and toward a conception of knowledge of nature as grounded in a special kind of revelatory experience. To return to the initial point of contact among Fernel, Paracelsus, and Reuchlin mentioned above, the central notion of a nature organized and energized by occult properties—rather than by merely manifest elemental causes—raises questions of how one has access to those secret and hidden properties, virtues, or causes in nature. For Fernel, the intimate relationship between the originary divine, or Adamic, language and the hidden causes of things provides a key to human inquiry into nature through medicine and natural philosophy. Insofar as the redeemed Christian optimistically enjoys access to the Word of God through exegesis of its surviving traces in the languages of man, Fernel's project of bookish, textual learning can succeed in uncovering the occult properties of natural things.

Reuchlin's project shrouds itself in greater ambiguity. On the one hand, Hebrew, the divine language, wields enormous authority as source of all wisdom: terrestrial, celestial, and divine. But, although tantalizingly within reach, the secrets of divine wisdom are accessible only through a forbidding and secret art, one open only to a very select few: the cabalistic art. Moreover, though this art engages in a special kind of exegetical hermeneutics, the power and force of the cabalistic art, the power of the "wonder-working word," lies more in the enactment of the "word" grasped and enunciated properly through cabalistic knowledge than in the mere "reading" of its meaning through semiotic exegesis. [8] That is to say, the performative, ritualistic dimension of Cabala is paramount in Reuchlin, adding a dimension that reinforces the notion of *scientia* as rooted in divine revelation made manifest in man's illuminated experience of texts and nature.[9]

What I mean to suggest is that human access to knowledge of occult properties within nature follows from the nature of human access to the *verbum Dei* in Reuchlin's scheme. Although mankind enjoys the presence of the divine Hebrew language in the postlapsarian epoch, true access to the Word of God comes only through faith and the illumined enactment of the wondrous power of the Word in ritual, prayer, incantation, and the like. In like manner, access to the occult properties of nature—hence, to true *scientia*—comes only to those who, filled with the piety of faith, see through the pure and transparent light of the divine the most secret dimensions of things in their engagement with nature.

As we shall see, Reuchlin's emphasis upon the agency of occult properties in nature and upon man's ability, through faith, to apprehend the intimate workings of nature suggests aspects of a hermeneutical enterprise that Paracelsus and his followers would soon develop.

---

8. Zika (1976–77:236) notes Reuchlin's debt to cabalist Joseph Gikatila, the "pupil of the seminal thirteenth-century Kabbalist, Abraham Abulafia." On Abulafia, see Idel (1989).

9. For Reuchlin's performative view of the Cabala, I have relied upon Zika (1976–77), who sees this emphasis as the key to understanding relations between Reuchlin and Erasmus and the former's conception of reform. As Zika argues, "The esoteric and elitist character of Reuchlin's knowledge, the permeation of mystery, the emphasis on ceremony and ritual, the overflowing of knowledge as miraculous power, the validation of philosophy and religion by wonders and miracles, the analogy between Christian prayer and ancient magic—all these run directly counter to the style and content of the Erasmian reform programme, the *philosophia Christi*" (1976–77:240–41). In essence, Reuchlin moves sharply away from a merely exegetical and textual conception of the Cabala. Zika helpfully cites the case of Paolo Rici's defense of the Cabala, stating that "Rici defined the essential task of Kabbalah as the allegorical interpretation of the Bible. At the same time, he warned against practical Kabbalah with its angelic invocations and letter combinations and claimed that in the *De Arte Cabalistica* Reuchlin had misrepresented Kabbalah. In other words, Rici clearly deplored the theurgical techniques involved in Reuchlin's practical Kabbalah, and it is not difficult to see why Erasmus should have found his approach to Hebrew learning more palatable" (1976–77:243). For details, see Zika.

Paracelsus

Etymology is dead, and with it, the Age of Bookish culture! Paracelsus, as the epigraph at the beginning of this chapter attests,[10] was profoundly wary of the textual, exegetical practices of his contemporaries.[11] This is perhaps best known with respect to his alleged flamboyant iconoclasm: his burning of books by such ancient authorities as Aristotle, Galen, and Avicenna. Paracelsus' entire enterprise has thus been seen as a long overdue revolt against the verbal wranglings of Scholastic dialectic and learned Galenic disputation. But Paracelsus was no less contemptuous of other projects to revive learning, like the humanist movements of the Italian and Northern Renaissance, by returning to its sources in the texts of antiquity. Intensifying the fideistic and antirationalist impulses of a Reuchlin, Paracelsus met with supreme impatience notions that one could ferret out the hidden properties of things—nature's secrets—merely by opening the pages of dusty tomes. No matter how learned, how pious, how clever one might be in tracking down the traces of the originary Adamic tongue in the languages of man, one is bound to trip and stumble over mere words.

For man has not the wit to win at the game of etymology, of an exegetical hermeneutics that, like Fernel's, takes texts and languages as its primary referent and dominant world. In such games played with languages, it matters not whether man is cat or mouse: both victory and loss produce a species of oblivion. One's play with language is barren; rather than revealing the "being" in things, such play induces a pervasive forgetfulness—a withdrawal from the presence of things. It is, then, as the epigraph states, truly of no use.

Yet this disdain for etymology, for the play of and with languages, does not mean that the problem of language is alien to the discourse and practices of Paracelsus and his followers. Far from it! Paracelsus' nature is itself a "book," every page of which bears the unmistakable imprint—"signatures"—of the divine. No mere metaphor, Paracelsus' trope signifies a profound engagement with the *verbum Dei* and with the problem of how humans may gain access to it. What is radical and

10. I would like to thank Peter Schöttler for checking my translation of the epigraph, whose source is Paracelsus (1520:3).

11. It is no doubt a coincidence—a mere artifact of serious scholarship—that this passage by Paracelsus, a passage away from both dialectic and exegetical hermeneutics of nature, constitutes the very first line of the very first treatise that opens the very first volume of Sudhoff's canonical edition of Paracelsus' collected medicophilosophical works! On Paracelsus (in addition to studies cited elsewhere), see Braun (1981); Dilg-Frank (1981); Fussler (1986); Kammerer (1971, 1980); and Schipperges (1988).

iconoclastic about Paracelsus is not his book burning, disdain of authorities, or even legendary (if unlikely) drunken "professional" behavior,
but rather his disdain for the languages of man as vehicle for engaging
the radiant and saving Word. In short, Paracelsus' radical turn toward
nature as text marked him as iconoclastic.

This turn was, of course, negotiated through discourse: its course
(em)plotted by a narrative about God, man, nature, and—no less!—
language. Here, then, are some of its essentials. Man, says Paracelsus,
"received from God in Paradise the privilege of ruling over and dominating all other creatures, and not of obeying them."[12] Although relin-
quishing this "right" as a consequence of the Fall (Paracelsus 1537:378),
humans still retain freedom even in face of an unruly nature. Paracelsus wishes to make something of this freedom, and to do so he must
unveil both humankind's strengths and its weaknesses. Postlapsarian,
Christian "man's" assets and deficits may be reckoned by attending to
Adam's prelapsarian powers and the legacy of the Fall. Here Paracelsus
invokes the *ars signata* as link between paradisical and fallen Adam.
This *ars signata* "teaches the true names to give to all things."[13] Under-
scoring its significance, Paracelsus tells us that

> Adam our first father had complete knowledge and perfect understand-
> ing of these names. For directly after the Creation he gave to all things
> their own proper and specific names. He gave to each of the animals,—
> and also to the trees, roots, stones, ores, metals, waters and to the fruits of
> the earth, water, air and fire—its own special name. And as Adam then
> christened them with their names, so was God pleased to ordain them.
> For their names were based upon real foundations, not upon their pleas-
> ant appearances, but rather upon a predestined art, namely the signatory
> art [*kunst signata*]. And for that reason, Adam was the first practitioner of
> this art of signs [*signator*].[14]

12. Paracelsus (1537:378): "So in doch got im paradeis privilegirt hat, also das er uber
alle andere geschöpf sol ein herr und kein knecht sein . . ." (translation Paracelsus
1976:1.174). The Waite English edition is based upon a Latin translation of Paracelsian
works published in Geneva in 1658. Where necessary, I have modified this English version, below, with my own translation from the German. Although Sudhoff (Paracelsus
1537:XXXI–XXXIII) casts doubt upon the authenticity of the *De signatura rerum naturalium*, for our purposes it not important whether this text was written by Paracelsus or
by a follower. Bianchi (1987:70–71), suggests: "Chiunque sia l'autore di questo testo,
Paracelso o, come è molto probabile, un suo sconosciuto discepolo, è indubbio che esso si
inserisce senza sforzo nel quadro delle concezioni fin qui esaminate e riflette, almeno
nella sua problematica generale, un punto di vista genuinamente paracelsiano."

13. Paracelsus (1537:397): "die kunst signata leret die rechten namen geben allen
dingen." Paracelsus (1976:188).

14. Paracelsus (1537:397): "Die hat Adam unser erster vater volkomlich gewusst und
erkantnus gehabt. Dan gleich nach der schöpfung hat er allen dingen eim iedwedern

Although Paracelsus attributes to Hebrew (obviously in light of the belief that Adam spoke that language in the Garden of Paradise) the ability to capture the nature of things through their Hebrew names,[15] that capacity of language does not concern him. Rather, what remains significant for Paracelsus is the conviction that Adam's ability to give all creatures their proper names was dependent upon our first parent's capacity to penetrate the mere appearances presented by natural things to uncover their real natures. Indeed, Paracelsian natural philosophy turns upon a vision of man, not as dwelling in the house of language, but as restless sojourner in the world of things. For postlapsarian man must become, like Adam, a *signator:* a practitioner of the *ars signata,* one who operates within and on nature, fathoming signs and manipulating the hidden virtues of things.

This, Paracelsus hints, is what the wise man (*der weis man*) does.[16] His wisdom comes, not from his intellect, not from his ability to read texts like a slavish scholar, but from his ability to transform himself into a new Adam through his direct engagement with nature. The nature of that engagement ultimately depends upon Paracelsus' understanding of the relationship of the *verbum Dei,* the divine and creative Word, to the world and to man. That relationship is at the very heart of what Paracelsus calls the *ars signata,* which turns upon the Paracelsian notion of *signatura*—the signatures of natural things.

Thus the Paracelsian account prompts us to ask how it is that Adam, as the first practitioner of this "art," can give things their real names. Let us examine the passage concerning Adam's naming of all creatures

---

seinen besondern namen geben, den tieren einem ieden besondern namen, also den beumen einem ieden seinen besondern namen, den kreutern ire besondere underschitliche namen, den wurzlen ire besondere namen, also auch den steinen, erzen, metallen, wassern und andern früchten der erden, des wassers, lufts und feurs eim ieden sein namen. Und wie er sie nun tauft und inen namen gab, also gefiel es got wol, dan es geschach aus dem rechten grunt, nit aus seinem gut gedunken, sonder aus einer praedestinirten kunst, nemlich aus der kunst signata, darumb er der erst signator gewesen." Paracelsus (1976:188). I have substantially modified the English provided in Waite's edition.

15. Paracelsus (1537:397–98): "Wiewol nit minder ist, das aus hebraischer sprach auch die rechten namen herfliessen und erfunden werden, einem ietwedern nach seiner art und eigenschaft. Dan was für namen aus hebraischer sprach geben werden zeigen mit an desselbigen tugent, kraft und eigenschaft."

16. Paracelsus (1537:378). Paracelsus contrasts the wise man to the bestial man (*einen viehischen menschen*). One rules the stars, the other is ruled by them. Implicit here is the notion that fallen man, though he has lost Adam's dominion over nature, may still win back some of his lost control. To do so, he must revive and master the Adamic *kunst signata.*

more closely. Our Paracelsian author states that Adam's naming reflects his "complete knowledge and perfect understanding." What is the source of this understanding? Here we encounter a knot in the Paracelsian narrative, for there seem to coexist two contesting themes. On the one hand, Adam appears to choose names freely, if aptly; on the other hand, the names themselves are grounded in what the story tells us is a "predestined art." This "knot" can, I believe, only be loosened, not untied. Indeed, the tension it embodies is, in my view, essential to the Paracelsians' vision of man in the created universe. While emphasizing the creatureliness of humans—their dependence upon God and, indeed, the fallen nature that clouds the unaided human intellect— Paracelsus and his followers for the most part draw back from reducing humans to creatures utterly devoid of the freedom to act, to control nature, and to influence their own destiny.

Adam—prelapsarian man—freely chooses the names that he gives to all creatures in the Garden of Eden. Yet this freedom does not mean that the names they are given are in any sense arbitrary. Language, or at least the Adamic language, is not a product of use, a human social construct, in Paracelsus' view. Free from all sin, Adam's language bears a close relationship to the *verbum Dei*. God has not directly imposed this language upon Adam, either by talking with Adam in that language, or by infusing knowledge of it directly upon his mind through divine illumination. Rather, God provides Adam with a tool that, given his pure and unfallen nature, he is then able to use correctly. This tool—the signatory art (*kunst signata*)—is not itself knowledge of the true names of things. Rather, its proper use requires that Adam turn directly to nature—to God's creatures—where he may then read those "signs" that mark each natural thing as a natural kind. That is to say, Adam's perfect understanding of the names of things depends upon his ability to grasp, through direct experience, signs that reveal to the able practitioner of the *ars signata* the innermost secrets, hidden properties, occult virtues, and hence true natures of things. By freely and accurately reading such signs, Adam exercises his privilege as unfallen *imago Dei* and microcosm to rule over all other creatures. He also comes to enjoy knowledge of those very elements of things that are based, or patterned, upon the divine plan or ideas that are the blueprint for all creatures. As a result, the names spoken by Adam—the Adamic language—is one that "pleases" God as the true, correct names that God then, in His power and freedom, ordains as such. The Adamic names thus bear a direct and univocal relationship to the Word of God.

Postlapsarian man has lost full knowledge of the signatory art that brought Adam perfect understanding of things. But while his intelligence is clouded, man's freedom enables him to choose the path of wisdom or of bestiality. Choice of the former path leads humans to the *ars signata* and thence to nature's secrets. How is this path opened to humans in Paracelsus' view?

Not through language. Not through mere textual exegesis and etymologies. Paracelsus will not abide humans playing cat and mouse with language and texts. The way to the Word, to the effects of the divine *logos* upon nature, is through nature itself. But humans— postlapsarian man—must approach nature properly, in the proper spirit. Man's freedom—limited though real—must seek the proper path and encounter nature piously and, hence, with the aid of God. For "Adam" represents not only humankind in paradise; rather, "Adam," as Kurt Goldammer insists, "c'est l'homme au sens théologique, c'est-à-dire tout d'abord un être *perdu* et *reprouvé*, un *prisonnier du péché*" (1973a:248). This prisoner of sin nonetheless retains a certain liberty and can consequently avail himself of whatever help God provides him in his fallen state. And it is precisely the divinely implanted light of nature (*lumen naturae*) that enables humans, Goldammer asserts, "à surmonter le *status corruptionis*" (1973a:248).

But it is not such an easy task, this overcoming of man's fallen nature.[17] While the key may be the light of nature, access to that light requires preparation. Given that preparation, however, nature and its secrets lie open before all of mankind:

> La "*lumen naturale*" est le mot clé pour l'ensemble de la théorie de la connaissance et la méthodologie scientifique de Paracelse. C'est elle qui relie Chrétiens et Non-Chrétiens, car elle a une portée qui dépasse et surmonte le domaine de la révélation chrétienne. Les païens, en effet, qui connaissent dans la lumière de la Nature sont donc aussi illuminés du Saint Esprit.[18]

Man must prepare himself for God's illumination in order that the light of nature may awaken his intellect to the correct interpretation of

---

17. Goldammer (1973a) provides a complex understanding of Paracelsus' divided, tension-filled view of man, at once limited and filled with possibilities. See 248–51 and 258.

18. Goldammer (1973:242). Goldammer notes Paracelsus' conflicting tendencies: at once separating nature from revelation and drawing natural and religious knowledge together. In the end, he suggests, "Le dilemme paracelsien se clôt en une *synthèse paracelsienne* dans laquelle lui vient en aide, d'une façon très importante, la notion de *lumière de la nature* avec son double aspect philosophico-scientifique d'une part et théologique d'autre part" (1973:242).

the signs God's created nature has stamped upon things. Man must learn, through the *ars signata*, those practices that allow him to be bathed in the radiance of the *lumen naturale* in his direct engagements with nature.

In short, humans must turn, with piety, toward nature to encounter in the concreteness of lived experience both the grace of divine insight and the light with which nature reveals its own innermost secrets. In his conceptions of man, nature, experience, and God's "signatures," Paracelsus points the way to a hermeneutics of nature that promises to complete the narrative of Adam: of the "old," once privileged, and now fallen Adam and the "new" regenerate Adam to come.[19]

Why, for Paracelsus, is it essential that man, this prisoner of sin, turn to direct and concrete experience of nature? How can such experience open the human mind to the "light of nature"? The answer to these questions perhaps lies in what would have seemed a paradox to Paracelsus. For it was precisely what is most accessible, ready-at-hand, and visible that provides the key to unlocking the most intimate secrets of nature hidden to human reason. Aristotelians, Galenists, and other postlapsarian students of nature ignored what was under their very noses to pursue fictitious entities raised up by their clouded but prideful reason. While it is true for Paracelsus that the properties, powers, or natures that give rise to the activity of things in nature are occult, i.e., hidden from our direct apprehension, it is equally the case that nature has left evident "traces" of such animating properties in the visible texture of things themselves. Thus, as Paracelsus was so fond of saying, "It is the exterior thing alone that gives knowledge of the interior; otherwise no inner thing could come to be known."[20] Postlapsarian man has ignored the external signs of the internal nature of things; by his failure to engage things directly, he has failed to penetrate the very secrets of nature. His failure, then, is at least in part a failure to learn how humans must read the visible book of nature placed before them by our Creator.

Paracelsus' "doctrine of signatures" is then both an account of the relations between the inner and the outer aspects of concrete things in nature, and a hermeneutics that teaches fallen man how he is to read things, to read nature itself. The themes of nature's signatures and of

19. On the new Adam, Christ, God's Word, and the light of nature, see citations below to Paracelsus (1537–38: esp. 398). On Paracelsus and theology, in addition to Goldammer, see Rudolph (1980, 1981).

20. Paracelsus (1529–30:97): "Allein die eussern ding geben die erkantnus des inneren, sonst mag kein inner ding erkant werden."

our dependence for knowledge of the hidden, inner nature of things upon external "signs" that signatures (re)present to man pervade virtually all of Paracelsus' work. In different ways, he tells us that the understanding of signs is indispensable to the knowledge and manipulation of nature. "Nothing," he asserts, "exists without its signature." Indeed, there is nothing produced by nature that it has not marked by a sign of what is inside it.[21] But such signatures have completely fallen out of use, forgotten entirely by fallen, error-ridden man.[22] Hence it is imperative that "those who wish to depict natural things must grasp their signs and understand the same through their signatures."[23]

Paracelsus urges that humans become adept at reading signs. Nature imprints all things that it creates with an outward mark, a sign or signature. Signs enable one who knows how to read them to know what it is that a given thing—a plant, an animal, a mineral, even the stars—contains within it. This "inner" thing controls, shapes, and activates that which we recognize by its external form and properties in nature. But it is the internal, occult, and secret properties of a thing that constitute its very nature, that are the key to our understanding its "thingliness" as a creature of God, and that contain those very "virtues" that empower it to act according to its nature. Understanding of these properties—an understanding that may be grasped only experientially—enables man to tap into or extract the powers that lie hidden in nature and thence to control, transform, channel, and transmute them. Thus, Paracelsus envisions a society in which those who are privileged by their experience—by the expertise that grounds itself in intimate experience of nature—can use their expert knowledge to fathom the secrets of nature and use such hidden virtues to improve the lot of humans on earth. For example,

> The expert practitioner of the art of signs [*signator*] may recognize by means of the signature the virtue inhabiting each material being—that which is in herbs and in trees, in sensible and in insensible things. For consequently such expert signators discover a great many medicaments, remedies and other powers in natural things. And whosoever does not

21. Paracelsus (1525:86): "dan nichts ist on ein zeichen, das ist, nichts lesst die natur von ir gon, das sie nit bezeichnet das selbig, was in im ist." The entire page dwells upon the role of signs and signatures in Paracelsus' conception of nature.

22. Paracelsus (1525:86): "Die selbig signatur ist gar aus dem brauch komen und ir gar vergessen worden, aus dem dan gross irsal folget."

23. Paracelsus (1525:86): "Der do wil beschreiben die natürlichen ding, der muss die zeichen fürnemen und aus den zeichen das selbig erkennen." I have not been able to locate a copy of Quecke (1955); Goltz (1972) proves not nearly as relevant to my discussion as the title might suggest.

note the power and efficacy of a plant from its signature, that person does not comprehend what he writes. Indeed, he writes like a blind person who does not understand what he writes.[24]

Postlapsarian man, blinded by his fall and turned away from things to the contentious world of mere human words, does not comprehend what he writes and does not gain access to the hidden nature of things through his vain "sciences." The order prescribed by God in His creation does not permit such access except to those who practice a different "science"—one that relentlessly follows the tracks ordained by God in nature that links the external signs of things back to their inner virtues and actions.[25] This new Paracelsian "science"—the true knowledge of natural things—reads in the visible signatures of things their origins and meaning in the invisible and spiritual nature of their inner constitution.[26]

Visible and invisible, material and spiritual, concrete signature and symbolic order of beings: all find themselves inextricably woven together in the texture of God's creation. God's authorship of the Book of Nature, like Christ's voice speaking of divine mysteries in the languages of man, animates the mere material traces of the letters He has stamped upon this world. In fact, the very light of nature that Paracelsus extols throughout his works would be nothing, were it not for the fact that it came from God. God's Word is the source of our illumination through the light of nature, just as our very nourishment

24. Paracelsus (1537–38:173): "der signator mag durch das signatum erkennen die tugent im selbigen corpus, es sei in kreutern, beumen, entpfintlichen oder unentpfintlichen. dan also haben die signatores vil medicamina, remedia und andere vires in natürlichen dingen gefunden, und wer nicht aus der signatur die kraft der kreuter schreibet, der weiss nicht was er schreibt. er schreibet gleich als ein blinder, der weiss nit, was er schreibet."

25. See Bianchi (1987:62): "Il costituirsi del mondo visibile come un insieme di segni che rimandano all'invisibile viene a garantire, per il medico di Einsiedeln, la completa accessibilità della natura al conoscere dell'uomo." Chapter 2 is devoted to "Paracelso"; its second section pays special attention to the theme of "Inner e Eusser" (1987:44–63) in his works, while the third dwells upon the theme of "Segni e Signaturae" (1987:63–86).

26. Paracelsus (1537–38:177): "Also hat die natur verordnet, das die eussern zeichen die innern werk und tugent anzeigent, also hat es got gefallen, das nichts verborgen bleibe, sonder das durch die scientias geoffenbart würde, was in allen geschöpfen ligt. es möcht manchen verwundern, warumb got solches verordnet hat, das der mensch durch die kunst das verborgene sol erfaren? so ist es doch alein die ursach, das der leib des menschen sein ubung habe oder dergleichen, und das erfare, was got in die körper spiritualisch gelegt und verschaffen hat. darumb had er nicht geordnet, das die scientia unerkantlich sei, sonder alle heimlikeit seind in ir und alle tugent seind den scientiis underworfen in der erkantnus, durch sie zu erfaren was spiritualisch in allen dingen ligt."

and life itself comes "not from the earth, but from God through His Word."[27] Where the blind Galenist or Aristotelian sees but mere accidents—a mere cloak of passive matter to be cast aside in the profound search for *entia rationis*—the divinely illumined Paracelsian sees living symbols of nature's hidden truths. Thus Paracelsus rails against the mere "bookish" medical learning of a Galen, Avicenna, Mesue, or Rasis. The door they hold open to their so-called art of medicine leads to but a blind alley. Only nature itself offers humans true access to the secret healing powers in things. The true door opening upon the art of medicine is that produced by the "light of nature" alone. Against the false, prideful, and pagan books of man, one must oppose the only true book: "that which God Himself has given, written, devised, and established."[28] It is to this book alone that man must turn if he is to uncover, through the light of nature, the occult properties and secret virtues deep within natural things. Those who ignore God's book, who foresake the *lumen naturae*, are condemned to wander in the world as if in a labyrinth without hope of escape.[29]

Exalting not the fictions of human reason, but the concrete particularity of individual things themselves, the Paracelsian discovers in things the symbolic order of nature ordained by God. This order, moreover, constitutes itself as a universe of individual things joined together by the symbolic relations forged among the occult, secret properties and virtues animating each thing in nature. The individual, by virtue of its

27. Paracelsus (1537–38:396): "Also ist die speis und das leben nicht von der erden, sondern von got durch sein wort"; and "So das ist, so ist das natürliche liecht nichts, sondern es muss aus got gehen, dan so ist es genug" (1537–38:397). More generally, "Das acht capitel, probatio in coelestem philosophiam adeptam" (1537–38:395–99) provides a good starting point for the relationship Paracelsus envisioned between the divine Word, Christ, the new Adam, the light of nature, and man's attempt to understand nature.

28. Paracelsus (1537–38a:169): "Sagen sie mir, welches ist zur rechten tür hinein gangen in die erznei? durch den Avicennam, Galenum, Mesue, Rasim ac oder durch das liecht der natur? dan da sind zwen eingeng, ein ander eingang ist in den bemelten büchern, ein ander eingang ist in der natur. ob nun nit bilich sei, leser, das da ein ubersehen gehalten werde, welche tür der eingang sei, welche nit? nemlich die ist die rechte tür, die das liecht der natur ist, und die ander ist oben zum tach hinein gestiegen; dan sie stimmen nit zusamen. anders sind die codices scribentium, anders das lumen naturae; anders das lumen apotecariorum, anders lumen naturae. so sie nun nit eins wegs sind und doch der recht weg in dem einen ligen muss, acht ich, das buch sei das recht, das got selbst geben, geschriben, dictirt und gesezt hat. und die andern bücher nach irem bedünken, consilia, opiniones geben so vil sie mügen; der natur ist nichts genomen."

29. Paracelsus (1537–38a:170): "Was concordirt in das liecht der natur, das bestet und hat kraft. was aber in das nit concordirt, das ist ein labyrinthus der kein gewissen eingang noch ausgang hat."

innermost constitution as a being—a natural thing—participates in a universal network of occult powers. Thus the individuality and particularity of material things give way to a deeper spiritual unity: a holistic universe resonating with hidden powers.

Paracelsian alchemy, astronomy, and, capitally, medicine hold out the prospect for humans of tapping into this vast network of forces. The key to healing is thus in the signatures made manifest in nature to those who cultivate the *kunst signata*, the Adamic art of reading the Book of Nature. This art allows man to search for those hidden correspondences between afflicted organs (and their particular, individual disease entities) in the microcosm that is the human being and the specific minerals, plants, animals, and heavenly bodies that constitute the macrocosmic world of the universe.

Cultivating the *ars signata*, the art of reading signatures, requires mastery of a number of skills, such as chiromancy and physiognomy.[30] We need not concern ourselves with learning the details of these, and other, skills. Instead, we should note the quality of the "experience" that lies at the heart of this Adamic art and how, by following the quest set for us by God, the pious human engaged in divinely sanctioned experiential engagement with nature uncovers the *lumen naturale* and thus the source for all true knowledge (*scientia*):

> For God willed that all that He has created for the good of man and placed in his hand as his property not remain hidden, occult, or secret. And although he had made them occult, he has also, after all, left nothing [in nature] unmarked with an outwardly visible sign. Here His special predestination ought to be recognized. Just in the same way, men themselves, if they bury treasure, mark the place by the addition of some sure signs. . . . The old and wise Chaldeans and Greeks, if in time of war they feared siege and exile, buried their treasures, and only marked the place by proposing to themselves a certain fixed day, hour, and minute of the year. They waited until the sun or moon cast a shadow there, and in that spot they hid or buried their treasures; which [procedure] they thought of as a special art and occult wisdom. Hereafter, they called this art Sciomancy, that is, the art of shades or spirits, which they otherwise termed shadow [*umbra*]. From such an art many other arts have sprung up and many occult things have been revealed, all spirits and heavenly bodies came to be discerned: for such are cabalistic signs which in no way can deceive, and therefore should be given special regard.[31]

30. See, for example, Paracelsus (1537–38:15–77; 1537: book 9), "De signatura rerum naturalium," esp. 377–87.

31. Paracelsus (1537:393): "Dan alles was got erschaffen hat dem menschen zu gutem und als sein eigentumb in seine hent geben, wil er nit das es verborgen bleib. und ob ers

The Paracelsian *ars signata*—the foundation for Paracelsus' natural philosophy and medicine—is, in one sense, akin to Reuchlin's cabalistic art. Looking to the Book of Nature rather than to merely human books, to the *explicatio* of God's creative Word in things rather than to the presence of His Word in the dead, external form of mere letters, Paracelsus seeks to uncover the symbols embedded by God in nature and to read them as ciphers of God's presence in the world. We can, I think, hear this Paracelsian message resonate in the language of that infamous English occultist of the seventeenth century, Robert Fludd:

> The variety of the Species upon the earth, did radically proceed from the very act of creation, when the word *Fiat* was spoken, and immediately the will of the speaker was accomplished by his Son, which, by the way of emanation, was sent into the world to do the will of his Father. And there are some that will not shrink to say, that all the Species or kinds of creatures, were expressed in and by the 22. Hebrew letters, not those externall ones which are vulgarly painted out with Ink or Art, which are but shadows; but the fiery formall and bright spirituall letters which were ingraven on the faces or superficies of the dark hyles, by the fiery word of the eternall Speaker in the beginning, and therefore they are tearmed originally *Elementa quasi Hylementa*, or Elements; as engraven in the forehead of the dark abysse or Hyle, and by reason of the essence of that divine Word, which received the mystery of the Typicall creation, and did trace it out after the Archetypicall patern, and delineated it in characters of formall fire the language which was framed out of it was called *Lingua Sancta, a language* (I say) much spoken of by the learned *Rabbies* of our age, but little known or understood by them, and yet of an infinite importance, for the true enucleation as well of sacred Mysteries, as of all true Cabalisticall

---

gleich verborgen, so hat ers doch nicht unbezeichnet gelassen mit auswendigen sichtbarlichen zeichen, das dan ein sondere praedistination gewesen. zu gleicher weis als einer, der ein schaz eingrebt, in auch nicht unbezeichnet lasst mit auswendigen zeichen, damit er in selbs wider finden könne. darumb sezt er darauf oft ein markstein, ein biltseul, ein beumlein, capeln &c oder ander dergleichen ding. die alten weisen Chaldeer und Griechen, wann sie in kriegsgeferlikeiten (da sie besorget sie möchten vertriben werden) schez ein und undergraben, haben den ort anderst nit bezeichnet, dan das sie im iar ein gewissen tag, stunt und minuten für sich genomen und achtung gehabt, wo die son oder der mon seinen schatten hingeworfen, daselbst hin haben sie den schaz eingegraben, welches sie für ein sondere kunst und heimlikeit gehalten. dise kunst haben sie hernach sciomantiam geheissen, das ist schattenkunst, welche man sonst umbram nennet. aus solchen umbrabilibus vil künste iren ursprung haben und vil verborgner ding offenbart, alle geister und siderische corpora erkent werden, dan solches sein cabalistische signata, die in keinen weg betriegen können darauf sondere achtung zu geben ist." Paracelsus (1976:1.185). I would like to thank Barbara Hahn for help with my translation of this passage.

abstrusities. But to proceed: According to the tenor of the divine Word, and his formall characters, the effects whereof passed unalterable into the world, each species or kind was framed. . . . [32]

It is by reading this "text" of nature—symbolically, "cabalistically"— that the practitioner of the art of signatures can recover the true Adamic names of things.[33] Experience, and practice of this art, bring humans access to the *verbum Dei*.[34]

Oswald Croll

The socioreligious implications of Paracelsus's concept of alchemy were profound and revolutionary. Not only was the peasant-artisan elevated to the status of an alchemist, he was allotted a positive rôle in a great drama which was nothing less than the redemption of the world. Just as Christ redeemed man the microcosm, who had fallen from grace through the sin of Adam, so man in his turn would redeem the whole of nature, which had fallen with him, by separating the pure from the impure and refocusing the virtues and spiritual powers of nature on himself, the center of the great world. Thus the whole of nature would be redeemed—nature through man and man through Christ. This theology of the priesthood of the laborer was at the center of Paracelsus's social and religious challenge to his times.[35]

Owen Hannaway eloquently reminds us that Paracelsus' vision of man as a second Adam was no mere theological abstraction. His anthropology, no less than his medicine, was "revolutionary" in its potential import and concrete in both its artisanal models and its articulation of the prospective role of the physician-alchemist. At the center of this revolutionary ethos was the ideal of man as imperfect, but free, agent of personal and collective regeneration. Paracelsus embeds this ideal, or, rather, resurrects it, from within the traditions of Judeo-Christian narratives of the Fall and Christ's redemption of man. The very history of

32. Fludd (1659:161). On Fludd, see Huffman (1988).
33. Paracelsus (1537–38:92): "Und die kunst signata lernet die rechten namen geben einem ieglichen, wie im angeboren ist." See also, Bianchi who cites Paracelsus (1537– 38a:204), on the importance of magic and the Cabala to his art: "Tramite essa e le sue *species*, la *gaballia* e la *gabalistica*, tutto cio che vi è di segreto e misterioso nella natura ('alle heimlikeit in verborgner natur') si fa aperto e manifesto (*offenbar*)" (1987:67).
34. On the notion of experience as central to Paracelsian practice, see Paracelsus (1537–38a: esp. 190–95, "Das sechst capitel: Von dem buch der arznei, so experientia heisst, wie der arzt dasselbig erfaren sol").
35. Hannaway (1975:44–45). Hannaway cites Goldammer (1948–52, 1953). See also Fussler (1986).

Paracelsianism as a natural philosophy, then, unfolds through successive recountings of this basic story. (It is a story that opens itself to "revolutionary" turns; we need not, however, commit ourselves to this revolutionary trope and its historiographical implications!) Those who recast the Paracelsian narrative increasingly negotiate a dramatically changing socioreligious terrain in early modern Europe that defines their own local positions as agents and subjects. When compared with their origins in Paracelsian thought, such positions can appear highly idiosyncratic.

Far from being an exception, Oswald Croll (c. 1560–1609) is a sterling case in point. Where Paracelsus' natural philosophy sprang from a Catholic, if heterodox, anthropology, Croll, in Hannaway's words, transmuted his forebear's corpus into "a Calvinist Paracelsianism."[36] According to Hannaway's account, Croll's Christology differed from that of Paracelsus, who saw Christ in human terms as the "second Adam." Instead, "Christ," for Croll, is "the Word Incarnate"; through Christ as the divine Word, "man is reborn in the light of grace" (Hannaway 1975:46–47, 50). Furthermore, man's knowledge of nature— knowledge that Paracelsus had regarded as rooted in the "light of nature" and as an inherent part of man's destiny that he must strive after in reclaiming his Adamic legacy—was strictly subordinated in Croll's view: "Man's works in and of themselves have no value for salvation; they are only justified through Christ in the light of grace. Thus there can be no independent path to God in the light of nature, nor can man achieve any redemptive works in nature. . . . it is only through the Word of Christ in grace that we can gain access to the Word of the Father in nature" (Hannaway 1975:50). At the risk of disturbing some of the ground cultivated by Hannaway, I plan to explore how Croll understood the relationship between the Word of God, man, and nature in the following pages, and to locate hermeneutic practices available to Paracelsians.

Croll repeats a by now familiar theme: the creativity of the divine Word. Starting from the "DIVINE NOTHING, or invisible Cabalisticall Poynt," God's creative act educed all creatures, all things, "out of the invisible Darknesse." How? Out of this primordial darkness Croll insists that things "were called out to the visible Light by the WORD speaking."[37] Moreover, it was "by vertue of that omnipotent *Word*," that God

---

36. Hannaway (1975:47). For differences between Croll and Calvin note Hannaway's discussion (1975:53–57).

37. Croll (1675:70). The text is a translation of "The Admonitory Preface . . . To the Most Illustrious Prince Christian Anhaltin," which, in the original Latin, prefaced Croll

implanted "a certaine specifix vertue concealed" in "Seeds" in each and every individual thing in the created universe (Croll 1657:77; see also 34–35, 70, 77).

The agency of the Word in the origins of the world—the world inhabited by humans—can thus be taken for granted. What perhaps is of more interest, and certainly of great pertinence to our exploration of the hermeneutic strategies and practices of Paracelsianism, is how the Word—and, derivatively, words, signs, and symbols—operate in Croll's cosmos and in human apprehension and manipulation of it.

Croll insists that "the WORD of God . . . is that which healeth all things." It is this divine Word that operates in all through all medicines used by the physician for healing. In glossing what might appear to be no more than a pious recognition of God's majesty and man's dependence upon His infinite love and mercy, Croll is still more emphatic. He insists that the very power of medicines is occult, or hidden, and that the Word engenders the virtue or force exhibited by them. This force is something "beyond any naturall actions," so that, in Croll's view, the only conclusion one can draw is that "hearbs are not the medicines, but a signe onely of the *Word* signified" (1657:83).

The text of nature explicitly requires thoroughgoing semiotic analysis if humans are to learn how to use things in nature for their own and society's benefit. Croll's semiology distinguishes between mere signs encountered in nature and the efficacious "Word" such signs represent. What appear to be potent natural substances are, at best, mere vehicles for divinely originating occult properties. Indeed, Croll goes to great lengths to articulate the relationship between divine Word and created thing in his semiological and ontological analysis of the nature of medicine as a healing art.

Croll's analysis employs various tropes, particularly metaphors, that have strong affiliations with the tradition of radically alternative medicine stemming from Paracelsus himself. Such metaphors typically introduce or reinforce key pairs of terms that are meant to suggest binary oppositions, although, in the hands of a figure like Paracelsus, their very opposition may at times be forcefully contested, or, less dramati-

---

(1609). Prince Christian of Anhalt was Croll's patron and "the Calvinist champion of central Europe" (Hannaway 1975:56, where further bibliography can be found). I cite this text in part because of Hannaway's use of it, and in part because of its close connection with English developments concerning Paracelsianism, science, and language in mid-seventeenth-century England. Volume 2 will analyze developments in England from roughly 1640 to 1670. I shall reserve any comments on Henry Pinnell's interesting introduction or "Apology" for my treatment of this latter period.

cally, subtly subverted by unresolved tensions in the text. I have in mind, for example, such Paracelsian dyads as visible/invisible and inner/outer that we have encountered previously.

Thus Croll argues that "physick"—medicine—has a "two-fold" nature that man learn to appreciate. First, there is a "visible" physick "which the Father has created." These are what in the twentieth century one might call "natural" substances—herbs, for example. According to Croll, such visible medicines "ought not to be administered before there be a separation of the pure from the impure" (1657:83). This suggestion conforms to the broad theoretical strictures and practice of Paracelsus' alchemy or chemical medicine, in the sense that it recognizes in visible substances active properties whose healing powers can be released only through the purifying flame of the alchemist's fire. However, the precise semiological relationship between sign and signified entailed by Croll's use of this metaphoric structure in distinguishing two kinds of physick gives, as we shall see later, a somewhat different complexion to the Paracelsian discourse evoked by the use of such terms.

Opposed to this "visible" physick was an "Invisible, from the Son by the *Word*" (Croll 1657:83). Yet, oddly enough, that does not mean that for Croll there are two different medicines—two different healing agents—in a given substance: one visible, the other invisible. What heals is but one, namely, the Word: "the Physitian cureth by means which are the Hearbs in which the medicine is, the Hearb is not the medicine, for that is invisibly hid in God himselfe" (1657:83–84). If this is the case, what sense does it make for Croll to state that there are *visible* physicks?

The answer to this question comes only at the end of an excursion Croll takes into the significance—better yet, power—of words, names, characters, and signs. For the very invisibility of that which cures, its origins in "God himselfe," appears to Croll intimately connected to the ability of "the *Prophets* and True *Cabalists*" to cure "with a word onely" (1657:84). The connection Croll sees between invisible cures and words-as-cures (apart from the invisibility of the spoken word itself)[38] centers upon the intimate relationship of words to God:

---

38. The significance of the invisibility of the spoken word should not be overstressed, for even spoken words can be regarded as mere carnal, material embodiments, even distortions, of a pure, spiritual sense. In this sense, then, not just any word will do as agent of healing. Carnal words must also undergo, tropologically, a fiery purification, just as the visible dross of medicines must yield to the pure healing virtue within.

> God is a living God, the NAME also of the living God is lively, and so the
> Letters of the living Name are also lively: God liveth for himselfe, his
> Name liveth because of him, the Letters live by reason of the Name; as
> God hath life in himselfe, so hath he given to his Name to have life in it
> selfe, and the Name also to the Letters. (1657:84)

Having spoken just moments before of the creative power of God's
Word and its efficacy manifest as a force within medicines, Croll
transfers the power of God as *logos* to a particular, concrete word.
The "name of God" does not merely reflect the power and liveliness of the divine Word—of God—itself; rather, it itself is alive, and
shares its divine vitality with the very letters out of which it is composed. Croll here stands on the threshold of dangerous territory, one
with hidden risks threatening at any moment to erupt into the chaos
of full-blown heterodoxy. Secure in his convictions, Croll does not
hesitate to proceed: "Great things have been affected by True *Magicians* . . . those accurate searchers out of Nature, by a *Word written*
and *Characters* or *Signes*, framed at a certain time according to the
power of Heaven, far from all superstition" (1657:84). The power of
the Word—a power that belongs to God and that God manifests in
the marvelous operations of his creatures—now belongs to the Divine Name, and through that transfer of power it belongs, too, to the
spoken and written letters in which God's name is known directly to
humans. "Man," who is both *imago Dei* and corruptible flesh, has
access via his own use of corporeal words, characters, and signs to
the incorruptible, incorporeal power of heaven, a spiritual power inhabiting language. In short, man as prophet, cabalist, magician evokes
the very power of the *verbum Dei* through his use of language. Access
to the Word of God is, for Croll, somehow immediate, extensive, and
wondrously empowering.

The paradox of a Calvinist physician evoking the ability of man—
fallen man!—to share directly in the inexhaustible power of the divine
should not escape us. Although he attempts to domesticate this practice, to ground it not in man's extraordinary powers, but in the power
or virtue "which God or Nature hath ordained to such a Name or Character" (1657:84), it remains a heterodox, potentially dangerous notion
for a seventeenth-century Calvinist to espouse. Indeed, Croll's invocation of the sixteenth-century occultist and heterodox thinker Agrippa
von Nettesheim in support of his conception of the magical power of
names, characters, and signs was itself a daring tactical move. While
the allusion to Agrippa purports to support the divine origin of the

power of language—the notion that such power "descends" from God and that its exercise depends upon "his permission"[39]—Agrippa's views were highly suspect.

Yet, despite the potential for heterodoxy contained in the invocation of Agrippa, magic, and the divine power of words, Croll's appeal to the use of words is part of a larger hermeneutic strategy, one tied to his understanding of language and consequently to his semiotic analysis of nature. In this view, both language and the created order of nature have their origins in the divine Word, the creative *logos*. Man's turn to language through magic and the Cabala is therefore analogous to, and congruent with, his turn to nature, and especially to the invisible Word in visible things.

In either case, it is the divine Word that sustains us and supports life: "The WORD of *God*, the First begotten of every Creature, is truly our *Dayly Bread* for which our Saviour commanded us to pray; it is the supercaelestiall Mummy, the supernaturall Balsome comforting poor Mortalls more then Mans own Mummy or naturall Balsom" (Croll 1657:86). Paracelsus, too, invokes the power of God's Word and the dependence of our ordinary, quotidian staple, bread, upon that Word for the nourishing effects it has upon man's body. For Paracelsus, nature is God's Word: it is a book written by God. Croll, as I have suggested earlier, draws upon these familiar Paracelsian metaphors to articulate his own version of occult, chemical medicine and natural philosophy. But, while evoking Paracelsus and his discourse, Croll simultaneously exploits what are tensions and paradoxes in Paracelsus, pressing them in directions that alter the tenor of Paracelsian semiotics and hermeneutic practice. The two "chemical philosophers" share a vocabulary that may appear, at first, to elide any differences:

This *Word* then is the true medicine that healeth all things, but is not known to every one, nor can every Scholler treat and write of it though plunged over head and eares in the dusty learning of School-Divinity: our friend *Theophrastus Paracelsus* a Disciple of the Mosaicall and Living Phylosophy hath written of the Secrets of Nature and the Wonders of God, to wit, of the *Word* of *God INCARNATE* which may be found in the Creatures, and is the Physick and Staffe of our Life; by this *Word*, *FIAT*, the seed of the whole world, were Heaven and Earth created, and this is that which is efficacious in all the Creatures, and to which the Creatures are justly in subjection as to their own soule. (1657:86–87)

39. Croll (1657:84–85). On Agrippa, see Cigliana (1985); Jaeckle (1945); Keefer (1988); Nauert (1965); Zambelli (1966, 1976).

What this passage masks is precisely the turn Croll has taken away from Paracelsus' understanding of the Word and its relationship to things, to the visible and invisible—or inner and outer—and to the healing powers inherent in nature. For Paracelcus as for Croll, the *verbum Dei* constitutes the world in the very act of creation. The world is, in a strong sense, an expression of God's Word. In this sense, too, both Croll and Paracelsus agree that the powers manifest by the occult properties of natural things have their origin in the Word of God. Beyond this point, the two begin to diverge as Croll raises certain Paracelsian polarities into outright oppositions.

Let us take, for example, the polarities inner/outer and visible/ invisible. While Paracelsus contrasts the outer, visible, and material form of things in nature to their inner, invisible, and spiritual essence, their hidden, or occult, nature, the relationship between these polarities is not a simple one of opposition. Not only is the material form a visible stamp, or signature, signifying an inner spiritual essence (a notion that Croll ventriloquizes in his own fashion and that marks his discourse as Paracelsian), it is for Paracelsus something more as well. For the outward form or signature of a thing for Paracelsus bears an essential relationship to its inner, spiritual core. Inner and outer are linked and inseparable: matter is spiritualized, and spirit is materialized in the very constitution of a unique individual thing. This interpenetration and inseparability of spirit and matter, inner and outer, the visible and invisible accounts for the "thingliness" and specificity of material objects in God's created universe.

Croll views these polarities much more ambivalently, so that they tend, for him, to approach the status of binary opposites. This tendency explains, I think, the rather self-contesting and inconsistent account Croll gives of some basic features of Paracelsus' ideas about matter and spirit. But rather than turn to such examples, I want for now to dwell on how Croll's reading of these Paracelsian tropes—the above and other polarities—affects the semiological relationship between sign and signified, between visible/material signature and occult property, and how this entails a different reading of the nature of God's Word and its connection to natural things. This reading, in turn, allows us to see why Croll places such extraordinary emphasis upon the power of words, characters, and signs.

While Paracelsus' use of the tropes inner/outer and visible/invisible accommodates his "system" to a predominantly immanentist view of God, matter, and spirit, Croll reads this same metaphorical structure differently. In his reception of Paracelsian ideas Croll reinscribes these metaphors within different discursive boundaries. Indeed, key Paracelsian metaphors serve as a medium of exchange allowing for a transfer of meaning *to* an essentially exemplarist Neoplatonic discourse of the creative *verbum Dei* (found in a medieval author like Ramon Lull and reflected in Paracelsus' own discursive system) *from* an essentially Protestant, biblical discourse of the Word as source and expression of God's absolute power (*potentia absoluta*).

As sites of discursive instability and change, tropes like inner/outer and visible/invisible allowed for the evolution of Paracelsianism as a specifically European phenomenon of the sixteenth and seventeenth centuries. Such tropes enabled basic Paracelsian themes and practices to adapt themselves to changing religious, social, political, and cultural matrices by providing concrete linguistic locations where distinct domains of practice and discourse intersected and, in theory, might interact. Such interaction might occur when, as in the case of Croll, the religious and even sociopolitical meaning of fundamental Protestant-Calvinist tropes was transferred to Paracelsian medicophilosophical discourse, thus reinscribing those shared tropes within a new system of meaning.

More specifically, Croll's use of metaphors of the visible and invisible, of the inner and outer, led to a subtle reorientation of Paracelsus' "originary" medical discourse through an exchange of meanings with Protestant discourses. In Paracelsus' use, these metaphors, as just explained, carry a dominant meaning that lends to things in the world a complex nature as neither matter nor spirit, but rather as either spiritualized matter or materialized spirit. Inner and outer, visible and invisible, are inseparable dimensions of created, real things. Yet even this brief formulation suggests a nascent contrast between matter and spirit that remained a possibility within Paracelsian thought and leaves its trace in Paracelsus' discourse as a pronounced tension between immanentist and transcendent views of God, matter, and occult causes.

Precisely this tension was exploited by Croll. For in Protestant, specifically Calvinist, discourse, the metaphoric system of meaning constructed by tropes like the inner/outer (man) and the visible/invisible construed them not as inseparable and harmonious components of a whole, but quite dramatically to the contrary, as near binary opposites, as contingently unified, but warring, entities. Croll's active reception of Paracelsus' occult medical and natural philosophical discourse exploited both the divergence and tension of metaphoric meaning associated with these Paracelsian tropes, thus allowing the latter to occasion a metaphoric exchange of meaning between two different discursive formations. As a result, not only was the meaning of the original Paracelsian tropes altered, but, in turn, Croll provoked a new emplotment of fundamental Paracelsian practices and underlying ontological and semiotic relationships.

Such exchanges, reinscriptions, and emplotments leave their traces in Croll's text, making it at once familiar in its family resemblance to its progenitor, Paracelsus' texts, and strangely, though almost imperceptibly, alien. Croll's account of "physick," of signs, characters, and words, and, ultimately, of the Word of God most centrally display such familiarity and difference.

As I have emphasized, Croll's invocation of an outer, visible, material form of things contrasted to an inner, invisible, spiritual essence is characteristically Paracelsian, particularly in its consequent articulation of a doctrine of signatures. Like the visible letters forming the outer sign that we recognize as a word, the external form of a thing is a sign—or signature—of an invisible, hidden spiritual essence/meaning. That of which a signature is a sign has this dual characteristic: it is at once an essence or specific occult power of an individual created thing and it embeds the individual, concrete created thing in a system of meaning that transcends its apparently discrete, isolated status in the world. A signature therefore points to an inner, invisible, underlying virtue of a thing, and to a symbolic order in which the thing simultaneously exists. In both ontological and semiological senses, a signature conceals and reveals the "force" of an object in the world.

Croll's reinscription of Paracelsus' foundational metaphors retains this fundamental structural pattern of Paracelsian medicine and natural philosophy, but introjects a highly consequential shift in how such signs function as signatures. For this shift entails a reordering of relations between signatures and the powers and meanings they signify. Such shifts result from the transfer, or exchange, of meanings mediated by the very tropes, or metaphors, undergoing reinscription. In this exchange, an immanentist and exemplarist discursive regime comes into contact with a transcendent and voluntarist discourse, giving rise to a mixed discursive currency circulating within a post-Reformation cultural economy. To put the matter baldly, within this post-Reformation cultural economy, signatures tend to function extrinsically and contingently for Croll, rather than intrinsically and necessarily as they predominantly did for Paracelsus.

God's absolute power, or—as in Gary Deason's formulation—his sovereignty, ensures the contingency of all matter and, hence, of all *material signs* in nature. That same divine majesty also empowers God's Word and its spiritual offspring, over matter and mere things, as effective agents in the world. Hence, in the economy of Croll's Protestant polity, the visible becomes opposed to the invisible, the outer to the inner. Agency—whether natural, moral, or spiritual—becomes metaphorically associated with the invisible as opposed to the visible, the inner as opposed to the outer, the transcendent as opposed to the immanent. As Deason usefully summarizes:

> Luther chided physicians and philosophers for ascribing procreation to "a matching mixture of qualities which are active in predisposed matter." "Aristotle," he claimed, "prates in vain that man and the sun bring man into existence. Although the heat of the sun warms our bodies, nevertheless the cause of their coming into existence is something far different, namely the Word of God." Similarly, Calvin believed that "fruitfulness proceeds from nothing else but the agency of God. . . ."
>
> In his discussion of Providence in the *Institutes*, Calvin formulated a systematic view of God's relation to the natural world. He made clear that God's activity in nature is ever-present and that nothing in nature can be attributed to natural causes alone. . . . Under no circumstances can nature be seen as an independent entity running under its own power toward inherent ends. . . .
>
> As instruments of God's work, natural things do not have an inherent activity or end. Although they may have received a certain nature or property at creation, this constitutes only a "tendency" that is ineffective apart from the Word of God. (1986:176)

Deason's comments about Luther, Calvin, and God's absolute power as manifest in His Word serve as fundamental premises to his historical argument regarding the relationship between "Reformation theology and the mechanistic conception of nature." Indeed, he asserts that the reformers' "understanding of natural things as passive recipients of divine power was entirely consistent with the mechanical philosophy" (1986:175). I do not wish to dispute this claim—in fact, it will figure prominently in my discussion of seventeenth-century English science in Volume 2. But I do want to suggest that the Lutheran and Calvinist viewpoint summarized by Deason allows for a number of permutations, and hermeneutic practices, in studying the Book of Nature. The strategies associated with mechanism in the seventeenth century are by no means an exclusive or privileged consequence of Protestant theology. In the present context, I want to stress that this theological viewpoint is also consistent with a view of nature pervaded by divine spiritual powers: that is, with a kind of mystical view of the cosmos as bearing signatures of the divine written upon the Book of Nature.

How do Croll and Paracelsianism fit into this picture of the relationship between nature and Reformation theology? How does Croll's metaphorical reinscription of Paracelsian thought evince the presence of a Protestant, especially Calvinist, semiotics of nature? How is Croll's understanding of signatures as signs, and of the presence of the Word of God in nature, evidence for a hermeneutic strategy for reading the "text" of nature that is at once fundamentally Paracelsian and yet a contesting variant at odds with aspects of Paracelsus' own practice?

The nature envisioned by Croll, like that of Paracelsus, is alive with hidden powers of a nonmechanistic character. These powers have their origins in God's creative Word, and, for Paracelsus no less than for Croll, they may well act as instruments of God's pervasive and all-powerful will. In these respects, both occult physicians place paramount importance upon the ascendancy of God's will over His intellect. But, for Paracelsus, this voluntarist strain does not bring with it a related tendency to dissociate strongly the active, nonmechanistic powers found in nature from concrete material bodies. Rather, as exemplifications of God's inner, archetypal ideas embodied in material forms through the creative agency of His Word, such hidden, inner, invisible, and spiritual powers become inextricably joined to matter, forming a kind of inherently active, and monistic, spiritualized matter.

By contrast, Croll's theory of matter, and fundamental ontology, are far more ambivalent and even problematic. Without adopting a mechanical model of nature, Croll appears strongly attracted to a "Protestant" theology of nature that regards discrete material objects as inherently passive. That is to say, for Croll the agency and vitality of nature—that which makes it alive with occult virtues, hidden potentialities, and transformative powers—do not spring from any essential property inherent to matter itself.

Croll's Calvinist belief in God's sovereignty underwrites a subtly variant reading of Paracelsian notions of matter, activity, and occult properties. In turn, belief in God's sovereignty reflects Croll's adoption of a "radical" Calvinist narrative of redemption via the agency of the Word. In this narrative, it is God's Word alone that is creative, active, and productive, just as God's Word alone can redeem the fall both of man and of nature. Hence, the relation of the Word to nature, the meaning of God's authorship of the Book of Nature, underlie not only Croll's Paracelsian views of matter, but also the very process by which Croll's

chemical philosopher links signs—the *signatura rerum* of Paracelsus—
to the active, occult agents they signify in this Paracelsian hermeneutics
of nature.

What, in short, has happened is that the metaphoric shift induced by
the interaction of Paracelsus' tropological system of meaning with that
of Croll's Calvinist religious discourse has displaced and redefined the
inner and invisible agent of natural processes in Paracelsian thought.
What, in Paracelsus, was an occult "spiritual" core inseparable from the
specificity of individual things themselves becomes displaced through
a metaphorical exchange of meaning to God as divine Word animating
a passive nature. In Croll's brand of Paracelsian medicine, it is the Word
that is at the center of things; the Word that is efficacious and that heals,
transmutes, transforms. Croll's metaphoric reinscription of Paracelsus'
thought thus attests to the new dominance of a Protestant narrative of
creation and redemption. Within this narrative frame, man as physician, alchemist, and natural philosopher adopts a different role as investigator of the divine book of Nature than Paracelsus had articulated.
Practices of reading nature and interpreting the Word of God inscribed
in this visible, material book—ultimately the regime for practicing the
very arts of medicine themselves—experience significant alterations
through Croll's interventions.

Let us return to Croll's discussions of the efficacy of "words," and of
visible and invisible "physick," for concrete instances of such alterations. As you may recall, Croll distinguishes between visible and invisible medicines, and yet seems to collapse this distinction by claiming
that what heals is but one thing, the Word. Juxtaposed with this discussion of medicines is Croll's daring assertions that "man" can gain access to the pure incorporeal power of the divine through words themselves. Wedding Neoplatonic language theory and Agrippa's word
magic with Paracelsianism, Croll suggests a dual, if split, path to healing: "Medicines are visible bodies; Words are invisible bodies: whether
the Hearb or Word healeth, it is by God the Naturall Vertue thereof, to
wit, by the Spirit of God made One with Nature by his Word FIAT"
(1657:85; see also 172). Croll's conception of healing and the Word, I
would suggest, turns upon the kind of "Calvinist" reinscription of the
metaphorics of Paracelsus' occult medical-philosophical discourse
spoken above. Careful reading of this, and related, passages in Croll's
text can serve to buttress this suggestion. The metaphoric polarity
visible/invisible is much in evidence in this passage. At first sight, it
may appear that the polarity does not entail a strict, binary opposition.
After all, both visible bodies—herbs that serve as medicines—and in-

visible words do exhibit effective activity as healing agents. Both the visible and the invisible are active. But this is only apparently the case.

For whether humans heals by means of medicines or words, the virtue responsible for the efficacy of healing derives from God, implanted in nature by God's Word. Here the play of metaphors is dense and tangled. For the tropes visible/invisible assume meanings that shift and subvert, or affirm, only in relation to other tropes. Thus, while herbal medicines are visible and words are invisible, the contrast one might expect between them never materializes, precisely because *both* are described as "bodies." Implicitly, then, the meanings associated with both classes of curing agents must be understood within the metaphoric system suggested by another binary opposition: the carnal versus the spiritual. Herbs and words are carnal things; the former is visible, the latter is not. Yet the words spoken by cabalistical physicians, or "true magicians," bear the tones and shapes of carnal, human sounds. The true healing force of words, then, rests not in this carnal, outer (to evoke another tropic register) word, but in the inner, spiritual Word of which it is both a sign and a contingent vehicle:

> And the True Cabalist . . . he doth above Nature, DEALLY or like God accomplish in a moment by firme confidence and strong faith, the very GATE of miracles in that *Only* Divine Name ISHUH in which all things are reckoned up and contained, that is he doth performe it in the WONDERFULL WORD by the Mind, Faith and Prayer, to wit, prayers made in Spirit and in Truth. The New Birth is the Field of Caelestiall Physick which healeth with a word without Externall means: that one operation is in respect of God as the Artificer, and in respect of Man as the Instrument. (Croll 1657:87–88)

Therefore, as agents in the world, both medicines and words are in themselves passive. The powers attributed to them act through them by virtue of the contingent fact that God, through His Word, permits such spiritual, divine powers to operate in conjunction with the proper use of such outward, carnal signs. God's power is made known to humans only in the guise of such signs. And human use of such carnal "agents" constitutes humankind's only means of access to divine powers.

(These considerations account, I believe, for why Croll can suggest that "physick is two-fold"—visible and invisible—and yet claim that only the Word heals. For, in the case of visible medicines, Croll conforms to Paracelsian theory: herbs, minerals, and animal substances are visible, but their healing virtues are constituted by the invisible, occult properties internal to such created things. But because Croll's metaphors create slippage from the inner, spiritual, and occult to the Word as agent of healing, he consequently raises up an independent category of invisible cures that are directly associated with this Word. The category of visible medicines thus makes sense to Croll not because their material, visible properties are healing, but, rather, in a semiological sense because the very term "visible" points to an important *differentium:* herbal, as opposed to verbal, cures are packaged in visible containers.)

This reading of Croll's Paracelsian therapeutics of medicines and words resonates, I think, with his tropologically ordered appeals to the Word of God within his larger medical and philosophical discourse. Here, it may help to note another tropic polarity assumed by Croll: the created versus the increated. The increated is, of course, God. But, more explicitly, Croll refers to the "Word" as "the increated Mercy of God." This increated mercy or Word is our "Saviour," Christ; it is also the Word spoken by our Father the Creator and the Word through which Christ performed miracles and healed the suffering sick. The "increated Mercy of God" is also that "by which are all created things, from which all simples flow, which also with the Father dayly worketh all in all" (1657:85). Hence, the increated is at once opposed to the created and in a sense linked to it.

The increated is opposed to the created as the eternal and all-powerful is opposed to the temporal, mortal, and contingent. And yet the very finitude of the creature links the latter to the increated source. Thus, in his wisdom and mercy, God has given humans not only his increated Word or Mercy, but also "the *Created Word*, or the incarnate Mercy." That is to say, God has made his Word accessible to man—for his use and preservation—in a carnal, external, and visible form. Through such incarnate mercies, through the use of merciful drugs clothed in the material forms of God's created order, the Word has provided "physitians" with the means to effect "great cures." And these cures "are done by the efficacy of the *Triune* and *Divine Word onely,* which healeth and preserveth all things" (Croll 1657:85).

The centrality of God and God's Word to Croll's vision of the physician and to his version of Paracelsian chemical medicine and occult natural philosophy can not be denied. The meaning of this central theme, and its import for the hermeneutic strategies and practices of Paracelsianism, demand further exploration. Here Croll seems to adopt the Protestant/Calvinist emphasis upon God's sovereignty noted earlier by Deason:

What vertue and operation soever ther is in the Creatures, as well in the great [i.e., the "macrocosm"] as in the little world ["microcosm," i.e., man], all that for certaine is wrought of God incarnate in his explicit and manifest bond of one Spirit filling all things inseperably gathered into one, which Spirit therefore is the only fulnes of the whole world, and may well be called *The Fulnesse*. Nothing is made out of God, for in him all things live, are moved and doe subsist. (1657:85–86)

As the marginal gloss to this passage notes, "grace exceedeth Nature, and the thing signified excelleth the signe." Croll, following his Protestant convictions, extols the power of God's grace over nature, thus suggesting that man's reading of this divine text of nature must proceed beyond the mere carnal signs that lie before him. But man's world is the world of carnal—or incarnate—signs. How is he to break through the bonds of this carnal world, to uncover that world of grace where "the thing signified excelleth the signe?" How is "man"—fallen, weak, and prideful man—to become a reader of such signs, to invest the world of carnal appearances and human fictions with the eternality and meaning of the divine?

The key to unlocking the door to such divine wisdom for Croll—a key also to differences between Paracelsus and Croll—is his affirmation of the central role of God's activity in man's knowledge of the Book of Nature. Medicine, Croll's "Physick," is "the most excellent" among the different forms of knowledge and inquiry, in no small measure because "it taketh its rise from Theology or the Light of Grace, and endeth in the Light of Nature" (1657:90). Behind this formulation of medicine's excellence, behind Croll's understanding of the semiotics of nature— the relations between signs and things signified—and fundamental to his own special variant of Paracelsian hermeneutic strategies for reading the Book of Nature lies Croll's retelling of Protestant narratives of man's fall and redemption. Out of such recast religious stories, Croll fashioned a vision of "man's" relationship to God and nature that enabled him to see in Paracelsian chemical medicine God's hand and the key to God's purification of man and nature.

The vision of medicine and nature that Croll articulates explicitly draws upon a narrative recasting of the relationships between God, man, nature, and language for legitimation and for defining the very methods human's should use to unlock nature's secrets. Thus, Croll announces to his reader that the physician

should be born out of the light of Grace and Nature of the inward and invisible Man, the internall Angell, the Light of Nature, which like a

sound Doctour teacheth and instructeth men, as the Holy Spirit taught the
Apostles in fiery tongues: It is perfected and brought to light by practice,
not established by Humane, but by the institution of God and Nature; for
it is not founded upon any Humane figments, but upon Nature, upon
which God hath written with his own sacred finger in sublunary things,
but especialy in perfect Mettalls; God therefore is the true Foundation
thereof. (1657:22–23)

As with Paracelsus, the gift of knowledge—the knowledge that the
physician needs to heal others—can be found in the text of nature. The
"light of nature" instructs humans in the secrets of this text, but access
to the "light of nature" for Croll depends upon God's mercy. Access to
the divine in nature depends upon man's first receiving the "light of
Grace" infused by God upon the "inward and invisible Man." It is this
theological light that allows "man" to "see," or better yet to "read,"
God's visible text, the Book of Nature. Hence, as the allusion to the
New Testament gift of tongues at Pentecost suggests,[40] God and nature
allow the illumined man access to the divine Word—the divine language "written" by God upon nature "with his own sacred finger"—as
a means, presumably the only means, of repairing the disastrous effects of Babel upon him.

Fallen man is cut off from God; but, through illumination, he can
begin to read His Book of Nature directly. Croll here contrasts the fictive, and by implication misleading, nature of human discourse (mere
"Humane figments") with the perfection of "things." "Things" in nature are the bearers of divine truth, stamped as they are by God's own
hand with traces of His wisdom. The pious, illumined man—the man
aided by God's grace—can through "practice," through direct engagement with God's created order, bring to light what God has written in
the symbolism of things in His Book of Nature. The tropes of "light"
and "language" that animate Croll's discourse announce the presence
of the "Word of God" in nature itself. To this "Word" in nature the
physician, who would translate it into human discourse and, more important, into the activity of healing, must turn.

Nevertheless, Croll's insistence upon the central mediating role of
the light of grace introduces a profound shift in the metaphorics of
Paracelsianism. This shift moves Croll's discourse away from the authenticity of Paracelsus' *kunst signata* as the art that "man" in his free-

40. Due to considerations of space, I have avoided discussion of the story of the Pentecost and its bearing upon Renaissance narrative reconstructions. For some very intelligent and telling indications of the importance of such narratives for late Renaissance
science, especially natural history, see Céard (1980).

dom can choose over fallen reason as the way to become the new
Adam. Direct engagement with nature became for Paracelsus an act of
piety, an act in which man earns his access to the light of nature and
learns its innermost, occult and divine, secrets. Human effort does not,
in itself, bring such light and such rewards for Croll. Rather, fallen man
is thoroughly dependent upon God and the light brought by His grace
for his access to nature. As we shall see, the narrative frame surrounding Croll's brand of Paracelsianism legitimates some of the practices
that Paracelsus advocated, but either supplements or reinscribes them
in a system of meaning more fully in keeping with the Calvinist emphasis upon the absolute power of the Word of God noted above.

Croll makes it clear that what Adam lost can be restored only through
God's gift which man, in turn, must be prepared to receive. "Adam," to
be sure, "was full of the wisdome and the perfect knowledge of all
Naturall things" (1657:217). But where Paracelsus stresses Adam's simple familiarity as a *signator* with nature and nature's signs as source of
his profound wisdom, Croll insists upon primitive man's dependence
upon God's revelation of His wisdom to him. Thus, those who lived in
Adam's day enjoyed long life, in part "by the help of Secrets and by
Wisdome which was revealed but to a few, and by speciall knowledge
which God gave them in this particular" (1657:218). The source of
knowledge, of wisdom, and, later, of man's redemption is God. Resis-
tance to such mercies is the result of sin, of man's fall.

Because of sin, the path to knowledge, wisdom, and redemption is
not—cannot be—easy. Humans must strive hard and long, through
great trials, to achieve the wholeness enjoyed by Adam before the Fall.
Nor is this simply a repetition of Paracelsus' stress upon the human
need to labor and experience nature directly in order to unlock her
secrets and become a new Adam. Nor, alternatively, is it a foreshadow-
ing of Bacon's *Great Instauration*, in which work, labor, and, in short,
mimicry of the pure, unfallen Adam's mastery over nature earned
"man" restoration to his former perfection. The difficult, painful path
sketched out by Croll does not lead to man's uncovering nature and
winning his own salvation, but rather represents a necessary suffering
and atonement that prepares man for God's mercy:

> For this onely is the Kings high-way, not onely to come all the desired
> Secrets of Nature, but that which is the chiefest thing of all it leadeth even
> to the very workmaster of the universe, by which ONE infinite OCEAN of
> all Divine GOODNES through Regeneration (alterity being swallowed up
> of unity) in the Sabath of Sabbaths or when the eternal Jubile is come for

which we were created, we do by consent of divine Clemency, attaine the
scope and true mark in the full fruition whereof we shall hereafter be
delighted just like a miserable Exile and pilgrim (tossed up and down
through various hazards, hardships, streights, and miserable sufferings)
restord againe to his rightful Country: for he deserveth not sweet, who
hath not tasted of bitter things: There is no recovering or returning to what
we have lost but by the Crosse and Death: Nor will God have mortall Man
who now is wandring from him that he should come to immortall blessed-
nesse and glory in a delicate journey, but through the Fire of Temptation
and Tribulation, with a sad and sharpe death, because the Coronation and
wiping away of all teares is after the victory, when we have overcome all
our enemies, eternall Life will recompence greater wars and wrestlings.
(1657:193–94)

Croll adapts Paracelsus' vision of man as active, pious laborer in the
garden of nature to a harsher, Calvinist image of man as wayward "exile and pilgrim" whose journey must not be "delicate." The very practices that Paracelsus extols as liberating and flowing from man's free
choice of his Adamic role as *signator*, Croll views as the necessary toils
and sufferings of fallen man awaiting the saving grace of his Lord and
Creator. Man as alchemist enacts the death of mortal matter, subjects
impure things to tribulation and suffering, wresting purity from the
raging fires of temptation so that, in the end, man and nature may be
redeemed by God.

Knowledge, human destiny, and the practices expected of the occult
physician legitimate themselves through webs of meaning spun by the
very story Croll recounts. In that story, prelapsarian, paradisical man
provides the model for the regeneration of fallen man. Adam's
knowledge—the source of his divine understanding and mark of his
perfection and destiny with and in God—was itself *God's gift, not* the
product of his own experiences as the first practitioner of the *kunst
signata:*

For there is no knowledge . . . that will abide in the soule for ever, but is
subject to forgetfulnesse and will vanish, but that onely which is inwardly
received by Essentiall knowledge in the secret understanding: which Essential intrinsecall knowledge is not from flesh and blood, nor from the
multitude of Books and reading, nor from the abundants of Experience
and old age, nor in the inticings of mans Word or wisdome, and wrangling
of reason, but the mind of man is perfected and compleated by a passive
reception of Divine things; not by study and paines, but by patience and
submission. (1657:48–49)

Paracelsus' active, "artful" Adam becomes Croll's submissive recipient of divine things. Though "man," in Piconian fashion, finds himself a microcosm of the universe, containing all things, and hence all knowledge, within himself, man nonetheless depends upon God's gift to bring knowledge to light. Indeed, fallen man must receive that light *passively, submissively* from God again in order to awaken from carnal slumber into lost Adamic knowledge:

> Therefore all Naturall and Spirituall good things were, and are in man at first, but as by sin that Divine Character was darkned in us, so sin being satisfied for, & done away, that Character shines out againe more and more . . . we are onely to awake out of our slumbring and snorting, who through sin have fallen asleep in the gifts that God hath bestowed upon us, so that we can neither see or perceive and believe that these good things are at present in us. (1657:49)

How does this awakening occur? Not by man's will, not by his own unaided efforts, but rather through the light of God's grace:

> When we therefore know our selves aright according to both kinds of Light, according to the Spirit and Nature, then by Gods help we enter into the gate that is opened in us, and we open to God who stands and knocks at the door of our heart, living according to the will of God, we have all things necessary as well for wisdom as for life, both for present and ever hereafter. From this diligent contemplation & knowledge of a mans self, the true knowledge also of God doth immediately arise. . . . (1657:50)

These two lights are inextricably linked to one another in Croll's understanding of matters. But it is the light of the spirit, or of grace, that is prior and primary. So that, for Croll, all knowledge of the world depends ultimately upon God:

> And as by the knowledge of the visible world we come to the knowledge of the invisible Workman; so & from Christ visible or the life of Christ we learn to know the Father, for he is the way to the Father: And as none can come to the Son unlesse he hear and learn from the Father, so none can rightly know the frame of the world but he that is taught of God. (1657:51)

Such grounding of knowledge of nature in the light revealed to man by God's grace results from Croll's metaphorical redescription and transposition of Paracelsus' occult medicine and natural philosophy. Though still recognizable as a variant of Paracelsianism, Croll's chemical medicine displays a profound ambivalence toward the experiential, practical engagement with nature that was a hallmark of Paracelsus' "system." By adopting a metaphorics of illumination that reveals an entire register of associations largely repressed, if not absent, from the discourse of Paracelsus, Croll has both redefined Paracelsian semiotics of nature and reworked the very foundation of relationships that gave meaning to its notion of signatures.

Where Paracelsus foregrounded a hermeneutics of reading the external signatures of things through direct experience of nature, Croll transformed the hermeneutical act of "reading" into an essentially "spiritual" encounter with divine meaning. This is not to say that an encounter with nature is not, for Croll, of fundamental importance. But the significance of engaging nature undergoes a subtle shift. For Paracelsus, only direct experience of nature afforded humans access to the inner recesses and occult properties of things. Reason could not penetrate the secrets of nature because human reason is itself fallen and thus divorced from the very architectonic divine ideas that served as a blueprint for the world. Only as *signator*—as reader of God's signatures stamped upon His Book of Nature—can humans uncover nature's secrets, the traces of the divine Word.

Moreover, this activity of seeking God's traces and of readings His signatures in things valorized a range of empirical practices—from collecting plant specimens, intact if not in situ, to exploring mines and examining minerals and metals—as themselves essential to Paracelsus' hermeneutics. For it was only such practices that afforded human access to the link between the outer, visible signature impressed upon things by God and its inner, occult, and "spiritual" nature. Nature was indeed a system of signs for Paracelsus, and it was the very uniqueness of each symbol that God had stamped upon nature that required such practical engagement with nature. The point for Paracelsus is that the link between sign and signified, between outer and inner, was absolutely real, impenetrable to mere reason, and capable of being grasped by humans only concretely and practically.

However, it was precisely this necessary link between sign and signifier that Croll's Calvinist theology of God's sovereignty called into question. While constructing his medical-philosophical discourse around familiar Paracelsian metaphors and dyadic relationships, such as the inner and outer, the metaphoric register of the Calvinist text within which Croll reinscribed these dyads stretched, if not severed, the concrete link between sign and signified. Mere matter—visible and carnal matter—bears no necessary and essential connection to the spiritual force and reality that lie hidden within the invisible, divine core of things. The fact that such occult properties are enveloped within particular carnal forms is a mere contingent fact, a consequence of God's absolute and arbitrary power.

Hence, for Croll, there is no way in which humans through practical engagement with nature alone—through the exclusive operation of the light of nature—can fathom the inner, spiritual nature of things. The signatures that God has impressed upon nature are arbitrary, and contingent upon His will. Attempting to read such signatures directly from nature alone is like attempting to break an indecipherable code without access to the hidden key. In Croll's Paracelsian universe only God possesses the key that will unlock the arbitrary code linking an external signature with its internal spiritual force. This arbitrary conjunction of signature and spirit is further emphasized by Croll's insistence, noted above, that it is the Word itself that acts directly in and through material things to effect cures. The carnal form assumed by medicines may be unavoidable, and, certainly, contingently allows humans to identify effective medications, but in itself neither divulges useful information, nor effects cures. It is only retrospectively that we can grasp its contingent link to the indwelling healing power of the Word.

No wonder then that Croll's Paracelsianism places the light of grace at the center of chemical medicine and occult natural philosophy. Reading God's signatures in the Book of Nature necessarily becomes for Croll an essentially spiritual act. As such, the role of experience is profoundly ambivalent in his hermeneutic system. Rather than flowing from practice and experience, Croll's ability to decipher the divine signatures depends upon an act of grace—upon God's act, God's unconstrained mercy in choosing to illuminate man's intellect so that the contingent link between sign and signifier is, of a sudden, revealed. This illumination does not depend upon human effort; we can not force God's hand.[41] Rather than a result of work, of pious human labor in the fields of nature, "man's" illumination and consequent ability to read the *signatura rerum* are a result of his faith.

Yet, and here is where we may wish to read his stance with respect to experience as ambivalent, Croll in Paracelsian fashion does evoke images of man laboring away on earth as part of his role as Paracelsian physician and philosopher.

> Nature therefore, as it is now, gives us nothing that is pure in the world, but hath mixed all things with many impurities, that as by the spur of

---

41. Croll (1657:188): "let that man use what Arts soever he can, yet shall he never get any thing against the Will of God; for the Spirit proceedeth from Grace, who inspireth whom he will."

necessity . . . we should begin to learn the knowledge of Chymistry from our cradles, that so long as we are shut out of Paradice into the subburbs of this world, we ought to till and manure the EARTH . . . and that we should labour to get our bread, and other necessary things for this present life, as Natures Labourers, not lazily, but in the sweat of our browes, that by this means, by laying the Crosse upon us which we should bear with patience, it might stir up our industry in this LAND of LABOUR to attain the fruits of Terrene and Caelestiall Wisdome. . . . (1657:96)

Paracelsus had clearly seen such practical activity both as efficacious in itself—as how one discovers the semiotic link between sign and signi- fied in nature—and as an act of supreme piety: a prayerful offering to God in the medieval tradition, *laborare est orare*. By contrast, Croll implic- itly rejects the implications of such justifications of experience: work takes on the character of a necessary punishment for sin, a conse- quence of the Fall, the sad harvest reaped by Adam's transgression. Rather than a means of direct engagement with the divine in nature, such work as the alchemical labors at the furnace is necessary to purify man and nature.

Croll's text eloquently voices this ambivalence about experience, while articulating a theocentric vision of occult wisdom that, in the end, subverts some of Paracelsus' cherished iconoclasm, ironically rais- ing him and his own *human* texts and language into latter-day icons. Again, it is the centrality of the Word and of divine illumination— the "Light of Grace" rather than the "Light of Nature"—that under- writes these currents in Croll's Paracelsian discourse. God's interven- tion, not human freedom and experience, brings knowledge of nature's secrets.

Of course, much of Croll's rhetoric remains consistent with that of Paracelsus. Both decry the proud, bookish concern with mere human languages that in their views characterized the tradition of the Schools, though Croll typically juxtaposes such criticism with praise of the divine spirit that evokes images, not of a Paracelsian folkish piety, but of a religious vanguard, an elite of the spirit. Such an elite owes its existence, of course, to "the glorious God, who according to his good pleasure may inspire whom he will, and deny it to whom he please," and, again, "who alone of his speciall mercy graciously be- stoweth this singular gift of the Holy Spirit . . . both to whom and when he seeth good" (1657:187, 188). Reception of this gift, of God's inspiration and the illumination of grace, prepares mortal humans to receive the light of nature and thus to uncover perfect knowledge of those secrets that lie hidden in the Book of Nature. It is this gift that distinguishes the tedious ratiocinations of the scholar from the true chemical physician and philosopher:

> For when that which is perfect is come, the time of Revocation and Regeneration drawing nigh, every imperfect thing will of necessity come to nought: For where Titles, Degrees and glistering Names make men proud, their is no humility, no life of Christ, no holy Spirit. . . . Now then the Lord enlighten all the lovers of Truth with his Holy Spirit, and graciously deliver them from the chaines of Utter Darknesse and incessant janglings of Putatitious and conceited Schollers. (1657:112–13)

In opposition to this bookish Scholastic philosophy of nature and medicine, Croll extols the unsurpassed wisdom of Paracelsus, "the true Monarch of Physicke, and first Physitian of man, who alone since *Noe's* time hath written of the Internall Astrall Man and the service which God created him for" (1657:131)

Implicitly, then, Paracelsus constitutes our modern-day link to the lost wisdom of Noah, the pre-Babylonic reign of Adamic wisdom that survived man's fall in the as yet uncorrupted language of Adam. Yet this link was made possible only by Christ's sacrifice, by the Word of God made incarnate whose suffering on the cross has conquered sin and whose resurrection overcomes the darkness of death itself. Through Christ man is redeemed, and the fountainhead of wisdom, power, and light that Adam enjoyed in paradise and in which his progeny dimly participated before the Flood and curse of Babel is now once again within reach. Thus it is that

> The Intellectuall Soul of Man, that divine Light flowing out of that spiracle of God and Divine springs pertaineth to the Invisible Phylosophy, whose foundation is CHRIST: Our study therefore and profession of Phylosophy should be Christian-like . . . Nor are we onely to know all Nature externally and internally, but we are also to make it our onely businesse, that according to the Fundamentall knowledge of the same by the supernall help of the *Light of Grace* we may together with Christ and all the Elect possesse that Eternal Life unto which God hath created us[.] this is true Theologicall Phylosophy: Wherefore the New Birth is first to be sought for, and then all other Naturall things will be added without much labor. (1657:131–32)

By placing Paracelsus within this Christian master narrative of creation, fall, and redemption through Christ, a narrative within which Paracelsus situated his own discourse somewhat differently, Croll appears to echo his master and legitimize his status as heir. But Croll's Calvinist retelling of this narrative, instead, represents a telling tactical move calculated to suppress the eruptive meaning of Paracelsus' own metaphorics of illumination, the visible and invisible, and the inner and outer, in favor of a far more transcendent and voluntarist metaphorical reinscription of meaning. This move allows Croll to undercut Paracelsus' emphasis upon work, craftlike labor, and practical empirical engagement with nature as pious and productive activity. In its stead, the product of occult Paracelsian inquiry—knowledge of the secrets of nature—owes its generation in Croll's scheme to God alone: labor is necessary only as punishment for sin.

Thus, Croll's real model for postlapsarian, Calvinist man in his quest for redemption and mastery (or, rather, ministry) of nature is not Adam-*signator* at home in nature, but rather a dual image of Christ and his Apostles. The suffering Christ, laboring to carry his cross, as man must labor and suffer with the weight of sin, promises man an end to his trials on earth. But it is the Apostles who provide Croll with an image of the true Paracelsian. Croll situates them "in the third school of perfect Men, that *Mentall* or *Intellectuall* school of Pentecost, in which the Prophets, and Apostles, and all truly learned men walking in the Life and steps of Christ, have been taught and learned without labour and toyle" (1657:136). The spirit of the Pentecost must rest within the heart of man, within the true and truly learned Paracelsian. The new dispensation of Christ brings to man—to the inner, spiritual man, the man of wisdom and piety who, like the Apostles, has been chosen by God to receive His saving Grace and, in Calvinist fashion, to join the ranks of the elect—release from the consequences of sin. Most especially, as archetypically occurred with the Apostles at the Pentecost, Christ's victory brings release from the curse of Babel, the confusion of tongues that was one last cause of the clouding of man's intellect in the wake of the Fall. From confusion and subsequent descent into Babylonic chaos and alterity, God's spirit—his saving, incarnate Word— prepares to enter man's heart and to transform the utter alienation and otherness of nature and society into blessed unity.

Not through "labour" and "toyle" does man win release from his carnal bondage and Babylonic exile. Rather, one enters the school of the Pentecost by opening one's heart to the spirit. For "the heart of a Regenerate man is Gods *Eden* or Garden of Pleasure, wherein he dwelleth; For God made the World and Man that he might dwell in them as in his own proper house or Temple" (1657:137). The message Croll reads in "the true Monarch of Physicke," Paracelsus, is one of regeneration, that is to say, the regeneration of nature through the fiery spirit of the flame and the regeneration of man through the brightly burning light of the spirit. Such regeneration leads man back to his lost paradise, to his Adamic Eden, the garden of the earthly paradise. Coming full circle, the saving grace of the Word operating through the "Fulnesse" of the illuminating spirit returns fallen man to his Adamic paradise and to the *verbum Dei* in which Adam found the true names and secret wisdom of all things. All this is possible for men if

> they were indeed dead to themselves, even to the whole Animall Man, who is nothing but EARTH, and were supprest by the *Sabbath* and oblivion of Temporall things, and entred into themselves with *David, Psal.* 40.1. patiently waiting for God our master who dwelleth in his holy Temple, in the Abysse of the heart or inward parts of our Soule, *Psal.* 5.7. speaking in us by his spirit, and that they should not hinder him who is willing and desirous to inlighten our mind, and to work all our works in us, which is the utmost happinesse and Blessednesse of Man, and the very determinate and appointed End of the Cabala or secret wisdome. (1657:137)

Croll's apostolic and Pentecostal vision of man—of the true Paracelsian— presses him to see this secret wisdom as a gift of the Holy Spirit, like the New Testament gift of tongues. The occult properties of a Reuchlin, Fernel, or Paracelsus are for Croll the utterly marvelous and contingent presence of the Word in nature, in unique individual things. They are, then, analogous to, if not isomorphic with, the true names and divine words that make up the creative Word of God—the language of nature. Thus, for Croll, the path to the secret wisdom enjoyed by the true Paracelsian is twofold: through the visible thing, and the invisible word. Both lead to the presence of the invisible or increated Word in nature, through the illuminating light of grace. Hence, Croll's Paracelsianism ironically restores to the study of nature the centrality of *words:* there is, indeed, a divine Cabala revealed to man and bearing God's secret wisdom.

In this drift occasioned by the metaphorical slippage Croll introduces into his brand of Paracelsianism, one experiences a sea change; Paracelsus' disdain for mere words is transmuted into longing for the Word and even for true words. With this recuperation of words comes perhaps the ultimate Paracelsian irony: the recuperation of man-made texts, texts of Paracelsus and his followers! Croll raises Paracelsus' texts to the status of sacred, magical repositories of eternal truth and wisdom:

> He concealed his mysteries under vulgar and various names; therefore we must not take the similitudes for the truth it selfe, or that which is intended by them: For there are few that understand the Physicall Secrets,

that is, the hidden power of *God*, or the Magicall WORDS in *Paracelsus;* therefore they need and require a Delian swimmer, a most acute and sharp wit, a Magicall Understanding, even that purified eye of the Mind, which can pry into and search out their sentences and secret mysteries. (1657:150)

Indeed, Croll extends this aura of divine wisdom even to such later followers of Paracelsus as Valentine Weigel. The manuscripts of such Paracelsians are "engraven monuments" that are "no lesse then divine witnesses thereof unto eternity." Further, by "the good pleasure of the Divine Will, the minds of those that read them" with "the assistance of divine Grace" can come to uncover secret mysteries as well (1657:136).

While Paracelsus turned from books to things, Croll has chosen to cast an eye back upon the texts written by regenerate men, even as he keeps the gaze of his other eye fixed upon the divine Book of Nature. Words and things are both creations of the Word. Still, like Paracelsus, Croll looks to nature for symbols, for the imprint or mark that God has impressed upon nature in the form of his divine signatures. Indeed, for Croll, man has "a symbolicall operation and conversation" with the "threefold world" (1657:63). Thus, Paracelsus and the later Paracelsians turn from a Fernelian enterprise of interpreting books to interpreting the symbols embedded in nature itself. This turn was not a difficult one to negotiate. For it involved—as we have seen before—what was a metaphorically frictionless transposition from books as texts to nature as a text: "things" became the "pages" upon which God stamped his mysterious, veiled symbols in creating his Book of Nature for the pious Paracelsian to read. While sharing a conviction with Fernel that the proper strategy for unlocking the secrets of nature is an exegetical one, we witness in the Paracelsians the full flowering of a symbolical exegetical hermeneutics of nature.

Within such a hermeneutical strategy, the practice of symbolic exegesis varies considerably, as the above analysis suggests. For Paracelsus, man need concern himself only with the text of nature, where he will find an intimate and concrete link between the visible signatures of things and inner "spiritual" properties that remain hidden to weak human reason. For Croll, as we have taken pains to stress, inner and outer are but contingently linked by the power of God's divine will. Man is still a reader of signatures, the divine symbols stamped upon the Book of Nature, but requires God's direct illumination to transform that book into a text that he can read. This radical Calvinist reading of signs in nature—a reading that makes the link between sign and signified arbitrary and unfathomable except as a divine mystery and mercy—opens

Paracelsian discourse to a spectrum of social and political practice grounded in religious ideology during the late sixteenth and seventeenth centuries. Indeed, as we shall see in Volume 2, the eruption of religious enthusiasm in mid-seventeenth-century England can be seen as exploiting the possibilities inherent in variant forms of Paracelsian doctrines of language and illumination.

6

From Symbolic Exegesis to Deinscriptive Hermeneutics—Natural History and Galileo's Mathematization of Nature as Descriptive Practices

Narratives of Unity and Diversity in Early Modern Science

> De là la forme du project encyclopédique, tel qu'il apparaît à la fin du XVIe siècle ou dans les premières années du siècle suivant: non pas refléter ce qu'on sait dans l'élément neutre du langage—l'usage de l'alphabet comme ordre encyclopédique arbitraire, mais efficace, n'apparaîtra que dans la seconde moitié du XVIIe siècle—mais reconstituer par l'enchaînement des mots et par leur disposition dans l'espace l'ordre meme du monde. . . . De toute façon un tel entrelacement du langage et des choses, dans un espace qui leur serait commun, suppose un privilège absolu de l'écriture.
>
> —Foucault

The interweaving of language and things in the divine text of nature gave rise to a spectrum of strategies for reading nature as an ordered and meaningful text during the sixteenth century. Of these strategies, Fernel's exegetical approach to the kernel of pristine, hidden wisdom locked within the outer husk of ancient medical and natural philosophical texts and written in the fallen prose of post-Babylonic languages, and Paracelsus' symbolic exegesis of the divine signatures written upon things, constitute two divergent examples. Yet while Fernel places central importance upon words and Paracelsus by contrast upon things, both nevertheless share a belief in the creative power of the divine Word and in the text of nature as an expression of this Word. For both, then, the "Word of God" constitutes the language of nature and

167

the archetype for the universe, access to which is essential for true knowledge of nature. For Fernel, humans may gain access through words, while for Paracelsus such access can come only through the symbols that God has imprinted upon things. With Croll's adaptation of Paracelsianism to a Pentecostal vision of man saved from the curse of Babel by the intervention of God's spirit, the authority of symbols and, to a far lesser extent, of words within the hermeneutic strategy of exegesis is preserved, but at a potential cost. For, in Croll's Protestant discourse, "words," "symbols," and "things" are but contingently linked by the power of God's will to the *verbum Dei*, to the spiritual indwelling in nature.

This chapter is transitional. Its subject is precisely the making and unmaking of links between words, symbols, and things, on the one hand, and the "Word of God," as both archetypal model of the world and authorizing ground for a hermeneutics of nature, on the other. This chapter does not follow a strict chronological order; nor does it claim to find a neatly chronological "development" *from* symbolic exegesis *to* deinscriptive hermeneutics. It is concerned more with the opening up of possibilities for engaging the natural world and with how such possibilities came to occupy niches that might expand or contract in local sites operating within a larger ecology of competing discursive practices. The first part of the chapter will explore some examples of the kind of hermeneutic strategy for reading the Book of Nature that I have called symbolic exegesis. These examples exhibit a "mixed" engagement with texts and with things as symbols. They intentionally aim to illustrate what lies within the spectrum of exegetical hermeneutics defined by the extremes of Fernel and Paracelsus. Relying heavily on recent scholarly studies, I draw examples largely from natural history.

Natural history still remains one of the more neglected aspects of the Scientific Revolution. Yet in early modern Europe natural history arguably finds itself situated more directly within the crosscurrents of major discursive formations and cultural transformations than has until recently been recognized. The often explicitly narrative and historical character of natural history provides important ground for discussing the relations between language—especially language theory—and science in this period. Indeed, as we shall see, the tempering of the curse of Babel with the promise evoked by the Pentecost offers telling insight into the transformations within natural history and science more generally during this period.

Finally, we turn to that icon of the new science embattled by Aristotelian orthodoxy, Galileo, to broach the complex question of how mathematics might enter into a legitimate understanding of nature. The question of legitimacy and interpretive authority is at the heart of Galileo's intellectual and socio-institutional struggles. His advocacy of mathematics as the proper language of nature, I shall argue, explicitly undermines an exegetical—whether textual or symbolic—strategy for reading nature and endorses a kind of deinscriptive hermeneutics. Within this de-in-scriptive strategy, Galileo nonetheless recuperates a privileged authorial stance authorizing a synoptic and totalizing understanding of nature.

Renaissance Natural History

Thomas Laqueur has recently drawn attention to divergent readings of nature that, in his account, structure the discourse of the body and sexuality in Western civilization. While the story he tells specifically concerns the shifting dominance of one-sex and two-sex models within our culture,[1] his use of materials from the sixteenth and seventeenth centuries provides telling examples for us of the way in which the natural world was arrayed within a "common space" exposing the conjuncture of words and things and the authority of what Foucault calls *écriture*.

Consider an example from the sixteenth-century surgeon Ambroise Paré, here cited in its entirety:

> I saw a certain person (a shepherd) named Germain Garnier—some called him Germain Marie, because when he had been a girl he had been called Marie—a young man of average size, stocky, and very well put together, wearing a red, rather thick beard, who, until he was fifteen years of age, had been held to be a girl, given the fact that no mark of masculinity was visible in him, and furthermore that along with the girls he even dressed like a woman. Now having attained the aforestated age, as he was in the

1. Laqueur postulates the existence of two opposing historical regimes within the history of sexuality. From antiquity through the seventeenth century, a one-sex model dominated Western thinking. In this model, the anatomical dimorphism of male and female bodies—specifically, of the generative or "sexual" organs—so "evident" to modern modes of perception was masked (if not erased and made unavailable) for premodern eyes. As Laqueur succinctly states: "For thousands of years it had been a commonplace that women had the same genitals as men except that, as Nemesius, bishop of Emesa in the fourth century, put it: 'theirs are inside the body and not outside it' " (1990:4). The two-sex model, by contrast, dominates modern thinking about sexuality and reveals male and female anatomical dimorphism as the bedrock upon which was constructed the cultural calculus of difference that the modern world has naturalized. This decisive shift occurs in the eighteenth century. One limitation to Laqueur's model is its exaggeration of the dominance of the one-sex model up through the Renaissance (see below).

fields and was rather robustly chasing his swine, which were going into a wheat field, [and] finding a ditch, he wanted to cross over it, and having leaped, at that very moment the genitalia and the male rod came to be developed in him, having ruptured the ligaments by which previously they had been held enclosed and locked in (which did not happen to him without pain), and, weeping, he returned from the spot to his mother's house, saying that his guts had fallen out of his belly; and his mother was very astonished at this spectacle. And having brought together Physicians and Surgeons in order to get an opinion on this, they found that she was a man, and no longer a girl; and presently, after having reported to the Bishop—who was the now defunct Cardinal of Lenoncort—and by his authority, an assembly having been called, the shepherd received a man's name.[2]

This, no doubt, was a shocking occurrence. One can imagine the adolescent's terror, turning later to confusion. But, what perhaps astonishes most—and what remains astonishing even in the event that Paré's account is credulous—is the radically alien nature in which Paré, to our eyes, manages to interpret and domesticate this singular transformation.

The reason why women can degenerate into men is because women have as much hidden within the body as men have exposed outside; leaving aside, only, that women don't have so much heat, nor the ability to push out what by the coldness of their temperament is held as if bound to the interior. Wherefore if with time . . . the warmth is rendered more robust, vehement, and active, then it is not an unbelievable thing if the latter, chiefly aided by some violent movement, should be able to push out what was hidden within. Now since such a metamorphosis takes place in Nature for the alleged reasons and examples, we therefore never find in any true story that any man ever became a woman, because Nature tends always toward what is most perfect and not, on the contrary, to perform in such a way that what is perfect should become imperfect. (1982:32–33)

Here we find a transformative mechanism based upon an opposition of natural powers—"heat" and "cold"—linked to a scale of perfection. Through heat a woman may conceivably metamorphose into a man: to do so is to follow nature's own path from the imperfect to the perfect. But it is not this commonplace adaptation of Aristotelian natural philosophy, metaphysics, and theory of generation—with its assumption

2. Paré (1982:31–32). For the French text, see Paré (1971:29–30). See Laqueur (1990:126–28) for paraphrase and comments.

of woman's inferiority to man—that so astonishes.[3] Rather, Paré's almost matter-of-fact account of the eruption of male genitalia from a woman's body postulates a basic identity that radically challenges modern assumptions of sexual difference.

The fundamental comparability of sexual organs among males and females denies the dimorphism basic to much of the discourse of sexual difference in the modern world. For Paré and many of his contemporaries man lives within a universe that God has constructed hierarchically and analogically. One reading—by no means the only reading[4]—of that universe would accommodate difference to a larger and overarching system of correspondences and essential forms. Apparent anatomical contrasts marking sexual difference would, in fact, mask more fundamental analogies and correspondences that evoke a hidden identity. In the case of Paré's testimony to sudden sexual transformation, such underlying identity of "form" is dramatically "expressed" by nature.

It is this seemingly effortless accommodation of what we might regard as objective, given marks of natural difference that so astonishes Paré's modern reader. Where we see two different, if complementary, sexual organs, the penis and the uterus,[5] Paré was apt to see but two expressions of the same fundamental form, the one external and the other internal. Sexual difference did, of course, exist and was, in fact,

3. See Aristotle (1963). On Aristotelian theories of generation and sexual difference, see also Boylan (1984); Devereux and Pellegrin (1990); Jacquart and Thomasset (1988); Lloyd (1983); Preus (1970, 1977); and Cadden (1984, 1992).

4. See literature cited below in my discussion of natural history for the variety of ways in which the text of nature could be read in the late Renaissance and early modern period. For a more subtle discussion than Laqueur's of sexual difference during this period, see Maclean (1980: esp. 28–46). I am indebted to Maclean's book throughout my discussion. On the background to anatomy, correspondences, and sexual difference, see Jacquart and Thomasset (1988) and Cadden (1992).

5. Laqueur (1990:4 and passim) frequently claims that ancient, medieval, and Renaissance medical authorities, such as Galen, regarded the female vagina as comparable to the male penis; the former was an inverted, or internal, version of the latter. This scheme then associates the uterus with the male scrotum. As Maclean (1980:33) notes, Galen's parallel suggests the isomorphism of penis and uterus and of the ovaries (*testes mulierum*) and the male testes. Laqueur occasionally reports the uterus/penis analogy, often in connection with illustrations from early modern texts where the point is inescapable. The variety of opinion concerning anatomical parallels among male and female as well as the shifting *meaning* of anatomical comparability itself needs to be studied much more carefully. In this respect, Maclean's brief account suggests greater nuance and opens numerous avenues for further research: Falloppio's revised analogy ("clitoris = penis") and its popularity among late-sixteenth-century medical authors; Du Laurens's "very coherent account of this medical dispute in 1593, concluding against comparability"; and the rejection of Aristotelian notions of the imperfection of woman "by 1600 by most doctors" (1980:44). See Maclean (1980:104) for citations of Falloppio's and Du Laurens's texts.

enshrined as a basic polarity of Renaissance culture, but its grounds were far more complex and culturally embedded than simple, and merely apparent, anatomical differences would allow. Instead, sexual difference exhibited the interdependence of the text of nature—written in the language of things, of which the "language" of anatomical forms was but an instance—and the divine text constituted by the *logos*, the creative "Word of God." Within this larger text of the Word of God anatomical differences acquired a kind of intelligibility that revealed an identity rooted in resemblance. Both male and female, penis and uterus, were material instantiations of the same spiritual archetype generated in the divine *mens* and given expression by the divine Word. Both sexes were the *imago Dei*. And yet, despite their fundamental similarity, or anatomical comparability, the very enactment or expression of the creative Word of God within the material world introduced differences.

Such differences turned, however, no upon visual distinctions, but rather upon differences in the capacities, powers, or virtues of similar forms such as male and female. While "form" gave evidence to the relationship of a particular entity to a divine archetype and hence its "resemblance" to other material forms, the differing "virtues" of similar forms established the particular place of individual entities within the analogically organized hierarchy of created things and hence each one's difference within this divine text. For example, celestial or quintessential spirits, the "spirit" or soul of man, the animal and vital spirits, the alchemical spirits of sublunar minerals, and so forth, share not only a "linguistic" resemblance but also a common form linking them with their archetype in the divine Holy Spirit. Hence, the commonplace resemblances, correspondences, and dynamic exchanges among them that pervade the literature on astrology and alchemy. Yet the specific powers and transformative virtues of each type of spirit differ dramatically, enabling one to array them—for the most part—along a verticle axis of perfection and imperfection that allows one to constitute a hierarchy exhibiting difference within resemblance.[6] In like manner, the analogical resemblance between male and female sexual organs, while undermining the significance of merely visual, anatomical difference, nonetheless exists simultaneously within a discursive system that encodes male and female with sets of powers that differentiate and stratify them as both natural and social kinds. Difference is

6. See Bono (1984, 1990). The latter appears, with appropriate, if slight, changes, as Chapter 4 of this book.

therefore not simply "natural." It is not a matter of simple anatomy. Rather, the difference encountered in Paré's text, and in other texts of the late Renaissance and seventeenth century, exhibits the inescapable interweaving of things and language in the text of nature. That is to say, the very order and differences displayed by nature suppose the presence of *écriture*—of a privileged text, taken to be of divine provenance, that emplots both the visible and the hidden. Meaning—specifically the meaning of resemblance and difference—cannot be *discovered* in things but must rather be *uncovered* in the stories within which things have been arrayed in the divine text of nature. Nature has its "true stories"; organic change is not simply a physical transformation, but also and more significantly a "metamorphosis"—a "poetic" reemplotment within the text of nature. Paré's shepherd receives "a man's name" to mark nature's marvelous and poetic reemplotment.

Historia naturalia, natural history, carries with it such poetic resonances in the Renaissance and early modern period. Natural history is a *historia*—nature's story or narrative, not just a retrospectively constructed narrative that arrays nature in all her variety before our mind's eye. Indeed, the very real differences in the significance of natural history from the sixteenth to the eighteenth century turn precisely on how seriously—or, to use Paula Findlen's (1990) apt expression, how playfully—natural historians of the period understood nature's artifice. Moreover, as God's text, the playfulness exhibited by nature could be seen as a reflection of the Word of God itself. Thus, natural historians might seek in that text, in its variety and playfulness, in its analogies, visual metaphors, and poetic, even allegorical resonances, clues to the unity of the creation and to the divine meaning of nature and of the microcosm, man.

Within this larger perspective, we should perhaps regard Paré's story, taken not as clinical observation but rather as nature's *historia*, as revealing a system of analogical resemblances and hierarchical differences pervading God's creation. Laqueur's story of the transition from one-sex to two-sex models of the body and sexuality within Western culture might then properly belong to a larger story of changing cultural narratives in the Renaissance and early modern period. More specifically, we can perhaps understand the transition he describes as a consequence of a much broader transformation in the relationships between language, symbols, and nature.[7] As post-Reformation narratives

7. See the excellent review of Laqueur by Gould (1991), who suggestively links this transition, and transformations in views of fossils, to the undoing of "the Neoplatonic theory of signs" with its correspondences and macrocosm-microcosm analogies.

of Adam, the Fall, and the confusion of tongues gradually distanced
the languages of man and the text of nature from God's authorizing
Word, the system of correspondences that underwrote the one-sex
theory gave way to the analysis of differences foundational to the two-
sex model—at least in its modern formulation.[8] If, as I believe it to be,
this is the case, then the roots of this transition are firmly planted in the
late sixteenth and early seventeenth centuries and not, as Laqueur assumes, in the eighteenth.[9]

Hence, the relationship of natural history to changing views of language and symbols presents a crucial problem for assessing fundamental changes in attitudes toward nature and science. Fortunately,
the recent and excellent work of William Ashworth, Jr., and Paula
Findlen—which, together with the work of other historians of science, has begun to transform our understanding of early modern
natural history—enables us to begin envisioning answers to these
questions.

Ashworth and Findlen provide all sorts of examples of how early
modern scientists simultaneously utilize both symbols purportedly
found within nature and texts drawn from a wide variety of traditions
to decipher, through a kind of exegetical practice, the meanings embedded by God in His Book of Nature. Indeed, Ashworth's (1990) term for
this late Renaissance approach to the natural world—his "emblematic
world view"—consciously evokes the duality of this textual/symbolic
and traditional/empirical style of scientific investigation. In what follows, I shall attempt first to summarize the salient features of Ash-

8. Although Laqueur fails to emphasize the point sufficiently, there was a two-sex
theory before the late seventeenth and eighteenth centuries. Such theories certainly
existed within a dominantly hierarchical conception of the universe, although without
necessarily embracing the full implications of an analogical system of correspondences
and resemblances. Ancient authors such as Hippocrates and Aristotle do not always
conform to the rather rigid distinctions Laqueur has set up in his book. Indeed, in
Aristotle's case Laqueur's emphasis upon *anatomical* comparability often obscures, in
my opinion, subtle, but central, distinctions that Aristotle makes between the fluids,
substances, virtues, and functions of male and female (Laqueur 1990:35–43). For
Hippocrates and ancient views of gender and sexuality see Hanson (1992); Dean-Jones
(1992); and Lloyd (1983).

9. This chronology fits well with Maclean's (1980) claims regarding the late-
sixteenth-century rejection of Aristotelian notions of woman's imperfection and of the
medico-anatomical comparability of male and female sexual organs (see notes, above).
This disagreement with Laqueur is more than a pedantic quibble; it suggests a different
interpretation of events and cultural changes that underscores the significance of language, cultural narratives, and cultural aspirations associated with human mastery and
science in the Scientific Revolution.

worth's understanding of this emblematic worldview, relying heavily upon his own examples as well as others taken directly from sixteenth- and seventeenth-century sources. I shall then provide an overview of Findlen's subtle analysis of natural history and the "playfulness of scientific discourse in early modern Europe," while calling attention as well to the innovative work of other scholars in this area. Finally, I shall focus upon Jean Céard's (1980) wonderfully suggestive discussion of Babel and the Pentecost in sixteenth-century thought as a point of departure for understanding the relationships between language theory and the shifting strategies of symbolic exegesis and deinscriptive hermeneutics in the early modern period.

Ashworth's basic historiographical point is simple: Renaissance natural historians were not "trying to write biological textbooks" (1990:305). The enterprise of natural history did not conceive its task as the accurate description of individual zoological and botanical specimens with reference only to such concerns as their relationship to a taxonomic system. Instead, the kind of knowledge that Renaissance natural historians sought belonged to a much broader cultural matrix. That knowledge became knowledge—or at least knowledge worth having—and acquired significance only in so far as it uncovered *meanings* in the text of nature that forged deeply resonant links with mankind's moral, cultural, cosmological, and symbolic universes. Indeed, underlying much of Renaissance natural history was precisely a conviction of the fundamental unity of what we might call natural and "symbolic" knowledge. Theirs, as Ashworth insists, was "a world where animals are just one aspect of an intricate language of metaphor, symbols, and emblems" (1990:305). In short, for Renaissance natural history the Book of Nature did not stand alone on its own shelf for humans to open and plunder for bits and pieces of information. Rather, it stood securely propped up by other volumes; only by disturbing them could man open God's visible text of things. The text of nature was profoundly intertextual in its nature. And the universe of Renaissance natural historians knew no divide between a world of nature and a world of signs.

As for the specific content revealed through proper reading of the Book of Nature, Ashworth provides an intriguing, if glancing, summary of the intertextual network within which nature find "herself" suspended for Renaissance authors. Renaissance natural historians not only drew upon "the classical literature on natural history," but in addition mined, and in some cases helped construct, a variety of other "literary" resources (1990:307). These scholars—for to be a natural historian meant to be a scholar and student of textual, linguistic, and visual traditions—seized upon the polysemous associations natural entities invited through their pregnant incubation in classical antiquities (such as "medals and coins"), in the tradition of Aesopic fables with its stories of animals, and in adages and epigrams.[10] But these sources were not alone. For the network spun by natural historians was interlaced with strands from hieroglyphic, mythological, and emblematic traditions as well.

These latter strands are as important to Ashworth's story as they are to ours. With the rediscovery of hieroglyphs through the text of Horapollo's *Hieroglyphica*, the Renaissance search for a pristine, Adamic language mirroring the *verbum Dei* found a possible model in a kind of language that was, erroneously, thought to be visual and symbolic.[11] As Ashworth points out, this discovery conjured up the ideal—so central to the tradition from Ficino to Boehme that we have been studying—of "a language in which understanding is conveyed immediately, much as God understands things, without the mediation of conventional language" (1990:307). For natural historians, then, animals became more than natural kinds, more than the rudimentary, spare objects of scientific inquiry and analysis. They were ciphers, bearers of meaning, and elements of a divine text: "Weasels, cranes, and lions became part of a visual language; they were symbols, but even more, they were Platonic ideas, whose meaning the mind could immediately perceive. Animals were living characters in the language of the Creator, and the naturalist who did not appreciate or understand this had failed to comprehend the pattern of the natural world" (1990:308).

Mythology, and the new Renaissance tradition of emblems, further deepened this sense of the Book of Nature as symbolic language whose text could be deciphered in relation to other texts. As Ashworth points out, the most important resource for Renaissance naturalists was Ovid's *Metamorphoses*. This text and its tradition enjoyed wide popularity during the sixteenth century, spawning such well-known major handbooks as Giraldi's *De deis gentium* (1548), Conti's *Mythologiae* (1551), Cartari's *Le imagini colla sposizione degli dei degli antichi* (1556), and Ripa's *Iconologia* (1593) (W. B. Ashworth 1990:309, 327). While such texts, and earlier ones starting with Boccaccio's *Genealogia deorum*,[12] are

10. W. B. Ashworth (1990:307, 308 [antique coins and Renaissance medals], 309 [Aesopic fables], 310 [adages and epigrams]).
11. W. B. Ashworth (1990:307, 326). See also my discussion of hieroglyphs in conjunction with Dee and Bacon in Chapter 7, below.
12. Boccaccio (1930). The importance of the mythological tradition in the Renaissance, and Boccaccio's centrality to it, have been studied by Seznec (1953).

fundamental resources for students of Renaissance literature and art, their importance for science and specifically natural history has been far from adequately exploited. The mythological tradition, as well as the tradition of emblem literature initiated by Andrea Alciati's *Emblematum liber* (1531) and quickly adapted to include "a visual image, a short motto, and a slightly longer epigram,"[13] provided Renaissance natural historians with a wealth of representations for the depiction of virtues, vices, and other moral, spiritual, philosophical, and cultural associations that enriched the symbolic meaning of things—particularly animals—in the divine Book of Nature. Moreover, as I stressed earlier, such sources reinforced the intimate connection between words, symbols, images, and things, thus leading to an alternative tradition to the Paracelsians' reliance on signatures.[14]

Ashworth's attempt to characterize this important tradition of Renaissance natural history leads him to coin the phrase "emblematic world view." But the actual use of emblems must, of course, be taken only as one of several possible components of this genre of natural history. In this sense, then, the term "emblematic" serves as a synecdoche for a range of symbolic approaches to natural history that employ "texts"—verbal and visual signs—in the investigation and interpretation of things in nature. The particular range of interpretive practices encompassed by Ashworth's term stand, then, as a subset of the larger repertoire of hermeneutic strategies that I have called symbolic exegesis. Together, these strategies inform the investigation of a broad range of naturalistic, medical, and scientific inquiry during the sixteenth and seventeenth centuries. Ashworth's vivid description of this emblematic worldview fits well with mainstream naturalists, such as his chief exemplars Gesner and Aldrovandi, though it runs counter to the expressed sentiments, and much of the practice, of Paracelsus:

> The essence of this view is the belief that every kind of thing in the cosmos has myriad hidden meanings and that knowledge consists of an attempt to comprehend as many of these as possible. To know the peacock, as Gesner wanted to know it, one must know not only what the peacock looks like but what its name means, in every language; what kind of pro-

13. W. B. Ashworth (1990: 311, 327–28). See also Seznec (1953:100ff.), and especially Praz (1939).

14. See Chapter 5, above. Paracelsian medicine and natural philosophy represent but one exemplar of the kind of hermeneutic strategy for reading and mastering the Book of Nature that I have called symbolic exegesis. In a review of Bianchi (1987), Paula Findlen (1990a) draws attention to the centrality of signs in the array of symbolic approaches to nature in Renaissance science.

verbial associations it has; what it symbolizes to both pagans and Christians; what other animals it has sympathies or affinities with; and any
other possible connection it might have with stars, plants, minerals, numbers, coins, or whatever. . . . The notion that a peacock should be studied
in isolation from the rest of the universe, and that inquiry should be limited to anatomy, physiology, and physical description, was a notion completely foreign to Renaissance thought. (1990:312)

The truth of things, their place in the vast analogical scheme of the
Book of Nature, would yield only to the gaze of the learned naturalist
fixed at once on things themselves and also on the record of verbal and
visual allusion that promised to unlock a pristine core of lost knowledge. It is tempting to regard all of this erudite activity as no more than
antiquarianism and rhetorical display, and to dismiss all but the kernal
of innovative, naturalistic description and visual representation as
mere dross. Ashworth, indeed, complains that, of his two chief repre-
sentatives of this emblematic natural history, Ulisse Aldrovandi (1522–
1605) is often belittled as "just another Gesner, except that he did not
know when to stop" (1990:313). But the point for Aldrovandi and his
disciples was that one could not stop with less than a complete anatomy of the physical and symbolic universe to which individual natural
kinds—the world of animals—belonged. As Ashworth points out, that
universe had expanded enormously since Gesner's work in the late
1550s. Aldrovandi had access to a vastly larger store of symbolic associations than had Gesner (1990:313–16). In particular, Renaissance hieroglyphic lore took off with publication of Piero Valeriano's *Hieroglyphica*
of 1556 and was incorporated into the growing emblem literature as
well. Ashworth (1990:314–15) points to the four-volume collection
(1593–1604) by Joachim Camerarius that included four hundred emblems, all incorporating an animal or plant, as a culmination of this
interest in nature and emblems that also served as a source for
Aldrovandi. The bottom line was that, while Gesner's (1555) *Historica
animalium* devoted eight pages to the peacock, Aldrovandi's text runs
to some thirty-one pages (1990:313–14): "With all these resources
Aldrovandi is able to spin a net of associations and similitudes that is
far more complex than anything that Gesner was able to achieve.
Aldrovandi's world needed thirteen volumes to contain it" (1990:316).

Such monumental work—not to say the monumental expense—
perhaps has its limits. Ashworth sees Joannes Jonston's *Historia naturalis* (1650–1653) as a "watershed publication" marking, if not the end
of monumental publishing projects, the end at least of emblematic natural history. Jonston's discussion of the peacock, Ashworth (1990:317–

18) notes, constitutes by Renaissance standards an insubstantial two pages, with little more than description. What is Ashworth's explanation for this dramatic change?

Ashworth notes that attempts to write natural histories of animals found in the new world had to confront the singular circumstance that "the animals of the new world had no known similitudes" (1990:318). Nature, or at least a part of it, seemed to have escaped the Renaissance network of symbols and meaning. This lacuna was one chink in the emblematic worldview's armor. Another, for Ashsorth, was the emergence of critical assessment of claims contained in the various fables, proverbs, myths, and emblems that surrounded particular animals in works by Gesner, Aldrovandi, and others. This critical perspective he traces to the new "antiquarianism and the quest for historical truth" of the seventeenth century (1990:319–22).

Unfortunately, Ashworth's argument regarding the role played by northern European antiquarianism, with its attempts to use artifacts as "vital historical clues," though highly suggestive, remains insufficiently developed. In particular, *if* there is a new and critical attitude toward historical evidence evinced by the use of artifacts (Ashworth expresses this attitude succinctly as "the artifact does not lie" [1990:321]), Ashworth does not give a convincing argument for its emergence. Of course, he is quite correct to point to the impressive development of "great museum collections" in the early seventeenth century as a significant event in the evolution of natural history, and undoubtedly this had an effect on fostering a comparative and critical spirit.[15] But one suspects that this critical spirit was not as fully developed, nor as internally consistent and uncontested, as Ashworth seems to imply. Certainly, in his list of antiquarian scholars one finds names, such as that of Jan Goropius Becanus, who remain wedded to the central importance of etymology for uncovering divine wisdom through attempts to rediscover and study the originary, Adamic language (Eros 1972).

This latter example points to the importance of a context that we have stressed throughout this book, that of language theory and cultural narratives in the Renaissance. While I think it is quite likely that changed attitudes toward natural history and both botanical and zoological specimens were intimately related to transformations in the evaluation and interpretation of human artifacts, the grounds for such

15. W. B. Ashworth (1990:221). On early modern museums, natural history, and scientific culture, see Moran (1991); Findlen (1989, 1989a, 1991); Impey and MacGregor (1985); and Lugli (1983).

changes are not explored by Ashworth. Such radical revisioning of—
or, more accurately, reengagement with[16]—nature and history implied, and I would argue necessitated, radical recasting of fundamental
narratives about man, nature, and language. Only within such a radically altered narrative frame could a discourse emerge that placed
stress upon the critical *differences* distinguishing animals and plants,
rather than on the correspondences and similitudes that bind them into
an analogical and symbolic network. The latter network supposed a
creation narrative in which God was the Judeo-Christian *logos* whose
Word gave visible expression to the archetypes or ideas of the divine
mind. Nature was then the image of God and even a divine *language*
that man must decipher. The question raised by the rise and fall of the
emblematic worldview recounted by Ashworth is, then, how did this
story lose so much of its cultural force as authorizing narrative legitimating scientific practices?

Two developments contributed to this loss of narrative authority.
First, the intimate and necessary connection between the Word of God
and nature came to be questioned and redefined. Second, by a gradual
delegitimization of narratives that figured humans as privileged readers of the text of nature, language began to lose its status as originary
source of knowledge. As a result, the project of seeking knowledge
directly through deciphering the symbols imprinted upon things, or
indirectly, through seeking the truth of things in language—in etymology[17] or exegetical hermeneutics—lost its firm foundations.

In many ways the second development prefigured and prepared the
way for the first, although it is not possible to insist upon any strict
chronology. Not only did these developments overlap with one another, the very discourses and strategies authorized by conflicting cultural narratives survived and coexisted for much of the late sixteenth
through at least the early eighteenth century. The point, however, is

16. The metaphor of vision has dominated Western conceptions of knowledge and
the process of knowing since Greek antiquity (see Derrida 1982). Discourses of scientific
change that traffic in such metaphors—"revisioning," "gestalt, or paradigm, shift,"
"changing worldview," "perspectival shift," etc.—implicitly embrace a global and totalizing model of science. To emphasize the local, partial, yet powerful nature of scientific
practice/discourse, I prefer to speak of "world engagements" rather than "worldviews."
Through various cultural mediations—technologies of "reading" and interpretive strategies with all of their embodied ways of knowing—we come to *engage* nature in particular
ways and construct our own sciences. For a parallel emphasis upon "engagement" in
science studies, see Rouse (1990).

17. As Marian Rothstein (1990:332) reminds us, "the very word 'etymology' [is] . . .
from *etumos*; true, real."

that Ashworth's account stops short of connecting the impact of the new "critical" antiquarianism with the groundwork prepared by both of these developments. He does, significantly, note that Francis Bacon's rejection of the emblematic worldview and its approach to natural history was rooted in his firm denial that nature is in any way a mirror, image, or reflection of God, thus suggesting an important role for the first development mentioned above.[18] But Bacon's significance in the demise of the emblematic worldview is not solidly connected by Ashworth to the role he claims for the new antiquarian spirit of criticism. And, more fundamentally, the importance of Renaissance cultural narratives—and their gradual delegitimization—figuring man as enjoying privileged access to the "Word of God" through his Adamic inheritance (which allowed him to read the text of nature) nowhere appears in Ashworth's account as a context integral to the emblematic worldview and its decline.

In order to underscore the significance of this narrative context for an understanding of the hermeneutic strategies employed by natural history in this period, let me briefly turn to a text cited by Ashworth, Edward Topsell's *The Historie of Fovre-Footed Beastes* (1607). For Ashworth, Topsell epitomizes the emblematic worldview: "his 'Epistle Dedicatory' is a hymn to animals as symbolic images" (1990:316). We should therefore find telling Topsell's preoccupation with the study of nature as integral to humanity's quest for salvation; a quest that, without the aid provided by the text of nature, may be doomed to failure by the curse of Babel.

> As an Interpretor in a strange country is necessary for a traveller that is ignorant of Languages (or else he should perish,) so is knowledge and learning to us poore Pilgrims in this our Perigrination, out of Paradice, unto Paradice; whereby confused BABELS tongues are againe reduced to their significant Dialects, not in the builders of BABELL to further and finish an earthly Tower, but in the builders of IERVSALEM, to bring them all to their owne Countrey Which they seeke, and to the desired rest of soules. (Sig. A3)

What is this knowledge and learning that serves as an interpreter, allowing us access to the language man needs in his journey away from his unfortunate fall out of a state of grace in the garden of paradise toward the paradise of the soul's repose with God in heaven? Topsell

18. W. B. Ashworth (1990:322–23). Ashworth's remarks on Bacon are terribly perceptive. For further discussion, see Chapter 7, which is devoted to Bacon and many of the issues raised here.

would have us answer that it is knowledge and learning concerning the beasts that will reveal the one language that can guide us on our journey toward salvation. Indeed, Topsell's text insistently implicates and emplots human pursuit of natural knowledge within the grand narrative of possession of the Adamic language and its loss through Adam's tragic fall and the subsequent confusion at Babel.

Through his fall and through Babel "man" has lost his way on his journey toward God and eternal salvation in paradise. But, although he has lost the original Adamic language, postlapsarian and post-Babylonic man is still heir to the Adamic legacy. That legacy—knowledge of the true names of all creatures and, hence, access to the creative "Word of God"—is his, if man will simply turn to the text of nature and read its divine message. Knowledge of that text—knowledge deepened by the etymological, mythological, and emblematic learning of ancient and modern authors—serves as our interpreter on the journey to paradise.[19] The knowledge, once lost, is yet again available to man:

> Their [i.e., the beasts] life and creation is Devine in respect of their maker, their naming divine, in respect that Adam out of the plenty of his own devine wisdome, gave them their several appellations, as it were out of a Fountaine of prophesie, foreshewing the nature of every kind in one elegant & significant denomination, which to the great losse of all his children was taken away, lost, & confounded at Babel. When I affirm that the knowledg of Beasts is Devine, I do meane no other thing then the right and perfect description of their names, figures, and natures, and this is in the Creator himself most Devine, & therefore such as is the fountain, such are the streames yssuing from the same into the minds of men. (Topsell 1607: Epistle Dedicatory)

The "names, figures, and natures" of animals are in God Himself; from God as source, they may flow "into the minds of men." Thus, the possibility of recapturing Adam's lost knowledge of the creatures, of repairing the effects of Babel and reconstituting a language that mirrors nature perfectly—that, in effect, closes the gap between the languages of man and the Word of God—is alive and very much present for Topsell. That possibility can be enacted through study of the very beasts them-

19. Thus, note Topsell's (1607: n.p.) insistent tone in setting forth the claims of the Book of Nature for bringing God's counsel to his people: "For how shall we be able to speake the whole Counsell of God unto his people, if we read unto them but one of his bookes, when he hath another in the worlde, which wee never study past the title or outside; although the great God have made them an Epistle Dedicatory to the whole race of mankind."

selves, for Topsell is convinced that God saved all animals from the
Flood by placing them on Noah's ark precisely to enable man access to
divine knowledge and the saving wisdom that may come with it.
"Surely," Topsell, speaking of the animals saved by Noah, proclaims,
"it was for that a man might gaine out of them much devine knowledge, such as is imprinted in them by nature, as a tipe or spark of that
great wisdome whereby they were created" (1607: Epistle Dedicatory).

God's very Word, His Scriptures, "compare the Divell to a Lyon; . . .
false prophets to Wolves; . . . Heretickes and false Preachers to Scorpi-
ons. . . ." Man needs the knowledge God has "imprinted in" animals if
he is not only to uncover, but also to comprehend, His saving Word.
For it is "cleare that every beast is a natural vision, which we ought to
see and understand, for the more cleare apprehension of the invisible
Maiesty of God." Thus it is that Topsell counsels direct investigation of
all animals found in the Book of Nature, and that he brings to this task
"what the writers of all ages have herein observed" and recorded in
stories, images, and emblems (1607: Epistle Dedicatory). God is author
of the universe, and humanity's task is to learn to read the divine text of
nature: "but this sheweth that Chronicle which was made by God himselfe, every living beast being a word, every kind being a sentence, and
al of them together a large history, containing admirable knowledge &
learning, which was, which is, which shall continue, (if not for ever)
yet to the world's end."[20] Topsell returns, then, to man and language.
For him, natural history is the divine text written in the language of
creatures that contains that "knowledge & learning" that Adam and his
offspring lost through the Fall and the curse of Babel. In the very making of Ashworth's "emblematic world view" we therefore find a *remak-
ing* of the grand Renaissance cultural narrative of man's fall and redemption. Natural history, as both the tool and the product of reading
God's text, enables humans access to their lost Adamic heritage, and
with it, access to the saving *verbum Dei*. Renaissance natural history, the
emblematic worldview, and much of what we have called the hermeneutics of symbolic exegesis have been emplotted in this story of man's
fall and salvation. The natural history of Gesner, Aldrovandi, Topsell,
and others is therefore part of that Renaissance quest for the wisdom of
the Adamic language lost in the Babylonic confusion.

This narrative emplotment thus provided legitimation and meaning
for the very practices of this dominant strain of natural history from
the late Renaissance into the seventeenth and even (in some, by then

---

20. Topsell (1607: Epistle Dedicatory). W. B. Ashworth (1990:316) quotes this passage.

atypical, cases) into the eighteenth century. Let us look for examples of natural history in Paula Findlen's recent work, where we shall also find clues for understanding the eventual delegitimation of emblematic natural history. Ultimately, this concern will bring us back to consideration of changes in Renaissance language theory—particularly changes in those very cultural narratives that shaped attitudes toward language—and in the scholarly study of languages that were a crucial development in the decline of the emblematic worldview and, more broadly, of symbolic exegesis.

Findlen's concern has been with natural curiosities that Renaissance naturalists and collectors increasingly viewed as "jokes of nature": *lusus naturae.* Her achievement has been to bring to the forefront of the cultural and social system of early modern science, particularly natural history, what appears to the modern as patently marginal. Jokes of nature—for example, nature's replication of natural forms, such as clouds, mountains, or faces, upon stones; shells in the shape of ears; the transformation of coral from a plantlike to a stonelike substance; fossils; or flowers that assume the shapes of other natural creatures— are no mere oddities, nor simply the aberrant imaginings of otherwise rational, even systematic, students of nature. Rather, they display fundamental assumptions about both nature and the practice of late Renaissance natural history.

Nature is fundamentally playful and creative; it is, in short, poetic. As poetic, nature displays all the imaginative and creative powers associated with fiction. Nature can mimic; it can take natural forms and make them metamorphose into other shapes; it is inherently active, transformative, plastic. Because of her at times willful disposition, nature's playfulness may seem to subvert the very categories and kinds into which creatures are ordered. This aspect of the *lusus naturae* marks something of the independence of nature from the usually stable order of God's creation. Nature, through creative powers instilled in her, spills over with variety and invention and threatens to disrupt the very order of things. Yet her playfulness also exhibits the workings of patterns in nature. Stones and plants mimic the forms of animals and humans; menstrual blood undergoes metamorphosis into toads. Consider even Paré's tale of sexual transformation: the "metamorphosis" that "takes place in Nature" exhibits the regularity and complementarity of nature's forms. Woman is but an internal, unexpressed, version of man. Such transformations and mirroring, then, display the analogical, metaphorical, and hierarchical structure of the divine system of nature.

This is not the place to reproduce the rich complexity and subtlety of Findlen's examples and argument, to indicate, for instance, how the playfulness of Renaissance natural history with its system of correspondences invites exploitation of fables, myths, and literature. The point is that this playfulness was authorized by the way in which nature, as a divine language of things, was constituted by the replication of divine archetypes, vertically within different strata, and horizontally among different "kingdoms and worlds" of nature (Findlen 1990:325). Hence, the decline of nature's playfulness, and of the emblematic worldview, stemmed from the delegitimation of this archetypal, symbolic vision of nature as divine language.

Within such a view, variety may be rampant. Yet the overarching system of correspondences and analogies provides a way of recuperating variety and difference, of transforming them into so many instances of a unified language of things. The very image of God as creator unfolding and "explicating" the unity of His being into the multiplicity of material things carries with it the complementary image of movement back toward unity. Unity becomes the very ground for the possibility of diversity; and the diversity of natural forms becomes the occasion for the quest for an originative unity.

It is precisely the coincidence of this double movement, from unity to diversity and back again to unity, that the cultural narratives authorizing Renaissance language theory and legitimating the hermeneutics of symbolic exegesis in the study of nature enforce in their own domains and reinforce among each other. Consequently, we should view with great interest those movements across the grain of unity-in-diversity that erupt at the margins of Renaissance linguistics and language theory and that begin to complicate the story of language, both manifesting and instigating revision of the narrative of man's Adamic legacy.

Jean Céard enables us to see just such roots of change in sixteenth-century transformations of the story of Babel and the Pentecost. Céard, well known as a student of sixteenth-century natural sciences,[21] expresses fascination with the contesting tendencies in Renaissance engagements with language and its status. Transfixed by the epic quest for the original Adamic language, the Renaissance nonetheless finds itself drawn to the sheer diversity of languages (Céard 1980:577). The story he describes illustrates for him how a guiding assumption such as the notion of an originary language can foster inquiry that, in turn,

21. See, for example, the magisterial work, Céard (1977).

transforms the very terms in which language, and the ideal of an original and universal language, are understood (Céard 1980:577).

Céard recounts the Renaissance confrontation with the legacy of Babel and consequent search for a perfect, universal, Adamic tongue. He discloses the grounds of sixteenth-century obsessions with Hebrew, with etymologies, and with the lure that cabalistic attention to the form itself of letters, words, language exercised over the imaginations of contemporary scholars (1980:578ff.). This story is familiar to us; earlier, in Chapter 3, we treated it in our discussions of the Word of God and the languages of man in Renaissance and Reformation narratives of pre- and postlapsarian man. Yet even this obsession constantly stumbled upon the effects of diversity within language, the inevitable presence of a multiplicity that contrasted with desired unity. The unity of written language must concede the multiplicity of the spoken, the sheer variousness of pronunciation (1980:579–80).

The question was, what to do about this variety within, and of, language? The recuperative power of Renaissance discourses is much in evidence in the challenge posed by the obvious variety of the languages of man. Every tendency exerted its force to find order, and therefore traces of a deeper unity, in the systematic study of language. Céard testifies to what we ourselves have witnessed in earlier chapters, the imposition of the mental grid of Renaissance naturalists—also students of diversity—upon linguistics in the sixteenth century.[22] In particular Céard cites, as an almost archetypal example of this recuperation of variety by a hermeneutics that enforced unity as the ground of all interpretation, the great sixteenth-century naturalist and embodiment of the "emblematic world view," Konrad Gesner. Gesner was, as we have seen, both a natural historian of the first rank and an early student of comparative linguistics. He was, as Céard notes, author of both a text in linguistics, his *Mithridates* of 1555 (Gesner 1974), significantly subtitled *De differentiis linguarum*, and the *Historica animalium* of the same year. According to Céard, "dans les deux livres, Gesner met en oeuvre la même conception de l'ordre et de la variété" (1980:581).

Linguistic diversity was, of course, a legacy of Babel. As such, one might dismiss the sheer variety of vernacular tongues as merely exhibiting the decadence and degeneration from an ideal language implicit in the notion of a historic confusion at Babel. Hence, one response to this

22. Céard (1980:581): "En somme, la variété n'est pas prolifération sans ordre. Dans son activité linguistique, la Renaissance use des mêmes outils mentaux que dans ses enquêtes naturalistes."

linguistic variety might well be to deem its study ignoble. For Gesner, Céard notes, this conclusion fit the case of "barbaric" tongues, particularly Hungarian, which had nothing in common with languages that enjoyed a more noble pedigree. Just as Gesner the naturalist lavished scrupulous attention upon the viviparous quadrupeds whose affinities to man lent them a certain dignity, Gesner the linguist evinced preference for those vernaculars, and ancient tongues, dignified by relationship to the sacred languages of Hebrew, Greek, and Latin, "celles dans lesquelles s'exprime le verbe de Dieu."[23] For Gesner, barbaric languages like Hungarian occupy the same ignoble place among languages in general as the lowly insects do among living things.[24]

Thus, while Gesner encouraged some attention to the variety of linguistic phenomena, his strongly hierarchical understanding of languages—flowing from his implicit adoption of a reading of the Babel narrative as suggesting decadence and degeneration from an ideal—placed limits upon the significance of such diversity. Other readings that looked to Babel as the origin of "linguarum varietas" were perhaps more receptive to the details of such variation and the intrinsic significance of their study (Céard 1980:581). Thus, for some the very diversity of languages might mask a deeper, hidden harmony (1980:585). As Claude Duret (1613, 1605), early seventeenth-century author of a massive tome on language as well as of a text of natural history, revealed according to Céard:

> Il y a "cinq differentes sortes d'escrire": les Asiatiques et les Africains écrivent de droite à gauche; les Européens, de gauche à droite; les Indiens, Chinois et Japonais, de haut en bas; les Mexicains, de bas en haut et, dans certains cas, en spirale. Il en conclut qu'ainsi sont exprimés "les secretz et mysteres de la croisee du Monde, et de la forme de la Croix, ensemble de la rotondité du Ciel et de la terre." (1980:585)

Here we see the continuation of that impulse Céard noted earlier: the tendency promoted by belief in an original, Adamic language to embrace linguistic variety. In Duret's case, such variety leads back to a more fundamental, if sometimes occult, unity. But, as Céard strikingly contends, that willingness to embrace linguistic diversity authorized by the myth of Babel's originary view of language also led to radical revaluation of the relationship between linguistic variety and universality in the late sixteenth and early seventeenth centuries. If the encoun-

23. Céard (1980:583; 581–84 for this discussion in general).
24. Céard (1980:583): "Au fond, parmi les langues, le hongrois tient la place qui, parmi les êtres vivants, est dévolue aux insectes."

ter with variety prompted this revaluation, what extended and further authorized it was another myth or, rather, another rereading of Renaissance cultural narratives that gave prominence to the events of the Pentecost as privileged, Christian context for repairing the effects of Babel.

First, linguistic variety was no longer simply a divine punishment and hindrance to man's reformation. It was not a sad phenomenon of postlapsarian civilization that humans should embrace only to overcome and ultimately reject by tracing their way back to the lost harmony and universality of the original language of Adam. Linguistic variety was, instead, a fact of human existence, and one within which humans learned to operate and to communicate. Céard presents a catalogue of opinions, each illustrating ways in which Renaissance thinkers came to domesticate linguistic diversity and to see in such variety opportunities for enobling "man's" status. Thus, for example, the Renaissance came to admire the polyglot, the person who forged bonds of universality, if not unity, with others through an appropriation of diverse tongues. Languages became a vehicle not only for communication, but also for acquiring the totality of human heritages—learning and wisdom.[25]

The surgeon Paré, encountered earlier, testifies to certain changes regarding language as well. While animals enjoy the ability to understand every individual member of their own species, man, according to Paré, displays his superiority to animals in his "dexterité d'apprendre toutes langues" (Paré 1841; Céard 1980:589). Céard sees in such views the beginnings of a new attitude toward language and the curse of Babel inflicted upon man: "Par cette 'dexterité' universelle, il échappe à la punition de Babel: l'unité perdue, il la reconstitue dans l'universalité" (1980:589). Clearly, this step in itself does not constitute the end of the emblematic worldview. Paré himself did not abandon a vision of nature as expression of the divine Word accessible to man. His approach to nature remains very much wedded to analogy and to the semiotics of resemblance and a symbolic order.

But such revaluations of linguistic diversity opened a discursive space for the articulation, or rather transformation, of powerful Renaissance cultural narratives. And such transformations could have the effect of delegitimizing the very narrative and discursive foundations underwriting the hermeneutics of symbolic exegesis—which includes the emblematic worldview—in the Renaissance. It is this aspect of Céard's

25. Céard (1980). See Céard (1980:585ff.) for discussion of new appreciations of linguistic diversity.

examination of language and the myths of Babel and the Pentecost that is of such importance to our own story. His point is a decisive one for the historical changes we have been considering:

> Le mythe d'une langue unique et universelle cède la place à un autre mythe: celui d'une aptitude à parler toutes les langues du monde, à disposer de toutes les richesses du verbe. La langue universelle, c'est le "thresor" de toutes les langues de cet univers. Dans cette abondance se refait la langue parfaite des origines. (1980:589–90)

Even here, however, we must resist the temptation to proclam a rupture in history. What we have is no more than an alternative articulation, a recast narrative whose terms are so familiar that they are always in danger of being reappropriated to a familiar disourse, reinscribed within the very narrative it attempts to discard. The alternative myth of speaking all tongues does lead to an embrace of diversity and variety unparalleled in the more traditional quest for the traces of the lost, Adamic language. And yet that "new" myth remains haunted by the shadowy presence of a perfect originative language.

What Céard has exposed in sixteenth-century thought is not so much a rupture within discourse as its tensions and opposing valences. That does not mean that nothing has changed. Quite to the contrary, the exposure and articulation of these tensions serve only to heighten new possibilities for language theory and for the hermeneutics of nature. My point is simply that these new possibilities did not vanquish the old; rather they existed side by side, even indeed within the same text and discourse. This state of affairs was in fact productive of a veritable explosion of interest in the diversity and variety of nature. On the one hand, a Gesner could appreciate the diversity of languages and the variety of natural forms as worthy of scholarly and scientific investigation. Nature could be laid bare as a text upon which God had imprinted a rich array of forms, a veritable profusion of natural dialects of the unifying and creative Word of God. For Gesner, such diversity was contained within the larger frame of belief in a single, originary divine language of man and creation. Gesner's language theory and his practice as natural historian fit within the mainstream of the emblematic worldview. His world, and that of his standard-bearers like Aldrovandi, was one that awaited symbolic exegesis.

On the other hand, one could also find in a Guillaume Rondelet a fascination with the variety of natural forms that resisted reinscription in the cultural text of a quest for an originative unity—a unity, not just of the specific forms themselves and the orders to which they belonged, but of the similitude of all forms that constitute the originative divine language of creation. It is precisely this symbolic, emblematic quality to the natural order, and to natural history, that Rondelet resists, and that has set him apart—along with the likes of Belon— among Renaissance naturalists in standard histories of the subject. Indeed, Ashworth's historiographic plea for the centrality of his emblematic worldview constitutes a reaction to what he sees as the anachronism of elevating the spare, descriptive natural history of a Rondelet to the standard against which Renaissance texts must be judged.[26] Rondelet should not, in fact, become such a standard. Nonetheless, the very century that produced a Gesner and an Aldrovandi also produced a Rondelet. They must, I think, be seen as emerging from the very possibilities for alternative discursive solutions to the problem of variety and unity that Céard's treatment of the myths of Babel and the Pentecost allows us to see in Renaissance language theory and natural history. Finally, we may regard a Paré as exhibiting a tolerance for linguistic diversity that finds itself easily recuperated by the belief in an originary *verbum Dei* authorizing a basically analogical, metaphorical, and hierarchical understanding of nature.

What Céard has uncovered is that opening up of a discursive space in the Renaissance that would, in time, widen into something like a chasm. That space allows for the articulation of a new linguistic ideal authorized by a subtly recast narrative of cultural origins that never quite breaks with past, and one might even say dominating, narratives. The ideal is that of a universal language. The ideal itself is but a trope: a projective desire for a whole that will bind together the many fragmented parts constituting the languages of man. In this sense, then, this new ideal generates itself out of a trope that it shares with its rival. But this new ideal threatens to destabilize discourse by reemploting that trope in a narrative that alters its fundamental meaning. For the trope of a universal language now looks, not to the past, to the unity of Adamic innocence and perfection for its authorization and its model, but instead to the future. The metaphoricity of the Adamic language now becomes transformed into the synecdoche of an ideal universal language. The universal language is not something that has been, but

---

26. W. B. Ashworth (1990:304–5). In his criticism of Foucault's often sweeping characterization of the sixteenth-century *episteme*, Huppert (1974) raises both Pierre Belon and Rondelet to the status of exemplars of French scientific/biological thought, thus deflecting attention away from the critical role that the emblematic or symbolic plays in the story of scientific culture in the late Renaissance. The source of the epigraph at the start of this chapter is Foucault (1966:53).

rather an ideal to be created by humans out of the multiplicity of linguistic phenomena constituting the very languages of man.

> La langue universelle est, non pas derrière nous, mais devant nous: c'est une langue à créer, par nos efforts, par notre travail. Au lieu de récuser la diversité en n'y cherchant que les vestiges de la langue première, embrassons la diversité pour lui conférer une nouvelle unité,—mais une unité à mesure humaine, une unité d'ordre qui fait de la diversité sa matière même. (Céard 1980:592)

Here Céard turns significantly to Rondelet. Rondelet, poised between the discursive possibilities offered in the tensions provoked by this Renaissance problematic of diversity and unity, reinscribes the very project of a natural history of God's text of nature within a new narrative of Babel. Citing Rondelet's *Histoire entiere des poissons*, Céard exposes with vivid clarity the choices now tentatively broached within his century:

> Rondelet loue la variété da la nature qui, loin di nuire à son unité, donne matière à l'ordre: un dans son dessein, multiple dans ses parties, l'ordre concilie l'unité et la variété; il écarte deux périls: celui d'une altérité si confuse que l'homme ne saurait y tracer son chemin, celui aussi d'une unité si totale, si ramassée que l'homme ne se perdrait pas moins dans cet océan de la similitude. A première vue, on attendrait que Babel soit l'image du péril de l'altérité; chez Rondelet, Babel est l'image du péril de l'unité totale: "La confusion des langues en Babel ne donna jamais tant de poene, que donneroit ceste similitude de toutes choses entre elles." On voit assez, par cet exemple remarquable, que l'accueil de la diversité ne signifie pas exactement une renonciation à quêter l'unité perdue, que la diversité est le champ même de la quête d'une nouvelle unité. Dans cette perspective, le mythe de Babel se transforme profondément. (1980:592–93)

By the late sixteenth century, then, one can find the cultural narrative within which theories of language are emplotted broadened and diversified to allow for new possibilities in the understanding of the relationship between the Word of God and the languages of man. The story of the confusion of tongues at Babel may now be read through "le miracle de la Pentecôte" (Céard 1980:593). The languages of man thereby gain the possibility of a new relationship to the *verbum Dei:* the very diversity that was once a sign of their corrupt, fallen status—a status to be overcome through supression of their individuality and alterity—now becomes the very material out of which a rich and proper universal language of things may be forged anew.

But the prospect of such a universal language, as we have noted above, was not easily disengaged from the old ideal of an originative unity. Only a distant hope, this new ideal did help to open up the very scope of natural history and the investigation of natural phenomena. By allowing that the perfection and universality of truth—of true knowledge of nature—were not to be *rediscovered* through exegesis of texts and the Book of Nature, but rather were something to be *made* by mankind through our efforts and work, this new ideal—however tentatively and partially articulated—looked toward a new hermeneutics of nature. Just as the miracle of the Pentecost embraced the diversity of tongues and of mankind in an image of God's Apostles engaged in the work of creating a new Christian unity, those pious disciples of the Lord who choose to work upon God's Book of Nature can create a new order by de-in-scribing God's plan within the variety of His earthly creatures. This new Pentecostal narrative of man's fall and redemption hence pointed toward emphases upon human industry and the observation of nature in all its diversity—and away from a symbolic ordering of nature—that were to become a hallmark of the new science of Bacon, Galileo, Mersenne, Descartes, and Boyle in the seventeenth century.

Returning for a moment to Ashworth and to Laqueur, the parallelism and, indeed, the intimate relationship between language theory and natural history in the sixteenth and seventeenth centuries, together with their mutual legitimation through the work done in recasting cultural narratives, must not be forgotten in attempts to explain shifting models of sexuality or models for the practice of natural history. For, as in the case of the encyclopedic and comparative study of language in the sixteenth century, the sheer diversity of natural things that the discourse of encyclopedic, emblematic natural history uncovered (including the flora and fauna of the New World) tended to overwhelm and undermine the discourse of resemblance, correspondence, and symbolism and its underlying ideal of an originative unity. And yet, as was also the case in the study of language, this sheer diversity—the sheer weight of things and facts—was not of itself enough to transform the discourse of natural history. Instead, diversity in language as well as in nature had first to be reinscribed within transformed narratives. Only such new, legitimating narratives could work to delegitimize the relationship between the *verbum Dei* and nature that authorized the ideal and discourse of originative unity. By recasting stories of Babel, and the Pentecost, "man" was now authorized to study diversity—both in language and in nature—within the context of a new relationship between the languages of man and the Word of God.

Galileo, Mathematics, and the Language of Nature:
De-in-scribing God's Book of Nature

The drift *away from* words and symbols—from symbolic and exegetial hermeneutic strategies like those deployed in the practices of the emblematic worldview—built upon newly recast cultural narratives to authorize new strategies for interpreting the text of nature. What these narratives and strategies had in common was a tendency to see that "text" as God's creation, but not as a simple reflection of His ideas and nature. God did not stamp some image of Himself upon things in nature, thus leaving a symbolic "trace" through which one could "read," in the hierarchical and analogical pattern of all such traces, the true image of nature *and* of God. Rather, God writes—or inscribes—His Book of Nature in the language of things. Although emanating from God, things are marked by their alterity, their utter difference from the divine being. Still, as a divine text, the Book of Nature exhibits an order and regularity that are inherent in the very language with which God has inscribed nature. As early modern authors turn toward this nonsymbolic text of nature, we find them constructing new hermeneutic practices aimed at *de-in-scribing* nature, that is, at *describing* the order and regularity of God's visible text.

We shall turn, in the two following chapters, to a number of examples of deinscriptive hermeneutics. For now, I want to bring closure to this chapter by briefly illustrating the kind of new discursive strategies that such revised narratives and theories of language could authorize by turning to Galileo and his mathematical vision of the text of nature.

That Galileo's vision was mathematical is so well known to historians of science and, I would wager, to virtually all who have encountered the "Scientific Revolution" that it hardly seems necessary to note the fact.[27] Who has not heard Galileo's famous pronouncement that the Book of Nature "is written in the language of mathematics?" But what Galileo says, what he purports to establish when invoking this language of mathematics, is far richer and less well known than his famous dictum:

27. So fundamental is Galileo and his mathematical approach to the Book of Nature to the historiography of the "Scientific Revolution" that the literature on Galileo is vast. I shall note here only a few classic and fundamental studies and texts: Biagioli (1993); Burtt (1954); Finocchiaro (1980, 1989); Koyré (1943, 1966); Maier (1949); McMullin (1967); Moody (1951); Randall (1961); Redondi (1987); Rossi (1971); Schmitt (1969); Shea (1977); Wallace (1981, 1984, 1986).

In Sarsi I seem to discern the firm belief that in philosophizing one must support oneself upon the opinion of some celebrated author. . . . Possibly he thinks that philosophy is a book of fiction by some writer, like the *Iliad* or *Orlando Furioso*, productions in which the least important thing is whether what is written there is true. Well, Sarsi, that is not how matters stand. Philosophy is written in this grand book, the universe, which stands continually open to our gaze. But the book cannot be understood unless one first learns to comprehend the language and read the letters in which it is composed. It is written in the language of mathematics, and its characters are triangles, circles, and other geometric figures without which it is humanly impossible to understand a single word of it; without these, one wanders about in a dark labyrinth.

. . . but now Sarsi expects my mind to be satisfied and set at rest by a little poetic flower that is not followed by any fruit at all. It is this that Guiducci rejected when he quite rightly said that nature takes no delight in poetry. That is a very true statement, even though Sarsi appears to disbelieve it and acts as if acquainted with neither nature nor poetry. He seems not to know that fables and fictions are in a way essential to poetry, which could not exist without them, while any sort of falsehood is so abhorrent to nature that it is as absent there as darkness is in light.[28]

Galileo's "language of mathematics" is the thread that enables him to find his way through the labyrinth of nature; it is the lamp that sheds light on the truth, or rather the lamp of truth itself. Ostensibly a defense against the appeal to authorities in the world of seventeenth-century natural philosophy, Galileo's words are rhetorically shrewd and significant. Rather than argue the merits of Galilean "philosophy" compared

28. *The Assayer* (Galilei 1957:237–38). For the original text, Galilei (1896:232): "Parmi, oltre a cio, di scorgere nel Sarsi ferma credenza, che nel filosofare sia necessario appoggiarsi all' opinioni di qualche celebre autore. . . ; e forse stima che la filosofia sia un libro e una fantasia d'un uomo, come l'Iliade e l'Orlando Furioso, libri ne' quali la mano importante cosa è che quello che vi è scritto sia vero. Sig. Sarsi, la cosa non istà cosi. La filosofia è scritta in questo grandissimo libro che continuamente ci sta aperto innanzi a gli occhi (io dico l'universo), ma non si puo intendere se prima non s'impara a intender la lingua, e conoscer i caratteri, ne' quali è scritto. Egli è scritto in lingua matematica, e i caratteri son triangoli, cerchi, ed altre figure geometriche, senza i quali mezi è impossibile a intenderne umanamente parola; senza questi è un aggirarsi vanamente per un oscuro laberinto." This is followed (1896:234) by "Ma che in una questione massima e difficilissima, qual è il volermi persuadere trovarsi realmente . . . la mente mia debba quietarsi e restar appagata d'un fioretto poetico, al quale non succede poi frutto veruno, questo è quello che il Sig. Mario rifiuta, e con ragione e con verità dice che la natura non si diletta di poesie: proposizion verissima, ben che il Sarsi mostri di non la credere, e finga di non conoscer o la natura o la poesia, e di non sapere che alla poesia sono in maniera necessarie le favole e finzioni, che senza quelle non puo essere; la quali bugie son poi tanto abborrite dalla natura, che non meno impossibil cosa è il ritrovarvene per una, che il trovar tenebre nella luce."

with traditional "philosophy," that is, the relative merits of the *texts* written by Galileo and by his Scholastic opponents, Galileo instead cleverly alters the very grounds of the comparison. Philosophy, for Galileo, is no longer a discourse fashioned by human beings. Indeed, his very rhetoric depends upon the insinuation that the so-called philosophy of his Scholastic opponents is but the fabrication of human writers, like the authors of "the *Iliad* or *Orlando Furioso*." Rather than a product of limited, and distorting, human imagination, Galileo declares that philosophy is not a text written by humans at all! Rather, it is a text "written in this grand book, the universe."

Galileo rhetorically withdraws himself from the scene of writing, from active agency in the production of the text of philosophy, in order to claim a kind of authority for the "philosophy" he espouses that transcends human institutions and the limitations of human fabrication. Instead, Galileo makes the odd, and bold, claim that philosophy—that is to say, true knowledge ("science") of nature—is already constituted as a text prior to its articulation by humans. To defend his own achievements and the truth of his new philosophy, Galileo chooses to efface his role as author. Instead, he transfers authorship to a nonhuman agency: to nature itself and, by implication, to the author of nature Himself, God.

Certainly, we have seen that the trope of the Book of Nature is a commonplace throughout the Renaissance and seventeenth century.[29] Isn't Galileo, then, simply invoking a familiar topos? I believe that we witness in Galileo's text the appropriation and transformation of this familiar metaphor.[30] For, where earlier natural philosophers figured nature as a divine book whose *meaning* man must learn to interpret correctly, Galileo figures nature as an "open" text that he can read directly, without the need for interpretation! Man does need a key to this text—mathematics—but the text itself requires no interpretation: it requires only decoding, or what I have called de-in-scription.

Let me clarify this distinction, for it is a distinction central to Galileo's own rhetoric in the passage quoted above. Galileo fundamentally as-

29. On the "Book of Nature," see Curtius (1963: esp. chap. 16, "The Book as Symbol"); Eisenstein (1979: esp. "Part Three: The Book of Nature Transformed"); and Garin (1961). I have not seen Rothacker (1979).

30. See Bono (1990a) for discussion of the role of metaphor in changing scientific discourses. As this book suggests, metaphoric change and exchange are intimately connected with the function of narrative in scientific discourse. My current theoretical work explores connections among metaphor, narrative, and the "ecology" of sociocultural discourses and practices in "scientific culture"; this work will culminate in a book (Bono, in progress).

serts that philosophy is *not* a text produced by human authors. Unlike the texts of natural philosophy, medicine, or even magic produced earlier in the Renaissance, no human author needs to untangle the meaning of the Book of Nature. While earlier students of nature figured that book as a polysemous text written in symbolic and divine language that required etymological, symbolic, allegorical, cosmological, or moral exegesis,[31] Galileo saw the Book of Nature differently. Whereas the polysemous text of this exegetical hermeneutics of nature required that individual human authors uncover hidden vectors of meaning that bound God's Book of Nature into a coherent and meaningful whole, Galileo figured the Book of Nature as a stubborn but self-revealing text that did not require human agency to construct its meaning. In short, in the old regime of an exegetical science of nature humans not only "read" the divine language stamped upon the Book of Nature but also exercised their imaginative and intellective powers to create texts— *books* of, for example, natural philosophy—to capture through interpretation the multilayered, interwoven, *textual* meaning of things in nature. "Science"—the entire spectrum of natural philosophy, medicine, and magic—was enmeshed in "bookish" culture. And those who sought knowledge of nature were called upon to reimagine, represent, comment upon, and explicate the analogical multiplication and unfolding (*explicatio*) of the Word in the text of nature.[32]

It is this regime of the Word and words in the understanding of nature that Galileo not only rejects, but ridicules. Indeed, his ridicule proves essential as, ironically, it rhetorically masks his own use of figurative thinking to produce the unlikely notion of a *philosophical* text without human author! So unlike the merely *fictive* texts of so-called philosophers—texts that indiscriminately mix nature with fables and poetry—is the real text of philosophy that all human imaginative agency is banished from it in the Galilean trope of philosophy *as* text of "this grand book," namely, the "universe." Here, in Galileo's assertion that "nature takes no delight in poetry," we find the complete reversal of the "emblematic world view."[33] Nature is not poetic, multilayered, polysemous, mysterious, arcane, and a congeries of resemblances and affinities—the stuff of fictive similes and metaphors. Such falsehoods are "abhorrent to nature" whose contours are in-

31. See such figures as Fernel, Paracelsus, Croll, Gesner, Aldrovandi, and the like discussed above.

32. The Latin term is, of course, Nicholas of Cusa's. For an introduction to Cusanus, see Cassirer (1963) and Watts (1982).

33. As courtier and client in search of patronage, Galileo was, of course, thoroughly enmeshed in the literary and poetic culture of Italy. See Biagioli (1990, 1993).

stead defined by the sharp, distinctly etched forms of "triangles, circles, and other geometric figures." The task of the true scientist gazing upon this open book of nature is thus utterly different from his exegetical counterpart's.

Here we discover the force of Galileo's metaphor of mathematics as the language of nature. For Galileo conceives that language as literal, not symbolic:[34] Nature *is* constructed mathematically, geometrically. As a book, nature is unlike any other known to man; it contains no ambiguities, no hidden and symbolic meanings, no fables or poetic senses to obscure the truth and, indeed, to mock truth by transforming it into the fleeting, subjective construct of scientist-as-exegete/interpreter. For the Book of Nature, unlike the Scriptures, is open to our gaze and not shrouded in figures and allegorical meanings meant to accommodate the simple and bar the ungodly.

Galileo does not *produce* the text of philosophy already written in the Book of Nature. Rather, as scientist, his task is to gaze upon that book and to decipher its characters. And once he has grasped its language—the fact that its characters are geometrical figures—Galileo merely uses the instruments at hand to deinscribe what God has legibly inscribed upon things in nature. Hence, Galileo's role—in his estimation—is not that of interpreter; he need not seek below the literal surface of the geometrical characters inscribed in nature for some deeper, hidden meaning. Nature—and philosophy, the "science"of nature—has only a literal sense to reveal to mankind: that is, the order and structure of things in their geometrical relationships to one another—the *how* of nature, not the mysterious, polysemous *why* of theology. Never mind that Galileo's very institution of this regime, this deinscriptive hermeneutics of nature, itself rests upon the figural: that nature is *like* geometry and mathematics; that geometry and mathematics are themselves man-made tools, metaphoric creations for the *measure*, for the *comparison*, of different things. Galileo's myth, his new narrative of man as reader of God's open book, requires that the very figural basis of his narrative be elided and presented, instead, as the literal configuration of nature itself (Wojciehowski 1990).

Perhaps this elided figure and unarticulated conviction are why

34. I want to acknowledge the important work of Professor Dolora Wojciehowski of the University of Texas at Austin, who kindly permitted me to see a chapter, "The Will to Read: Galileo and the Book of Nature," from her forthcoming book. She provides a careful and compelling analysis of Galileo's desire for a literal language of nature, and of his own thoroughgoing implication in figurative language. My own analysis complements hers, but Dr. Wojciehowski provides a far more detailed analysis, with judicious use of examples, than I can attempt here.

Galileo thought that he could demarcate for theologians the boundaries of the Word of God. The Book of Nature with its divinely inscribed language of mathematics produces, and creates in man, *only* a literal sense. The Book of Nature tells us only about nature; it reveals only *how* nature carries out God's commands. It tells us nothing about God's intentions or about heaven. For the latter, we need His other book, the Scriptures, whose language perforce must be of a different order: layered, symbolic, filled with metaphors figuring what must always remain mysterious and cannot be "open," but which man can only hope to grasp through interpretation, much like a poem or a fable:

> For the Holy Scripture and nature derive equally from the Godhead [*Verbo divino*], the former as the dictation of the Holy Spirit and the latter as the most obedient executrix of God's orders; moreover, to accommodate the understanding of the common people it is appropriate for Scripture to say many things that are different (in appearance and in regard to the literal meaning of the words) from the absolute truth; on the other hand, nature is inexorable and immutable, never violates the terms of the laws imposed upon her, and does not care whether or not her recondite reasons and ways of operation are disclosed to human understanding.[35]

Unlike Scriptures, the language of nature is not accommodated to man, according to Galileo. Nature's book, while open to mankind's gaze, requires work to understand, not the work of exegesis, of interpretation, but, in its stead, the work required to achieve linguistic competence in its foreign, mathematical tongue. Galileo would surely add the work of observation and experimentation which alone can turn the leaves of nature's book to expose its characters to our view. Galileo's new deinscriptive hermeneutics of nature, while effacing its own status as interpretive practice, enacts a desire to enshrine a new strategy for reading the text of nature literally, if laboriously.

35. Galileo Galilei, "Galileo's Letter to the Grand Duchess Christina (1615)," in Finocchiaro (1989:93). For the original text, Galilei (1895:316): "perchè, procedendo di pari dal Verbo divino la Scrittura Sacra e la natura, quella come dettatura dello Spirito Santo, e questa come osservantissima essecutrice de gli ordini di Dio; ed essendo, di piu, convenuto nelle Scritture, per accommodarsi all'intendimento dell' universale, dir molte cose diverse, in aspetto e quanto al nudo significato delle parole, dal vero assoluto; ma, all'incontro, essendo la natura inesorabile ed immutabile, e mai non trascendente i termini delle leggi impostegli, come quella che nulla cura che le sue recondite ragioni e modi d'operare sieno o non sieno eposti alla capacità degli uomini."

7

The Reform of Language and Science

Sir Francis Bacon's Adamic Instauration and the Alphabet of Nature

> But the more difficult and laborious the work is, the more ought it to be discharged of matters superfluous. . . . First then, away with antiquities, and citations or testimonies of authors; also with disputes and controversies and differing opinions; everything in short which is philological. . . . And for all that concerns ornaments of speech, similitudes, treasury of eloquence, and such like emptinesses, let it be utterly dismissed. Also let those things which are admitted be themselves set down briefly and concisely, so that they may be nothing less than words.
>
> —Sir Francis Bacon, *Parasceve*

> For God forbid that we should give out a dream of our own imagination for a pattern of the world; rather may he graciously grant to us to write an apocalypse or true vision of the footsteps of the Creator imprinted on his creatures [ac veram visionem vestigiorum et sigillorum creatoris super creaturas scribamus].
>
> —Sir Francis Bacon, *The Great Instauration*

Discovering the true "pattern of the world" qualifies as a "difficult and laborious" sort of "work" for Bacon—the kind of work that requires humans to turn away from philology, etymology, and the exegesis of language and symbols that inhabit his all too fecund imagination and ensnare his intellect with mere images: fictions. Bacon, of course, was not the first to try negotiating the turn from words to things (think only of Paracelsus), but few before him claimed to have as detailed a map to avoid plunging recklessly off course. The Lord Chancellor was nothing if not shrewd in his estimation of the causes of others' mishaps.

John Dee

Take Bacon's Elizabethan predecessor as court natural philosopher, John Dee (1527–1608), whose alleged cavorting with spirits sat not well at all at the Stuart court.[1] Dee, too, was obsessed with things. How Dee attempted to negotiate his way to things, however, illustrates the continuing lure of tradition and bookish culture and the power that the very trope of nature as akin to a language had in enforcing an exegetical hermeneutics of things in nature. It is precisely this lure and power that the Lord Chancellor's "Great Instauration" sought to resist.

Fundamental to Dee's quest for things was his conviction that the occult and hidden knowledge that man seeks is not confined to one spoken or written language, however "holy" that language might be. Rather, as God is the God not only "of the Jews, but of all peoples, nations, and languages," the source of such sought-after knowledge lies elsewhere and is available to all humans and nations.[2]

Dee calls this source of knowledge the "real cabbala," which—though it corresponds to no actual language—assumes for him the status of "our Holy Language" (Josten 1964:134/135). Indeed, this "sacred language," the real Cabala, is radically unlike the common Cabala known to the sixteenth century, "which rest[s] on well known letters that can be written by man." This common, merely verbal art is but a cabalistic grammar that dwells only on "that which is said" (*tou legomenou*). By contrast, Dee's "real cabbala" concerns itself exclusively with "that which is" (*tou ontos*).[3] Here Dee reveals himself as operating within the discursive system of Renaissance magic while simultaneously exposing one of its foundations—the intimate connection between a divinely originating language and things—and even working to subvert it. For the point of Dee's distinction between a "vulgar," or verbal, Cabala and a "real" one is precisely to indicate the inadequacy of any spoken or written medium as a vehicle for grasping, uncovering, and penetrating "that which is"—the very things that God has created. Thus Dee can vigorously assert that "the real cabbala, which was born

1. On Dee, see Clulee (1977, 1988); Deacon (1968); French (1972); Heilbron (1978); Knoespel (1987); and works cited below.

2. The fundamental text is Dee (1564). I shall use the text and translation found in Josten (1964:112–220); see Josten (1964:132/133). Note that even numbered pages are for the Latin text; odd numbered for the English translation.

3. Josten (1964:134/135): "REALEM nominaui CABALAM, sive *tou ontos:* Vt illam vulgarem alteram; Cabalisticam nomino GRAMMATICAM sive *tou legomenou;* quae, notïssimis Literis, ab Homine Scriptibilibus, insistit."

to us by the law of the creation (as Paul intimates), is also . . . more divine. . . ."[4]

The real Cabala gives us knowledge of things; indeed, it constitutes as an art a turn to things. For, as Dee's allusion to Saint Paul suggests, man does not need any other "*written* memorials" (*Scriptum Monimentum*) to uncover truths. He has, in their stead, something far better, far more original, indeed, the very "originals" of truth itself in "that from which the Creation has been [written] by God's own finger on all creatures."[5]

To be sure, Dee's subversion of Renaissance quests for a verbal Cabala, for a hermeneutics to unlock the sacred truths found in the traces of the *verbum Dei* in post-Babylonic languages, does not abandon a Neoplatonic vision of language as, ideally, reflection and "embodiment" (in the materiality of verbal, written or oral, signs) of the true nature of things. Dee, in short, recognizes a "mystical" dimension to language, even as he regards spoken and written languages as inferior to those things that God has written upon. But, for Dee, it is "the science of the alphabet" that "contains great mysteries."[6] This is so, Dee says, because "He, who is the only Author of all mysteries, has compared Himself to the first and last letter."[7]

Dee's negotiation of the turn from language to things, partial as it may appear to twentieth-century readers, hinges upon a metaphorical association in which the significance of letters as the constituent elements of words—that on one level exist as meaningful *marks*—becomes transferred to the "real" marks that constitute things themselves. Thus, the world itself can be understood as built up out of a constellation of marks—an "Alphabet of Nature"—that God has written upon things. God Himself is the first and last letter, but He has left behind others in nature as well: "How great, then, must be the myster-

4. Josten (1964:134/135): "Haec autem, quae Creationis nobis est Nata Lege, (vt Paulus innuit) REALIS CABALA, GRAMMATICA quoque quaedam Diuinior est."

5. Josten (1964:124/125): "quod ex CREATIONE, ipso Digito DEI, in omnibus est exaratum Creaturis." Josten translates the verb as "inscribed"; I have substituted in its place "written." The reason for this change is to avoid confusion, since in my analysis "inscribed" and "inscription" are paired with what I call the hermeneutics of "de-inscription." Dee's hermeneutics of nature remains what I term "exegetical" and "symbolic." What God writes on all creatures are, for Dee, marks, traces, or symbols of a special sort. These are written *on* things; things themselves are not written, or *inscribed*, by God. The language, or alphabet, of nature is still symbols, not things themselves.

6. Josten (1964:124/125): "Alphabetariam Literaturam, magna continere Mysteria. . . ."

7. Josten (1964:124/125): "Cum IPSE, qui omnium Mysteriorum Author est SOLVS, ad Primam & Vltimam, SEIPSVM Comparauit Literam."

ies of the intermediate [letters]? And it is not surprising that this [mystery] should be so constituted in letters; for all things visible and invisible, manifest and most occult things, emanating (through [the medium of] nature or art) from God Himself, are to be most diligently explored in our wanderings, so that thereby we may proclaim and celebrate His goodness, His wisdom, and His power" (Josten 1964:124/125).

While this trope of an alphabet of nature depends upon the cosmogonical vision of the universe brought into being by the *verbum Dei* that we have encountered so often before in the Renaissance, for Dee that vision underwrites the search, not for a lost language or Word, but for unspoken marks, letters, or characters that are, as it were, the physical traces that God has etched upon things. These marks constitute, as it were, an alphabet that contains nature's mysteries and that generates the phenomena associated with "that which is"—the things that surround us in the garden of God's creation.[8]

What Dee seeks, then, are the marks, "letters," or characters that God has written on things in nature. These divine traces represent symbolic keys to unlocking the mysterious meanings in nature and are also the enlivening sparks that enable nature to generate and regenerate itself. They are what make nature both intelligible and productive: poetic in both senses of the term. But for Dee this search, this "wandering" among things, becomes, in addition to a search for "that which is," a quest for the true alphabet that will enable humans to communicate the very essence and divine presence in things. For, as God's writing on things, these marks, letters, or characters are *both* "that which is" and the only true, sacred, divine language: what Dee therefore calls "our Sacred Language." The doubleness of natural things as both ontic reality and divine semiotic sign pervades Dee's *Monas Hieroglyphica*, so that the search for the truths of nature turns out to be also a quest for nature's hieroglyphics—literally, for its originary "sacred writing." Things in nature are thus necessarily also *messages*. Our quest must then entail discovering that "sacred art of writing" that will allow us both to decode and to communicate nature's divine and secret messages.

The very title and central figure of Dee's work subtly convey this doubled, layered sense in which nature—things—must be encountered by humans. The *Monas* is a figure (see Figure 7.1)—not simply an allegorical image enshrined within the iconographical traditions of

---

8. See Clulee (1988: esp. 89–95) for a careful and detailed account of Dee's search for such an alphabet of nature and its implications. I am indebted to Clulee's discussion.

Dee's Monad

Figure 7.1. Dee's Monas. From Michael T. Walton, "John Dee's *Monas Hieroglyphica:* Geometrical Cabala," *Ambix* 23 (1976): 120. Reproduced by kind permission of the Society for the History of Alchemy and Chemistry.

Western art and occultism, but a geometrically generated form. Its elements represent major cosmological components of the universe (Figure 7.2), but, as a whole, it most closely represents the symbol for the planet Mercury (Figure 7.3). Dee says of this "hieroglyphic monad,"

> And indeed the very rarest thing of all is that all this should be embodied in one single hieroglyphic symbol, notably that of Mercury (to which a pointed hook has been added). Mercury may rightly be styled by us the rebuilder and restorer of all astronomy [and] an astronomical messenger [who was sent to us] by our IEOVA so that we might either establish this sacred art of writing as the first founders of a new discipline, or by his counsel renew one that was entirely extinct and had been wholly wiped out from the memory of men. We have done this in such a way that the hieroglyphic interpretations fall into place most gently and, as it were, of their own accord. (Josten 1964:121–23/120–22)

Here the planet Mercury and its geometrical, hieroglyphical symbol double as both thing and message. For Mercury is also the god Hermes, the keeper of occult wisdom so well known to the Renaissance and the messenger god, the god of communication. This messenger, with its secret and all-powerful message, has been sent to us by the one, true God of the Jews and Christians according to Dee. This very messenger, Mercury, is therefore also the *message:* the root of that "hieroglyphic monad" that unlocks for Dee the geometrical/mathematical alphabet of nature. By "reading" this message and messenger correctly, then, Dee asserts that humans can uncover—or recover—the "sacred art of writing" known only to God.

Implicit in Dee's account of Mercury as hieroglyph and harbinger of a new kind of "writing" is a narrative that embeds Dee's quest in a larger

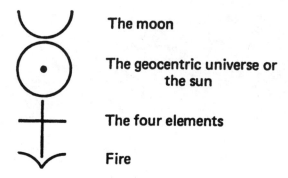

The moon

The geocentric universe or
the sun

The four elements

Fire

Figure 7.2. The major cosmological components of the universe as represented in Dee's Monas. From Michael T. Walton, "John Dee's *Monas Hieroglyphica:* Geometrical Cabala," *Ambix* 23 (1976): 120. Reproduced by kind permission of the Society for the History of Alchemy and Chemistry.

Mercury

Figure 7.3. The symbol for the planet Mercury. From Michael T. Walton, "John Dee's *Monas Hieroglyphica:* Geometrical Cabala," *Ambix* 23 (1976): 120. Reproduced by kind permission of the Society for the History of Alchemy and Chemistry.

history, both sacred and profane. Dee cryptically suggests that he and his reader "might either establish this sacred art of writing as the first founders of a new discipline, or . . . renew one that was entirely extinct and had been wholly wiped out from the memory of men." What can be at stake here? Why this uncertainty about his own, most central, enterprise?

Dee's ambivalence has to do, I think, with his own understanding of human cultural history and with its relationship to the divine history of creation, the Fall, and redemption. As we have seen, Dee remains profoundly skeptical of the claims of any human languages in these post-Babylonic times to universality. Even ancient and so-called "sacred" languages like Hebrew are inferior to the truly "sacred art of writing" practiced by the very finger of our Creator upon things. Implicit in

Dee's account is the suspicion that even "sacred" languages are but human in origin—at best, crude and corrupted reflections of an originary alphabet of nature in merely human, material forms of speech and written letters. Whatever we, fallen and corrupt mankind, have, it is not the divine "sacred art of writing." Therefore, the latter art is one that Dee, and his complicit reader, need to "establish" as something new to the history of mankind. Its traces, if every they existed in "historical" times, have been "wholly wiped out from the memory of men."

Dee's enterprise, his quest, and, through the "hieroglypic monad," his establishment of a "real cabbala" of things can therefore count as a new beginning for mankind: the introduction of a new art of decoding nature and communicating true knowledge to all those who seek to know God's sacred mysteries. Yet, at the same time, Dee holds out the possibility that this innovation is also a "renewal," a restoration. It is, to be sure, a restoration not of any literate and historical human tradition, but, rather, of an unspoken—or, if spoken, soon lost—Adamic encounter with the "real cabbala" of "that which is," the divine hieroglyphs of things themselves. As Dee says of "the hieroglyphic messenger": "He is (by the will of God) that most famous Mercury of the philosophers, the microcosm, and Adam."[9]

On balance, however, the degree to which Dee turns from any specific human language as possible repository of divine and natural wisdom (through its covert links to a pristine Adamic language reflecting the originary "Word of God") signals the unresolved tension in his discourse between innovative and reformist impulses. If Dee seeks the primitive unity of an Adamic earthly paradise, he nonetheless sees the very diversity of human languages and traditions as an impediment to, rather than a source of, that universal knowledge that postlapsarian man must recapture to fulfill his Adamic destiny. While caught in this ambivalent posture, Dee makes what may appear to be a radical move. He accepts, in short, the diversity of human tongues and nations; he largely abandons the attempt, through any "vulgar" verbal Cabala, to find universal meaning within any fallen discourse, within the finite and merely corporal signs of spoken or written human language. He moves instead to nature, to God's writing as the only source of a new and truly "sacred art of writing." Dee, as innovator and one of "the first founders of a new discipline," seeks to discover God's own hieroglyphics in order to create an art that can impose unity and universal knowl-

9. Josten (1964:164/165): "Et, (NVTV DEI,) iste est Philosophorum MERCVRIVS ille Celeberrimus, MICROCOSMVS, & ADAM."

edge upon the babble of humankind. The geometrical, mathematical forms God writes upon nature thus constitute an "alphabet of nature" uniting knowledge and the capacity for universal communication.[10] As both message and messenger, the hieroglyph transcends the diversity and contention of tongue and nation in the unity of a truly universal, "natural" eloquence:

> Though I call it hieroglyphic . . . there is [in it] an underlying clarity and strength almost mathematical, such as is rarely applied in [writings on] matters so rare. Or is it not rare, I ask, that the common astronomical symbols of the planets (instead of being dead, dumb, or, up to the present hour at least, quasi-barbaric signs) should have become characters imbued with immortal life and should now be able to express meanings most eloquently in any tongue and to any nation. (Josten 1964:120/121)

Here we glimpse the spirit of Pico, and before him of Ramon Lull, searching for a means to restore humankind to harmony and universality, as a basis for a universal conversion of Jew, infidel, and pagan. But Dee's quest remains curiously disengaged from any one tongue, from demonstration of the superiority of one sacred, spoken or written language. Hebrew, Greek, Latin all are too particular, too human to serve as God's universal alphabet of nature and as basis, therefore, for a universal knowledge that may wed man to man, nation to nation in the coming of a new age of Christ upon the earth.[11]

Some of these impulses anticipate our Lord Chancellor, Sir Francis Bacon. As with Bacon, Dee's turn to "that which is" fosters renewal of man's biblical quest for dominion and power over nature, although the language in which Dee envisions fulfillment of this quest is decidedly un-Baconian:

10. See Walton (1976:117): "Cabala's appeal to Dee, however, must have come not only from its general agreement with language and writing . . . but also from its similarity to an assumption set forth by Boetius that the world was created from numbers. As an Hermetic and a mathematician, Dee was committed to the virtues of numbers in both the natural and supernatural realms. In his 'Mathematical Preface,' he approvingly quoted the second chapter of Boetius' *De arithmetica:* 'All things . . . do appeare to be formed by the reason of numbers. For this was the principall example or patterne in the minde of the Creator.' " For details of Dee's cosmology and natural philosophy, which are beyond the scope of this study, see Walton's brief article and, especially, Clulee's (1988) excellent and comprehensive study.

11. See Chapter 2 above, for works on Pico. For Lull, see Avinyo (1925); Colomer (1961); Hillgarth (1972); Lohr (1968, 1990); Peers (1929); Pring-Mill (1955–56); Rossi (1960, 1961); Victor (1975); Yates (1982: esp. 9–77 ["The Art of Ramon Lull: An Approach to It through Lull's Theory of the Elements"] and 78–125 ["Ramon Lull and John Scotus Erigena"]) (both originally appeared in the *Journal of the Warburg and Courtauld Institutes* 17 [1954] and 23 [1960]).

(Once upon a time) four very famous men, philosophizing together, obtained by [their] work this true effect of it [*sc.* of the monad], whereupon, having for a long time been stunned by the very great wonder of the thing, then at length they devoted themselves entirely to singing and preaching the praises of the most good and great God, who had in this way granted them such great wisdom, power over other creatures, and large dominion. (Josten 1964:215/217)

Dee remains a Neoplatonist and occultist. He retains strong links to their underlying impulses. And he sees himself through the image of Christ, man's redeemer, as the very instrument of His spirit delivering an illumined message to man through the text of His *Monas*, his "Hieroglyphical Monad."[12] Hence, Dee's strivings, his "wanderings" through the visible landscape of this world, represent efforts not of a weak fallen man, but of a redeemed, divinely inspired intellect now raised to that incandescent peak of insight that allows him to grasp the most abstruse and hidden secrets of God's universe. Most strikingly, then, Dee authorizes a way of reading the Book of Nature as sign and symbol—as a hermeneutic act of symbolic exegesis—that, as we shall see, was foreign to Bacon's discourse and encounter with the world.

Sir Francis Bacon (1561–1626)

That "great instaurator," Bacon, like Dee before him, appears to the historian's retrospective gaze poised between reform and innovation. For Charles Whitney, the tension between tradition and innovation characterizes Bacon's project and underlies the emergence of a new critical—what he calls "metaleptic" (1986:79–90)—stance; Bacon is the instaurator of modernity. Closely related to Bacon's instaurating activities are, as Martin Elsky (1984,1989) has stressed, his confrontation with Renaissance traditions of hieroglyphs and language. Bacon's reform of language and his reform of science, I shall argue, are indispensible to each other—indeed, complementary aspects of a single project. Moreover, the problem of the relationship of language to the study of nature—so fundamental to Bacon's project—also proves central to understanding the tension between tradition and innovation in Bacon. Therefore, in my examination of the intersection of his reforms of language and of science below, I shall first consider Charles Whitney's recent analysis of Bacon. As I hope to show, Bacon's understand-

12. Josten (1964:200/201): "In Nomine IESV CHRISTI, pro nobis CRVCI affixi (cuius Spiritus celeriter haec per me Scribentis, Calamum tantum, esse Me, & Opto, & Spero.). . . ."

ing of language *authorized* a new kind of reading of the Book of Nature
that transformed beliefs about tradition and innovation and thus furnished him with both tools and values for a new legitimation of the
study of nature. I do *not* claim that Bacon's authorization preceded the
practices of working "scientists"; still less do I claim that Bacon himself
practiced what he preached in all instances. Nonetheless, Bacon's attempt to authorize "science" as cultural practice does rewrite the script
of science in Western thought, altering the role of the individual investigator and author in natural philosophy and simultaneously introducing
new characters—institutions, "fact," "method," and "experiment"—
into the story of science.

Bacon, of course, insisted on the importance of separating knowledge of nature from theology. More precisely, the fruits of inquiry into
natural phenomena for Bacon could provide no positive knowledge
about the divine mind and nature. This so-called Baconian separation
of religion and science has sometimes been read as a critical break from
a theocentric past, one that embraces secular and modern ideals of utility and of control of nature and human destiny.[13] With the seminal
studies of Paolo Rossi (1968,1970) and Charles Webster (1975), historians of science can invoke convincing contextual analyses of Bacon to
counter such tendencies toward a reductive secular explanation of his
program for delimiting the spheres of science and religion. Charles
Whitney's book, *Francis Bacon and Modernity*, attempts a rather different
reevaluation of Bacon's "modernity"—one that attends to multiple,
and often religious, valences of Bacon's language. "Beyond Bacon's
self-assurance" as vigorous proponent of science, technology, medicine, and social amelioration through useful, practical knowledge,
Whitney would have to look "to the struggle taking place in his work
between different discourses and ideologies" (1986:4).

More specifically, this struggle between discourses takes place
within the contexts of "classical and biblical prophetic" traditions in Bacon's works. As a consequence, Whitney suggests that Bacon's entire
project of a "Great Instauration" evokes "the oppositions . . . between
fulfillment and new beginning, and between piety and iconoclasm
(1986:5, 23–24). In other words, the religious language employed by
Bacon—the rhetoric associated with the Baconian vision of change—is
no mere ornament or concession to his audience. Rather, that language
pervades Bacon's analyses of the problems associated with inherited

13. For a brief discussion of some ways to read the separation of religion and science
in Bacon, see Whitney (1986:28–32).

schemes of knowledge and also structures the very articulation of choices to which Bacon points as remedies. Indeed, the prophetic language of "fulfillment" versus "new beginning" finds its most economical, and ambivalent, expression in Bacon's use of his key term, *instauratio*.

The religious tensions captured in this term underline the multiple valences associated with Bacon's attempt to reorder the study of nature in his own day. As Whitney suggests,

> The Great Instauration encompasses Christian ideas of reform related to biblical prophecy . . . as it gropes toward a secular idea of revolution and as it drafts a discourse of secular prophecy. On the one hand biblical prophecy and exegesis, and the visionary tradition of literature springing from them, are profoundly recuperative. . . . But in its call to commitment and action prophecy can be subversive and revolutionary. . . . Bacon's *Instauratio Magna* exemplifies the doubleness of prophecy. (1986:13)

This doubleness—to share for the moment the conviction voiced in Whitney's last sentence—animates much of Bacon's writings on natural philosophy, on the study of nature, and on language, imparting to them their sense of urgency and profound importance. What is at issue, however, is our historical understanding of the relationship between the recuperative and revolutionary dimensions of Bacon's project of an *instauratio magna*. Do the revolutionary impulses of Bacon's discourses subvert their recuperative, or reformatory, gestures, as Whitney seems inclined to argue?

To answer this question it will prove useful to examine Whitney's claims regarding the doubleness of the prophetic voice in Bacon. The prophetic context of *The Great Instauration* forces itself upon Bacon's reader at the outset. The title page (Figure 7.4) to the book skillfully combines image and text to suggest both fulfillment of ancient hopes and new beginnings signified by Bacon's choice of the polysemous term *instauratio* to characterize his project. At the center of the image we see an oceangoing ship venturing out beyond two pillars to the open sea. Below this image, the caption reads: "Multi pertransibunt & augebitur scientia." The latter, of course, is a quotation from Daniel 12.4, well known to Bacon scholars as biblical cornerstone for the Baconian and millenarian hope for an imminent advancement of learning.[14] Bacon's allusion suggests that the age of explorations is also an

14. Whitney (1986) translates this prophetic verse as "Many shall go to and fro and knowledge shall be increased." For an interpretation of Bacon that gives central prominence to this biblical verse and its millenarian readings, see Webster (1975).

Figure 7.4. The title page from *The Great Instauration*, by Francis Bacon. By permission of the Houghton Library, Harvard University.

age of religious fulfillment. Indeed, by venturing out beyond the Pillars of Hercules—the traditional limits to the known world of classical civilization—Europeans have begun to catch in their net a huge harvest of knowledge that awaits further exploration. By traveling to and fro and contributing by leaps and bounds to the increase of knowledge, late Renaissance Christian society appears to be the very instrument for fulfilling Daniel's prophecy, and, hence, for ushering in a new age of millenarian wisdom and religious renewal.

Whitney's analysis of Bacon's *instauratio magna* makes explicit the links between reform of learning and religious reform, between the age of exploration and the age of religious fulfillment. For, as he suggests, *instauratio* as a biblical term, with its implication of renewal, has dual religious and architectural meanings that converge in the image of "King Solomon's Temple of Jerusalem." The architectural "restoration or re-edification" of Solomon's Temple recounted in the Bible—itself an act of religious revival—for Bacon also invokes "the Christian process of spiritual edification toward salvation" (Whitney 1986:23). Hence, the Pillars of Hercules of the title-page image "also represent the temple of the world through which the ship of apocalyptic exploration passes, just as one passes through the twin pillars before Solomon's Temple" (1986:33). *Instauratio*, which as Whitney points out is the term used by the Latin Vulgate when characterizing "the many renovations of Solomon's temple," then suggests in Bacon's *Great Instauration* a rebuilding or "re-edification" of human knowledge that is, at one and the same time, also a spiritual re-edification of mankind (1986:24–25).

The confluence of dual architectural and spiritual tropes in Bacon's project for reforming human knowledge of nature is pervasive in his discourse, as Whitney points out:

> Bacon says first that man's knowledge has not been "built" (*aedificata*) properly, and is "like some magnificent pile without a foundation" (*tanquam moles aliqua magnifica sine fundamento*, 1.121). Therefore, he says, we must "make a universal Instauration of sciences and arts and all human knowledge, built anew on the proper foundations" (*fiat scientiarum et artium atque omnis humanae doctrinae in universum Instauratio, a debitis excitata fundamentis*, 1.121). To know is to construct a building in the mind; to innovate is to reconstruct, "instaure" that building and, through technology, the architecture of nature as well. (1986:24)

For the building that Bacon has in mind, that he wishes to see restored—man's intellect, or rather, his knowledge—evokes not only the physical and spiritual edifice of Solomon's Temple, but also that nearly perfect structure of natural knowledge, "Solomon's own lost work on natural history" (1986:24), for which Solomon's wisdom was legendary. Solomonic wisdom as the pursuit and rebuilding of natural knowledge gave both antique resonance and religious meaning to Bacon's project of a "Great Instauration" and prefigured the erection of that great seventeenth-century icon (and its associated "progeny") of scientific, religious, and political progress: "Solomon's House" of *The New Atlantis*, the Royal College of Physicians, and the Royal Society of London.[15] Indeed, Bacon's tropes gave rise to his own playful, but nonetheless serious, association of different social and intellectual realms through metaphoric exchanges among scientific, religious, and sociopolitical discourses. More particularly, Bacon appeals in typological language to Kind James I "as the instaurer of a new kind of Solomon's Temple, a temple of arts and sciences":[16]

> surely to the time of the wisest and most learned of kings belongs of right the regeneration and restoration [*Regeneratio ista et Instauratio*] of the sciences. Lastly, I have a request to make . . . namely, that you who resemble Solomon in so many things . . . would further follow his example in taking order for the collecting and perfecting of a Natural and Experimental History . . . such as philosophy may be built upon [*ad condendam philosophiam*].[17]

Bacon conceives his project of instauration as the fulfillment not only of an intellectual order, in which "that commerce between the mind of man and the nature of things . . . might by any means be restored to its perfect and original condition" (Bacon 1857–74:4.7) but also of a religious, social, and political order drawing upon ancient, and especially biblical, models. Such aspects of Bacon's program prompt Whitney to interpret his overall endeavor as, to some significant extent, conforming to a notion of change common to Christian Europe of the Middle Ages and Renaissance: the idea, or "ideology," of reform. As Whitney, following the classic work of Gerhart Ladner (1967), characterizes it, "In reform ideology, whether in an explicitly religious context or not, individuality, invention, and growth are maximized precisely through creative imitation and synthesis of prior models or forms. The reformer

---

15. On the connection between "Solomon's House" in Bacon's *New Atlantis* and the Royal College of Physicians and Royal Society of London in the later seventeenth century, see Hunter (1981); Rattansi (1964); Webster (1967, 1975).

16. The quotation is actually Whitney's own phrase (1986:25).

17. Bacon, Epistle Dedicatory to *The Great Instauration* (1857–74:4.2, 1.124[Latin]), as quoted in Whitney (1986:24).

carries on a tradition, but he also creates something distinctively new" (1986:27).

If reform remains rooted in tradition, if novelty arises through acts of imitation and recuperation of past models, then Bacon as reformer must ground his instauration upon a concrete, traditional model. As icon of an antique wisdom Bacon wishes to recuperate, Solomon serves as such a model. Moreover, the Temple of Solomon serves metonymically as a model for both the intellectual edifice and institutional structure that Bacon seeks to establish as a new monument to the advancement of learning.

Yet, in most respects, this Solomonic model is but hortatory. Bacon cannot point to an actual tradition of Solomonic natural history or philosophy. Indeed, the traditions of learning, knowledge, science, and natural history that Bacon finds concretely arrayed before him are, in his estimation, anything but Solomonic. As a foundation for reconstructing the temple of natural knowledge, such traditions are weak, crumbling, and irredeemable. Hence, Whitney finds intertwined with Bacon's reformist return to tradition and the language of restoration an iconoclastic impulse that adopts a metaleptic discourse—and "revolutionary stance" (1986:81)—of deconstruction and radical rebuilding from the very foundations of thought and knowledge. As he succinctly puts it, "The striking thing is that [Bacon] should . . . have considered first establishing his originality to be so essential to getting on with the project" (1986:89).

Whitney finds in the *Instauratio Magna* and its accompanying text, the *Novum Organum*, ample evidence for Bacon's revolutionary stance. Bacon, he quotes, sees "no hope except in a new birth [*regeneratione*] of science; that is, in raising it regularly up from experience and building it afresh. . . ."[18] Here, he claims, *regeneratio* does not mean " 'renovate' or even 'restore,' " but instead means " 'rebuild' . . . in the sense of 'replace,' i.e., to build a new building after a new design and after having completely removed the old building" (1986:91). Much of this rings true of Bacon's discourse. For Bacon does repeatedly express the point that reason and the arts and sciences that spring from it have little or no foundation; that, in consequence, the foundation upon which knowledge of nature has been erected must be destroyed in order to be rebuilt.

18. Bacon, *Novum Organum* (1857–74:4.94, 1.202), as quoted in Whitney (1986:90). Whitney's citation of the English page numbers and of the aphorism number is incorrect. The quotation is from aphorism 97.

The crux of Whitney's interpretation, however, is that such strong, iconoclastic statements and the powerful, leveling remedies they suggest definitively mark Bacon's discourse as moving "beyond the model of Christian reform altogether" (1986:92). While giving voice to such a model of reform as we have seen, Bacon, according to Whitney, adopts a rhetoric of revolutionary change that undercuts the former stance. Based upon an "assertion of absolute discontinuity with previous affairs" (1986:99), this "new" revolutionary rhetoric does not simply manifest itself in well-known and strategic Baconian assertions of the type that invite paraphrases such as "Bacon's instauration demands total purgation and a completely new beginning" (1986:92). Rather, it leads Whitney to the additional, and much stronger, claim that Bacon, in such iconoclastic moments, leaves behind "the model of Christian reform." Thus, Whitney can conclude that "In the *Instauratio Magna*, Bacon's claims of independence and originality represent . . . a fresh and independent starting point; as we have seen, he tells us that his resemblance to past philosophers is irrelevant, hardly the product of cumulative reform and gradual development. So much for walking in the old paths, let alone looking for them" (1986:95). But has Bacon, the reformer of human knowledge, really forsaken "the old paths"? Or has he looked for, and found, paths so old as merely to appear new?

Bacon's Adamic Instauration

Like religious reformers before him who eschewed the traditions of the Schools and Roman church in favor of an "originary" or "primitive" Christianity, or even, at times, in favor of an iconoclastic and radical return to a lost "egalitarian" conception of religion, Bacon's instauration rejects inherited traditions of rational knowledge, philosophy, and the "sciences" in favor of a still more primitive, virtually lost, model of knowledge.[19] His iconoclasm, indeed, required a complete and utter razing of the edifice of natural philosophy down to, and including, its very "foundations" in order to effect the radical *instauratio* needed by mankind. Yet, however total this rejection of past and present traditions might have been for Bacon and notwithstanding any inability of practice to live up to the austere demands of his substantive critique and rhetoric, Bacon's radical assertion of originality, of the need for a

19. On primitive models of human knowledge and radical strains of reformation, see Ficino, Pico, Paracelsus, and Agrippa and the literature cited in previous chapters. See also Ozment (1973); Webster (1982); G. H. Williams (1962); and, on Bacon, Webster (1975).

new beginning, was itself a call to embrace a misunderstood, if not forgotten, primitive model: the model of Adam before the Fall. In this respect, Bacon as advocate of science and of a new instrument or method (his *Novum Organum*) actively pursued the perhaps contesting, and surely tension-filled, ideals of tradition and innovation—better yet, reform and revolution—at one and the same time. Bacon's "revolutionary stance," if that coinage is not itself an anachronism, certainly carried no rejection of the ideals of Christian reform. But the model of reform, and its radical implementation, did entail a highly individual reception and recasting of that reformist ideal.

Such a model of reform was intimately connected with Bacon's rejection of Renaissance hermeneutics of nature, and, as we shall see later, of conceptions of language that served to underwrite what he considered to be worn-out—indeed, *foundationless*—approaches to knowledge of the natural world. As such, Bacon's recasting of the Christian ideal of reform is part of the larger story of language, cultural narratives, and hermeneutic strategies for grasping nature that has occupied so much of our attention in this book. More specifically, the turn *from* words, symbols, and unity that marks the exegetical hermeneutics of late Renaissance science, *toward* things, particularity, and diversity— what I have called a hermeneutics of *"de-in-scription,"* or description, of the Book of Nature—can be seen in many examples and in a range of subtle variations during the Scientific Revolution.[20] Increasing and evolving throughout the seventeenth century, this descriptive hermeneutics found influential, if not archetypal, expression in the work of Sir Francis Bacon.

The very basis for Bacon's new hermeneutics lies in his skillful and knowing recasting of the biblical story of Adam and the Fall, out of which he transformed Renaissance cultural narratives. For Bacon, the history of culture and of mankind became the story of the "Great Instauration": a narrative of human restoration to a kind of Adamic perfection. Within the boundaries of this narrative, language came to assume a special significance for Bacon, one that not only specified a certain relationship between "words" and "things," but, as a consequence, also prescribed how humans ought to interpret nature.

Bacon's "Great Instauration" represents a blueprint for repairing the effects of the Fall in man. To return to the prelapsarian state entails, for

20. I have discussed Harvey, contrasting his approach to nature with that of Fernel, in Chapter 4. I also discuss such different examples of "de-in-scriptive" hermeneutic strategies as Galileo, Mersenne, and Descartes elsewhere in this book.

Bacon, recapturing those powers Adam enjoyed in paradise. Hence, Bacon opens his *Great Instauration* by asking "whether that commerce between the mind of man and the nature of things" that marked Adam's pristine state "might by any means be restored to its perfect and original condition" (1857–74:4.7). Bacon's answer, of course, is that such means do exist; indeed, his "novum organum" purported to supply just such a means, a "new organ or instrument" for repairing human knowledge of nature and, hence, our own *fallen* nature.

Bacon's analysis of man's possibilities, of his future and that of society as a whole, drew upon a recasting of the story of Adam and his fall in the Book of Genesis. In his (c. 1603) work *Valerius Terminus*, Bacon (1859) draws connections between man's present state, his access to knowledge of God's creation, and his Adamic past. For Bacon, man's pursuit of knowledge in his present, postlapsarian state ought to draw its impetus from understanding of his Adamic past. The very character and nature of Adam's knowledge become the model, inspiration, and *hope* for fallen man's regeneration:

> And therefore it is not the pleasure of curiosity, nor the quiet of resolution, nor the raising of the spirit, nor victory of wit, nor faculty of speech, nor lucre of profession, nor ambition of honor or fame, nor inablement for business, that are the true ends of knowledge; some of these being more worthy than other, though all inferior and degenerate: but it is a restitution and reinvesting (in great part) of man to the sovereignty and power (for whensoever he shall be able to call the creatures by their true names he shall again command them) which he had in his first state of creation. (1859:222)

Adam's knowledge in the Garden of Eden brought with it power over nature, the ability to command creatures and hence exercise sovereignty over all that he surveyed. Bacon attributed this Adamic dominion to Adam's knowledge of the "true names" of all created things. In man's original state of innocence and grace, Bacon implies that Adam's knowledge was immediate—that there was a perfect conjunction between man's mind and nature, between "words" and "things." It is this Adamic legacy, the model of his knowledge and power, that Bacon would have man seek to restore.

Yet man's loss through his catastrophic fall makes this task of restoration daunting. That perfect, and perfectly *natural*, conjunction of words and things Adam experienced in the Garden of Eden has been lost. The nature of that loss and its meaning to man lie at the heart of Bacon's narrative of man's fall and subsequent history. Drawing upon a tradition of Protestant thought that regards man as weak, corrupt, and fundamentally transformed in his fallen state, Bacon refuses to grant humans direct access to the *verbum Dei*, understood as the divine language of creation and, hence, of nature. Man's traffic with words, his intellectual exchanges in the marketplace of ideas, are vain, idle, and unproductive. The link between words and things has been shattered by the legacy of the Fall, and to attempt to retrace or repair that chain by attending to the languages of man is a prideful, futile effort.

How, then, does Bacon propose to allow human access to knowledge? How can he hope for man to repair the effects of his fall? Bacon's answer dramatically redefines the relationship between the "Word of God" and the "languages of man," reorienting man's vision of himself as an agent in the world in ways that point toward not only a radically new hermeneutics of nature, but also a new social ethos and vision of the human community. At its core was Bacon's reformulation of the very story of man's cultural origins and history and of the place of the divine "Book of Nature" within that master cultural narrative.

Bacon begins by characterizing the basis for Adam's fall from his state of grace in the earthly paradise. Man's fall did not arise from a lust for power or dominion; in his own realm, man already enjoyed these as part of his divinely created nature. Rather, the Fall according to Bacon was the result of man's desire to intrude "into God's secrets and mysteries":

> Man on the other side, when he was tempted before he fell, had offered unto him this suggestion, "that he should be like unto God." But how? Not simply, but in this part, "knowing good and evil." For being in his creation invested with sovereignty of all inferior creatures, he was not needy of power or dominion; but again, being a spirit newly inclosed in a body of earth, he was fittest to be allured with appetite of light and liberty of knowledge; therefore this approaching and intruding into God's secrets and mysteries was rewarded with a further removing and estranging from God's presence. (1859:217)

That is to say, man's illegitimate desire and prideful attempts to fathom God's nature, and with it to gain divine knowledge of good and evil, precipitated his tragic fall.

The story of Genesis becomes for Bacon not so much a story of contesting divine and human wills, but one that defines for all human history man's nature and limitations. Man's nature, like that of Adam, is to exercise power and enjoy dominion over all creatures. Knowledge of creatures is licit; knowledge of God's nature is beyond man's ability. As a result of his fall, however, man lost that *immediate* knowledge of the nature of created things that Adam enjoyed. He lost knowledge of the true names of things, and with it easy dominion over all creatures. Moreover, since Adam's language was pure and uncorrupted, a divine gift that was forever lost, what remained in its place, mere human discourse, was faulty and inaccurate, if not serpentine and misleading. After the Fall, man could no longer rely upon his wit—his weak fallen mind—to fathom the nature of creatures through language alone. Words and things were abruptly alienated from one another.

What was left to man were his senses, wayward and in need of disciplined control as they were. Yet through those senses, man's mind could be restored to some measure of "commerce" with things themselves. Hence Bacon, while reading human history as a story of decline and prideful self-exaltation, nevertheless held out the prospect of regaining Adam's lost dominion over nature through careful application of his senses to the task of building up knowledge of things. What was needed was a clear apprehension of man's limits and, with that, curbing of his prideful, corrupt instinct to pursue illicit knowledge.

> For if any man shall think by view and inquiry into these sensible and material things, to attain to any light for the revealing of the nature or will of God, he shall dangerously abuse himself. It is true that the contemplation of the creatures of God hath for end (as to the natures of the creatures themselves) knowledge, but as to the nature of God, no knowledge, but wonder; which is nothing else but contemplation broken off, or losing itself. Nay further, as it was aptly said by one of Plato's school "the senses of man resembles the sun, which openeth and revealeth the terrestrial globe, but obscureth and concealeth the celestial;" so doth the sense discover natural things, but darken and shut up divine. (1859:218)

Bacon avoids the extreme pessimism of certain Reformation views of fallen man by redefining, through his narrative of human cultural history, the relationships between the Word of God, the languages of man, and the "Book of Nature." Bacon denies humans access to any originative, divine language: man cannot fathom the divine mind and will. In that sense, there is an unbridgeable divide between the *verbum Dei* and the languages of man. In another sense, God's Word speaks to man, most especially through the Bible, instilling in him a moral, spiritual sense that may help to guide his actions, fill him with piety, and open his heart to faith.

While thus denying to man any intellective grasp of the divine Word,

Bacon claims that God's Word is nonetheless visible in a form that humans can read and come to understand. For, according to Bacon, God's imprint or signature is to be found in nature itself, and man, if he takes care to exercise his talents properly, is capable of receiving and interpreting the traces of the divine Word written in the "Book of Nature":

> And lest any man should retain a scruple as if this thirst of knowledge were rather an humour of the mind than an emptiness or want in nature and an instinct from God, the same author defineth of it fully . . . declaring not obscurely that God hath framed the mind of man as a glass capable of the image of the universal world, joying to receive the signature thereof as the eye is of light, yea not only satisfied in beholding the variety of things and vicissitude of times, but raised also to find out and discern those ordinances and decrees which throughout all these changes are infallibly observed. And although the highest generality of motion or summary law of nature God should still reserve within his own curtain, yet many and noble are the inferior and secondary operations which are within man's sounding. (1859:220)

Bacon, in a sense, distinguishes between two aspects of the "Word of God." The first is revelatory of God's intentions for man, for his salvation and redemption. Man can fathom only the meaning this spiritual Word holds for him as a spiritual creature, the *imago Dei*, capable of salvation through God's will alone. This Word, then, affords man no direct knowledge of God's hidden and mysterious nature. The second aspect of the "Word of God" is His creative Word, that "word" that is productive of the created order itself—of "God's Works"! Man, Bacon argues, can know God's works; he can know, that is, the "inferior and secondary *operations*" God has inscribed in nature. Through knowledge of His works, man can "read" the book of nature, can uncover the divine signatures in nature, and, I would argue, begin to reconstruct this second aspect of the "Word of God"—the language of nature known to Adam in his earthly paradise. Man's access to this work, and Word, comes not through his intellect, not through exegesis of texts and nature, nor even through a privileged divine illumination, but rather through the work of the senses: the labor of experimentation.

> The access also to this work hath been by that port or passage, which the divine Majesty (who is unchangeable in his ways) doth infallibly continue and observe; that is the felicity wherewith he hath blessed an humility of mind, such as rather laboureth to spell and so by degrees to read in the volumes of his creatures, than to solicit and urge and as it were to invoke a man's own spirit to divine and give oracles unto him. For as in the inquiry of divine truth, the pride of man hath ever inclined to leave the

oracles of God's word and to vanish in the mixture of their own inventions; so in the self-same manner, in inquisition of nature they have ever left the oracles of God's works, and adored the deceiving and deformed imagery which the unequal mirrors of their own minds have represented unto them. Nay it is a point fit and necessary in the front and beginning of this work without hesitation or reservation to be professed, that it is no less true in this human kingdom of knowledge than in God's kingdom of heaven that no man shall enter into it "except he become first as a little child." (1859:223–24)

The relationship of the "Word of God" to the "languages of man" is not fixed in some ancient, recoverable past. Rather, it is prospective, looking toward the future, toward the moment when man's labor in the bounty of God's works will enable him to fashion names that are appropriate representations of things. Through such labors, man can hope to achieve something like the original Adamic language in which the creatures can again be called by their true names. Yet this ideal remains distant for Bacon. Language, the ordinary discourse of fallen man in history, is itself cut off from that Adamic past. Reconstituting the language of Adam is not, therefore, a matter of recovering its traces through the exegesis of mere words, nor of mystically grasping the names, analogies, and correspondences of things through an illuminated insight into the divine signatures traced in nature itself. Man—fallen, corrupt, imperfect man—has not the wit for such paths (fictitious as they, in any case, are) to the true names of things. Man has only the sweat of his brow, the labor of his hands, the testimony of his senses. But that, for Bacon, will be enough.

Bacon's narrative of man's fall, history, and eventual redemption generates a theory of language, a vision not of words as reliable signs of things, but rather of things as the only reliable criteria for framing words properly. With this theory of language, Bacon suggested a hermeneutics of nature that would compel science to interrogate nature relentlessly through experiment in order to force "her" to reveal "her" secrets.[21] We shall consider Bacon's theory of language and his herme-

21. From this vision of language, the Book of Nature, and man's agency in the world as interrogator of nature, Bacon fashioned a vision of society and culture as well. See, for example, Bacon (1859:221–22): "But yet evermore it must be remembered that the least part of knowledge passed to man by this so large a charter from God must be subject to that use for which God hath granted it; which is the benefit and relief of the state and society of man; for otherwise all manner of knowledge becometh malign and serpentine, and therefore as carrying the quality of the serpent's sting and malice it maketh the mind of man to swell; as the Scripture saith excellently, "knowledge bloweth up, but charity buildith up." And again the same author doth notably disavow both power and knowl-

neutics of nature more fully below. What should be clear from the discussion thus far, however, is that Bacon's advocacy of a new method and of a new form of human knowledge of nature remains firmly embedded in an ideal of Christian reform. By reconceiving the meaning of the formative cultural narrative of Adam in paradise and his subsequent fall, Bacon finds the purchase to recast radically that ideal of reform. As a result, reform becomes compatible for Bacon with a radical rejection of corrupt, fallen, prideful *human* traditions of knowledge— be they Aristotelian, Galenic, Paracelsian. Reform embraces the razing of old foundations and instauration of a radically new "temple of arts and sciences." But this temple, while without precedent in extant postlapsarian traditions, is not to be constructed without its own unique blueprint. Bacon's model—a "model" for Christian reform—is the pure, perfect, unfallen Adam.

This perspective carries consequences for our reading of Bacon. Hence, before turning to our consideration of language theory and hermeneutics in Bacon, I would like to return to Whitney's analysis of Bacon's innovations.

Whitney makes few references to Adam or to man's prelapsarian state in the course of his analysis of Bacon. As a result, he is led, I think, into some rather unusual readings of Bacon's texts by failing to attend to the narrative context that Bacon himself creates for his own discourse of instauration. The chief, and to my mind most troublesome, example concerns the supposed role of "absolute presence," "novelty," and "unmediated" induction in Whitney's account of Bacon. Whitney's analysis begins with the following claim:

> in the context of Bacon's revolutionary stance, an aspect of his vision of discovery emerges which, while part of his prophetic call, does not seem to be a part of the prophetic tradition of reform: novelty. The perceiving mind itself in effect has no past, no irrelevant thoughts or experiences; hence the encounter between subjective and objective fundamenta or foundations for the Temple of the World in the Mind takes place outside of time in a moment or moments of absolute presence. . . . (1986:105)

Let me reserve until later specific criticism of the reading of Bacon implied by this passage. There are interesting claims here, and I do not

---

edge such as is not dedicated to goodness or love. . . ." See, as well, *The Great Instauration.* For discussion of the religious, millenarian context of Baconian beliefs in the seventeenth century, see Webster (1975). Paula Findlen's (1990a) review of Bianchi (1987) is relevant to my discussion. Finally, for the tradition of "secrets of nature" and metaphors shaping scientific practice in Bacon and others, see Eamon (1990, 1990a, 1994).

want to suggest by my silence that I regard them as slight or trivial, however much I may disagree with them. Let me also add that the crux of my disagreement has in no way to do with the phrase "absolute presence." For some historians, invocation of this phrase may, by its evocation of a Heideggerian and Derridean perspective, raise questions of "anachronism" or touch deeply rooted fears of discourse and deconstruction invading the domain of concrete historical "experience" and "events." By contrast, I do think that Bacon's perspective on knowledge and nature, however much it may wish to distance itself from the "onto-theology" of traditional philosophy, does indeed come to depend upon an unacknowledged "absolute presence" as ground for its claims to reveal the facts and "forms" of nature through sense perception and the interrogation of nature through experimentation. My dispute with Whitney, as we shall see, has rather to do with his claims concerning how man comes to his knowledge of the "absolute presence" of/in things.

More specifically, I should like to examine and question Whitney's claims regarding unmediated perception and the emergence of novelty in Bacon's discourse. One crucial point for Whitney would seem to be that novelty, which I take it represents the irruption of "absolute presence" from "outside of time" into the temporal world of the scientific investigator, emerges in the mind in the act of perception. This is, at least, the sense I take from Whitney's appeal to Bacon's own words immediately following the passage quoted above about "absolute presence." The passage I am refering to is from the preface to the *Novum Organum* and is but one sentence long. The first part of the sentence reads, "the mental operation which follows the act of sense I for the most part reject. . . ." This statement is, of course, entirely in keeping with Bacon's well-known criticism of both the inborn and the adventitious tendencies of the mind to frame a distorted image of the world from sensations. The rest of the sentence provides the crucial "evidence" for Whitney's analysis: "and instead of it I open and lay out a new and certain path for the mind to proceed in, starting directly from simple sensuous perception."[22] Now Whitney, to judge from his subsequent discussion, evidently takes this passage to mean that novelty— the discovery of truths in nature—arises "directly from simple sensuous perception" according to Bacon. While this is certainly a plausible interpretation, the difficulty is that it does not address what the force of Bacon's use of the word "starting" might be in this sentence. My own

22. Bacon (1857–74:4.40), as quoted in Whitney (1986:105–6).

interpretation would place great weight upon this very point. Indeed, I believe that Bacon's reference to "a new and certain path for the mind" as *"starting"* from sensation arises from a specific narrative context regarding postlapsarian man that must occupy our attention.

Whitney's interpretation, however, ignores this narrative context, leading him to talk instead about an "unmediated aspect of induction" in Bacon's scheme and to see the latter as questing for "a pure, immediate relation of the mind to nature." What Whitney seems to claim, then, is that Bacon regards "sensuous perception" as the sound basis and unique source of true knowledge of nature, provided that the mind has been freed of all fancies imposed upon it by the imagination. It is the "absence of mediation"—the absence of mediating fictions or "dream[s] of our own imagination"—that enables man's "sensuous perception" immediate access to absolute presence.[23] Whitney's best evidence for such a reading of Bacon comes from a passage he quotes from the preface to *The Great Instauration.*

> For all those who before me have applied themselves to the invention of arts have but cast a glance or two upon facts and examples and experience, and straightway proceeded, as if invention were nothing more than an exercise of thought, to invoke their own spirits to give them oracles. I, on the contrary, dwelling purely and constantly among the facts of nature, withdraw my intellect from them no further than may suffice to let the images and rays of natural objects meet in a point, as they do in the sense of vision; whence it follows that the strength and excellency of the wit has but little to do in the matter. (Bacon 1857–74:4.19; Whitney 1986:108)

It is not at all obvious, however, that this, or other, passages that Whitney cites establish anything like an immediate perception or induction as an element of Bacon's vision. The above passage, for example, appears most directly concerned with the limits of man's "wit" or "spirit." Indeed, Bacon introduces this passage by stating that "wherein I have made any progress, the way has been opened to me by no other means that the true and legitimate humiliation of the human spirit" (1857–74:4.19). His point seems to be that previous "scientists" have all too quickly turned away from nature—from things—to invoke images, concepts, abstractions fashioned by their "wits" as explanations in place of the phenomena themselves. This prideful turn toward "wit" Bacon rejects, choosing instead to keep his gaze fixed upon "the facts of nature." Bacon does not tell us here how those facts of nature are

23. Whitney (1986:111). See also the second epigraph from Bacon at the beginning of the chapter (1857–74:4.32–33, 1.145). The source of the first epigraph is Bacon (1960:254).

grasped—whether their apprehension is direct and immediate through the senses. Just prior to the above statements, Bacon does suggest, quite to the contrary, that the senses alone are not sufficient to penetrate the secrets of nature:

> in like manner the discoveries which have been hitherto made in the arts and sciences are such as might be made by practice, meditation, observation, argumentation—for they lay near to the senses . . . but before we can reach the remoter and more hidden parts of nature, it is necessary that a more perfect use and application of the human mind and intellect be introduced. (1857–74:4.18)

Here I think that it is essential to insist upon the importance of "simple sensuous perception" as *just* the "starting" point for Bacon's "new and certain path for the mind." Bacon is profoundly aware of both the complexity and the intractability of nature—which hides "her" secrets from man—and of the limits of the senses as instruments for apprehending the truths of nature. Whitney's claim that such perception of truth "takes place outside of time in a moment or moments of absolute presence" rests on his assertion that for Bacon "the perceiving mind itself in effect has no past, no irrelevant thoughts or experiences." Clearly, by this latter statement Whitney means to suggest only that the *correctly* perceiving mind—the mind that has gained access to truth—is free of these liabilities according to Bacon. Only if this latter assertion is the case does it make sense to claim that induction, for Bacon, can be unmediated or, stronger still, that true knowledge is a result of the "pure, immediate relation of the mind to nature."

But it is precisely this assertion that the narrative context within which Bacon develops his discourse undermines. What Adam lost in his tragic fall was that very immediacy of knowledge that came from the perfect conformity of his mind to nature. The legacy of his fall, then, was not only man's alienation from things and from that creative "Word of God" reflected in the conjunction of words and things in the language of Adam, but also a necessary and inherent corruption of man's mind. Rather than reflecting God's created order perfectly, postlapsarian man's mind was fated to remain a distorted and unreliable mirror of nature. Man's fall—his descent into the flux and mutability of time and history—leaves him forever enmeshed in the contingencies and warped perspectives of his own unique place in time and space. He is then doubly disadvantaged. As an individual, his mind acts as a prism altering the world before it can be grasped in its purity. As a member of a historically situated community, he bears within himself the shared habits, language, and past that impose themselves upon perceived phenomena.[24]

Despite Whitney's implied claim, Bacon nowhere leads us to believe that man's mind ever acts other than to distort the "true" image of the world. Man, in short, never enjoys a moment "outside of time." Indeed, it is precisely because the mind *does* have a past that "simple sensuous perception" is *only* a starting point. Instead, the hope—or "mercies"—that Bacon does extend to man lies elsewhere. Moreover, that hope is one that is continuous with the story of postlapsarian man and his efforts to regain his lost, Adamic legacy as understood by Bacon.

The tendency of man's mind to distort the world is, in fact, a central legacy of the Fall. Bacon insistently reminds his readers to take this legacy seriously in the pursuit of true knowledge of nature. The woeful history of human traditions—a history of human error for Bacon[25]—results precisely from man's prideful belief that his fallen intellect is capable of an immediate apprehension of truths from sense perceptions. Indeed, Bacon chides the Aristotelians for this error.

> I . . . reject demonstration by syllogism, as acting too confusedly, and letting nature slip out of its hands. For although no one can doubt that things which agree in a middle term agree with one another (which is a proposition of mathematical certainty), yet it leaves an opening for deception; which is this. The syllogism consists of propositions; propositions of words; and words are the tokens and signs of notions. Now if the very notions of the mind (which are as the soul of words and the basis of the whole structure) be improperly and overhastily abstracted from facts, vague, not sufficiently definite, faulty in short in many ways, the whole edifice tumbles. (1857–74:4.24)

The entire syllogistic logic, with its major and minor premises connected by middle terms, relies wholly upon the veracity of definitions of the essences or forms of things. The latter derive not from proper inductions, but, on the contrary, from improper and all too hasty abstractions from sense perceptions.[26] Bacon's analysis of this and other

24. Bacon's famous "four Idols," analyzed in his *Novum organum*, offer just this sort of radical critique of traditional epistemology. See Shapiro (1983: esp. 61–66).

25. Bacon (1857–74:4.92ff.). See also Guibbory (1975).

26. Whitney (1986:111) refers to Bacon's discussion of the syllogism in *Advancement of Learning* (1857–74:3.392), where Bacon points out that "the proof" in the Aristotelian syllogism is "not immediate but by mean." Whitney takes this point to signify that induction, by contrast, is immediate according to Bacon, in the sense (I take it) that induction proceeds by the immediate relation of the mind, in its operation, to nature. My under-

sources of human error provides telling evidence against the "immediate relation of the mind to nature" and forms a cornerstone of his attempt to establish a foundation of hope for man's true knowledge of nature that is in accordance with the larger story of man's fall, redemption, and eventual regeneration of "instauration."

Let me then sketch in very general terms how Bacon moves from an identification of the weaknesses of human knowledge to such hope in the *Instauratio Magna*. As we have already seen, Bacon is wary of abstraction, of overhasty generalization from limited sense experience. But the answer is not simply more sense experience. On the contrary, "the senses deceive" (1857–74:4.26) and thus are not in themselves reliable sources of certain knowledge, leading Bacon to declare, "to the immediate and proper perception of the sense therefore I do not give much weight" (1857–74:4.26). In fact, Bacon further identifies the error of the "logicians"—or Scholastics—with their uncritical reliance on "knowledge" abstracted from the senses: "they receive as conclusive the *immediate* informations of the sense" (1857–74:4.25). Merely accreting more and more sense experience will not overcome its weaknesses which are inherent to human knowledge, "And again when the sense does apprehend a thing its apprehension is not much to be relied upon. For the testimony and information of the sense has reference always to man, not to the universe; and it is a great error to assert that the sense is the measure of things" (1857–74:4.26).

Thus, the real weaknesses of human knowledge are nearly intractable. They cannot simply be bracketed by, for example, attempting to remove the biases and distortions that individuals bring to sense per-

---

standing of Bacon is different. Bacon makes the point that the syllogism, in employing a middle term, separates the act of invention from the act of judgment. In other words, proof for the Aristotelian is discursive; it proceeds from propositions, through deductive chains. As such, proof can not be immediate, but is rather mediated by middle terms. On the other hand, since for Bacon "the invention of the mean is one thing, and the judgement of the consequence is another. . .," the mediated nature of judgment or proof for Aristotelians does not exclude the possibility that "invention" is immediate for them. In fact, underlying the Aristotelian syllogistic, for Bacon, is the claim to an immediate apprehension ("invention") or abstraction of essences from things, which in turn form the basis for definitions that constitute the premises and middle terms of syllogisms. It is this very claim of an immediate apprehension of things that Bacon takes as a prideful rush to abstraction among Aristotelians. By contrast, Bacon's induction avoids this prideful claim of immediate abstraction. For Bacon, induction does join both invention (or discovery) and judgment in one thing, or one operation of the mind; but this operation does not entail an immediate relation of mind to nature: invention, and knowledge, can—indeed must—be mediated for fallen man.

ception. Such obstacles to true and immediate perception are not simply contingent, but in addition "inherent in the very nature of the intellect, which is far more prone to error than the sense is." Indeed, "as an
uneven mirror distorts the rays of objects according to its own figure
and section, so the mind, when it receives impressions of objects
through the sense, cannot be trusted to report them truly, but in forming its notions mixes up its own nature with the nature of things."
What is more, this obstacle, or "idol," according to Bacon, "cannot be
eradicated at all" (1857–74:4.27).

How, then, can man hope to acquire true knowledge of nature and
hence reclaim his lost Adamic legacy? How, that is to say, can man
overcome such deeply ingrained and seemingly insurmountable consequences of his fall from the pure and immediate knowledge of things
he enjoyed in the Garden of Eden?

Bacon's answers are closely connected to his strategy of instauration.
In order to "restore to its perfect and original [that is, Adamic] condition" the "commerce between the mind of man and the nature of
things," postlapsarian man, as we saw earlier, must attempt to recapture the powers formerly exercised by man in paradise. Bacon accepts
man's fallen state, accepts his weak and distorting intellect, and accepts as well the inherent deceptions produced by the senses. Man's
great instauration must then start with these materials, including "simple sensuous perception." Thus, postlapsarian man's starting place on
his road toward perfect knowledge of and power over all the creatures
in nature is necessarily different from—even incommensurable with—
that of the unfallen Adam. Where Adam's nature allowed him immediate access to the knowledge that brought him dominion in paradise,
fallen man's access to such power over nature must necessarily be *mediate*. He must look, that is, to tools—to instruments that, fashioned to
his own limited powers, can enable mankind to exercise gradually increasing dominion over all creatures and over nature itself.

Such an aid Bacon proclaims his *novum organum*—his "new organon," or new organ/instrument—to be. In short, access to the hidden
secrets of God's creation could come to postlapsarian man only through
the work of the senses refined, checked, and transformed into effective operations upon nature through the labor of experimentation.
How is this transformation of human powers with respect to nature
possible through experimentation? First, Bacon acknowledges that
while man's senses deceive, they also "supply the means of discovering their own errors." Bacon proposes "to provide helps for the
sense—substitutes to supply its failures, rectifications to correct its

errors," namely, through experiments. Where the senses provide but incomplete, or even false, information, active interrogation of nature through experimentation will enable humans to ferret out the half-truths and lies by catching nature in its trap. Here, our Lord Chancellor surely found analogies with the law's interrogation of witnesses compelling. Though the senses be but unreliable witnesses to nature's mysteries, man has his own instruments to torture true confessions from her: only, however, through "such experiments . . . as are skilfully and artificially devised for the express purpose of determining the point in question." By introducing experimentation to the *active* study—not to say interrogation—of nature, Bacon thus feels confident in the capacity of the senses to play but a constrained, juridical role in natural philosophy: "I contrive that the office of the sense shall be only to judge of the experiment, and that the experiment itself shall judge of the thing."[27]

Man's labor in performing experiments befits his status as fallen progeny of Adam, sentenced to work in the fields of nature to earn his daily bread. Indeed, only through such labor can he hope to subdue and even conquer nature, wresting from such operations the bounty of her secret treasures. Yet, in order to engage fruitfully in such study of nature, man must begin to purge himself of those obstacles that pridefully separate him from such active searches and seizures. Only by ridding himself of the fantasies promoted by false philosophies, false methods, and false estimations of human reason can man turn humbly, but confidently, toward nature herself.[28]

27. Bacon (1857–74:4.26). See also Sargent (1989). The legal background to Bacon's thought and his reform of science and society is an enormously important subject that I cannot explore here. Since my concern in this volume is with the narrative reconfiguration of scientific discourse through the reworking of Renaissance cultural narratives, I have had to limit aspects of my analysis. The next step—microhistorical analysis of the cultural and sociopolitical contestation occuring within an "ecology" of competing *local* discourses and practices—will attempt to illustrate how such narrative legitimations of knowledge and practice become the foci for such *interdiscursive* contestation among contiguous and *overlapping* communities (see Bono, 1995). The present volume can only gesture in this direction, while such a focus will be close to the center of Volume 2. Bacon deserves such microhistorical treatment in which the boundaries between his "scientific" discourse, professional life, legal discourse, and the state are at issue. For now, see Cardwell (1990); Coquillette (1992); Martin (1988); Shapiro (1991); and Wheeler (1990).

28. Bacon (1857–74:4.27): "This doctrine then of the expurgation of the intellect to qualify it for dealing with truth, is comprised in three refutations: the refutation of the Philosophies; the refutation of the Demonstrations; and the refutation of the Natural Human Reason." See also Briggs (1989); Hattaway (1978); Howell (1961); L. Jardine (1974); McNamee (1971); and Perez-Ramos (1988).

This destructive—or, to evoke Bacon's telling architectural metaphor, deconstructive—moment, while necessary to replace the old temple with a "new temple of arts and sciences," must then lead to a constructive phase in which active search and experimentation provide necessary instruments mediating the discovery of new knowledge of nature. Hence, the hope that Bacon holds out as a beacon to mankind requires that

> Those however who aspire not to guess and divine, but to discover and know; who propose not to devise mimic and fabulous worlds of their own, but to examine and dissect the nature of this very world itself; must go to facts themselves for everything. Nor can the place of this labour and search and worldwide perambulation be supplied by any genius or meditation or argumentation; no, not if all men's wits could meet in one. This therefore we must have, or the business must be for ever abandoned. But up to this day such has been the condition of men in this matter, that it is no wonder if nature will not give herself into their hands. (1857–74:4.28)

Nature has not given herself "into their hands" precisely because men have used faulty methods—false instruments—to conduct "blind" inquiries on "bad materials for philosophy and the sciences." Instead, man must use the true instruments—experiments—available to him and turn them upon the proper objects of inquiry. In short, for Bacon, "the only hope therefore of any greater increase or progress lies in a reconstruction of the sciences" (1857–74:4.28): "Of this reconstruction the foundation must be laid in natural history, and that of a new kind and gathered on a new principle. . . . I mean it to be a history not only of nature free and at large . . . but much more of nature under constraint and vexed; that is to say, when by art and the hand of man she is forced out of her natural state, and squeezed and moulded" (1857–74:4.28–29). We have asked how it is that fallen man, with his deceptive senses and distorting mind, can hope to know the truths of nature and reclaim his lost Adamic dominion over all God's creatures. And we have seen Bacon's answer, at least in its rudiments as found in the *Instauratio Magna*, holding out the hope for human progress in the reconstruction of the sciences through experimentation and natural history. Yet, it remains for us to wonder, why? Why, for Bacon, must man choose this way? Why natural history? Our search for answers will lead us back to language, to the alphabet of nature, and to Bacon's apprehension of the relation of the languages of man to the "Word of God," with all of its implications for a hermeneutics of nature.

Bacon's Alphabet of Nature: Language Theory and Hermeneutics

Why natural history? The work of the senses, the labor of experimentation, was not for Bacon simply a punishment for Adam's original sin. Whatever Adam lost through the Fall, man can regain through those "mercies" that God endows him with. Natural history, employing experiments to probe "nature under constraint and vexed," was no arbitrary punishment, no tedious regimen imposed upon man, but rather a mercy that could lead man back to a promised land of plenty, charity, and community with mankind and with God.[29] Natural history was the way back to the *verbum Dei*, or rather, to its vestiges in God's works. Natural and experimental history is man's only window opening onto an unobstructed view of the vast panorama of God's Book of Nature.

Bacon, in his introduction to the *Historia naturalis et experimentalis* (1622)—the third part of the *Instauratio Magna*—frankly confesses that man has no alternative to the way of natural history and experiment except for repetition of that sorry state in which "every one philosophises out of the cells of his own imagination, as out of Plato's cave" (1857–74:5.131). Bacon proposes to reorient man's gaze from the darkness of his own imagination to the light of things themselves. In justifying natural history and experimentation as the proper *means* to accomplish this reorientation and as the necessary *mediator* between the light of things and man's mind, Bacon invokes a story of man's past, present, and possible future in language that goes to the heart of his project of instauration.

Let us carefully examine his words as clue to Bacon's hermeneutics of nature.

> For we copy the sin of our first parents while we suffer for it. They wished to be like God, but their posterity wish to be even greater. For we create worlds, we direct and domineer over nature, we will have it that all things

29. These themes are echoed throughout Bacon's works. See for example, *The Great Instauration*: "Wherefore, seeing that these things do not depend upon myself, at the outset of the work I most humbly and fervently pray to God the Father, God the Son, and God the Holy Ghost, that remembering the sorrows of mankind and the pilgrimage of this our life wherein we wear out days few and evil, they will vouchsafe through my hands to endow the human family with new mercies. This likewise I humbly pray. . . . Lastly, that knowledge being now discharged of that venom which the serpent infused into it, and which makes the mind of man to swell, we may not be wise above measure and sobriety, but cultivate truth in charity" (1857–74:4.20). See also (1857–74:4.33): "Wherefore if we labour in thy works with the sweat of our brows thou wilt make us partakers of thy vision and thy sabbath. Humbly we pray that this mind may be steadfast in us, and that through these our hands, and the hands of others to whom thou shalt give the same spirit, thou wilt vouchsafe to endow the human family with new mercies."

> *are* as in our folly we think they should be, not as seems fittest to the
> Divine wisdom, or as they are found to be in fact; and I know not whether
> we more distort the facts of nature or our own wits. . . . (1857–74:5.132)

The pairing of sin with prideful error, with, that is, the imposition of a
humanly imagined "pattern of the world" onto things, is a familiar
theme. But Bacon now goes on to elucidate the exact nature of this
prideful error and its consequences.

> But we clearly impress the stamp of our own image on the creatures and
> works of God, instead of carefully examining and recognizing in them the
> stamp of the Creator himself. Wherefore our dominion over creatures is a
> second time forfeited, not undeservedly; and whereas after the fall of man
> some power over the resistance of creatures was still left to him—the
> power of subduing and managing them by true and solid arts—yet this
> too through our insolence, and because we desire to be like God and to
> follow the dictates of our own reason, we in great part lose. (1857–
> 74:5.132)

Here Bacon contrasts man's impressing of his stamp, or mark, upon
things with God's imprinting of his trace (*sigilla*) in His works. The
former is clearly an extraneous—and therefore false—image that man
simply tries to impose on things, much like a mask that covers up the
true visage underneath. By contrast, Bacon suggests that the creatures and works of God have their own, true marks that are, in fact,
traces of the Creator that God has imprinted in things. Man ought to
be busy searching out these marks in things, and Bacon implies that
postlapsarian man's failure to exercise dominion over creatures results from his failure to carry out such a search. Bacon clearly thinks
that man has it in his power to regain such dominion, despite considerable liabilities imposed upon him by his fall. In fact, postlapsarian
man retained "true and solid arts" that afforded him some modicum
of control over otherwise brute and intractible creatures. But man's
pride and insolence has jeopardized these very arts, where he might
instead have extended both his knowledge and his power through
careful searches after the true marks—or faces—of things themselves.

There is something vaguely reminiscent of Paracelsus in Bacon's remarks. Both regard the exaltation of man's reason over nature as symptomatic of his fallen nature. Both allude to the survival, not of a pure
Adamic language, but of Adamic arts and of some capacity to exercise a
willful control over brute creatures. Both, finally, seek man's direct encounter with things in order to discover their divine marks. Beyond

this point, however, Bacon parts ways with Paracelsus.[30] For while the latter exults in the search for images and symbols impressed *on* things by God as signs, or signatures, of their occult sympathies and spiritual interconnections, Bacon seeks to discover marks of an entirely different sort. At root, Bacon proves insistently skeptical of images and symbols as anything other than the mere projection of the human imagination. Thus, his criticism of past traditions—his history of error—extends not just to reason (as did Paracelsus'), but more centrally to images and the imagination itself. It is the imagination that leads reason astray, that imposes its fictions upon the intellect, thus obscuring its vision.

Hence, Bacon's project of instauration rejects the Paracelsian mode of reading the Book of Nature as symbolic. Bacon's hermeneutics is not one of symbolic exegesis of external signatures imprinted on things by God.[31] Indeed, just as we have seen Bacon reject mere words as but limited, even pitiful, human constructs utterly divorced from the originative *verbum Dei*, he also rejects symbols—the "emblematic world view"—as another instance of prideful human imagination. In its stead, Bacon proposes an entirely different hermeneutics of nature: one that turns upon a quite different understanding of language, of God's authorship of the Book of Nature, of the relationship between words and things, and, finally, of the association between the "Word of God" and the "languages of man."

We can begin to uncover the nature of that Baconian hermeneutics of nature, and the necessity of natural and experimental history to

30. Looked at from the perspective of his matter theory and cosmology, however, Graham Rees has argued for other affinities with Paracelsianism. See Rees (1975, 1975a, 1977, 1977a). Related to Bacon's matter theory, see also Walker (1985) and Rees (1984).

31. Note also Bacon (1857–74:4.379–80): "Not that I share the idle notion of Paracelsus and the alchemists, that there are to be found in man's body certain correspondences and parallels which have respect to all the several species (as stars, planets, minerals) which are extant in the universe; foolishly and stupidly misapplying the ancient emblem (that man was a *microcosm* or epitome of the world) to the support of this fancy of theirs"; and also (1857–74:4.367): "For as for that natural magic which flutters about so many books, embracing certain credulous and superstitious traditions and observations concerning sympathies and antipathies, and hidden and specific properties, with experiments for the most part frivolous, and wonderful rather for the skill with which the thing is concealed and masked than for the thing itself; . . . this popular and degenerate natural magic has the same kind of effect on men as some soporific drugs, which not only lull to sleep, but also during sleep instil gentle and pleasing dreams. For first it lays the understanding asleep by singing of specific properties and hidden virtues, sent as from heaven and only to be learned from the whispers of tradition; which makes men no longer alive and awake for the pursuit and inquiry of real causes, but to rest content with these slothful and credulous opinions; and then it insinuates innumerable fictions, pleasant to the mind, and such as one would most desire,—like so many dreams."

Bacon's *instauratio*, by picking up his prefatory remarks where last we left them:

> If therefore there by any humility towards the Creator, any reverence for or disposition to magnify His works, any charity for man and anxiety to relieve his sorrows and necessities, any love of truth in nature, any hatred of darkness, any desire for the purification of the understanding, we must entreat men again and again to discard, or at least set apart for a while, these volatile and preposterous philosophies, which have preferred theses to hypotheses, led experience captive, and triumphed over the works of God; and to approach with humility and veneration to unroll the volume of Creation, to linger and meditate therein, and with minds washed clean from opinions to study it in purity and integrity. For this is that sound and language which went forth into all lands, and did not incur the confusion of Babel; this should men study to be perfect in, and becoming again as little children condescend to take the alphabet of it into their hands, and spare no pains to search and unravel the interpretation thereof, but to pursue it strenuously and persevere even unto death. (1857–74:5.132–33; 2.14–15)

So extravagant is this passage in its figuring of nature as language that we, in the twentieth century, may well find it tempting to think of its import as "merely metaphorical."[32] If, however, we stop to consider the seriousness with which Bacon characterizes God as author of nature (His "Work") and Adam as enunciator of the true names of creatures, we may resist this temptation. We can then ask how it is that God, in Bacon's view, authored that work and how man, in turn, may read it.

As Bacon's words earlier suggested, God leaves His traces in His works, imprinting His marks in the Book of Nature. What are these marks and traces? They are not, like the Paracelsians' signatures, symbols. They do not envelop things in complex webs of signification whose nodes, through resemblance, reverberate sympathetically, if occultly. They do not await human wit to reimagine them as interconnecting symbols. Rather, God's marks and traces *are* his creatures and works. "In them . . . the stamp of the Creator himself" must be "recognized" and "examined." God's works, in other words, are imprinted

---

32. Thus, in his excellent and important article, Elsky (1984:458) unduly slights the significant metaphoricity of such passages in Bacon when he says, "for Bacon the idea that letters are part of the order of creatures in the natural world has become nothing more than an attractive, 'magistral' metaphor." Rather, Bacon engages in the critical practice of resituating a related cluster of dominant and inherited metaphors within a radically reformed narrative context; the result is no less than a reconfiguration of discourses of nature, religion, humankind, and society. See Bono (1990a; 1995).

by things; His Book of Nature written not in symbols, but in things themselves.

Let us assume for a moment that this characterization of God's authorship of the Book of Nature through things adequately captures Bacon's understanding. What then does it mean for man to "read" that text? What does it mean to think of nature as a text written in a language of things? For one, it means that the "meaning" of that text is inherent, or immanent, to itself. The fundamental units that make up the text of nature do not acquire their intelligibility through metaphoric relations to other units and, ultimately, to other texts. The "text" of animate nature, for example, is not a "microcosm" whose meaning can be fathomed only through it resemblances to and sympathies with a larger macrocosmic world—ultimately, through its metaphoric relations to a common, foundational archetype. The language of things in which God has composed His Book of Nature instead embodies and enforces a metonymic order of meaning. Such an order grounds the intelligibility of nature in the contiguity and spatiotemporal relations between and among things. It assumes, then, a multiplicity and diversity of individual things subsumed under a set of ordered relations, rather than a unity of types made manifest in a diversity of exemplary instantiations.

Man must learn to read such a text in a new way, according to Bacon. For, unlike the imagined symbolic text of nature of the Paracelsians, or even the hierarchically conceived and metaphorically structured Book of Nature of the Aristotelians or Galenists, the true text authored by God in the language of things will not reveal its secrets to mere logicians or exegetes. Its order, structure, and causal relations can not be discovered verbally or metaphorically. Instead, Bacon's use of the metaphor of the Book of Nature suggests other routes to knowledge. The "volume of Creation" must be "unrolled," its leaves unfurled by man's active investigations to reveal the languages of things written upon it. This language is unlike human languages: it alone escaped "the confusion of Babel." To master that language, man must become again like a child and, starting with its mere rudiments without flying first to interpretations, learn that very "alphabet"—the alphabet of nature—through which that language can begin to acquire intelligibility for us.

Bacon's allusion to Babel and the curse of the confusion of tongues should strike us as unusual and unusually significant. It suggests that there is no Adamic language accessible to man, and certainly no access to the Word of God, in any usual sense in which the quest for such an originary, primitive language was conceived by the philologists and etymologists of the late Renaissance. Bacon's whole project of an *instauratio magna*, it seems to me, turns upon his insight that return to an Adamic state of Edenic perfection requires a turn to things. For things—the language of things in which God authored the text of His creation—were at the heart of that lost language of Adam. Adam's divinely ordained gift was his ability to apprehend the nature of things immediately, that is, without the mediation of observation, experimentation, and the sustained work of scientific investigation. His language, through this divine gift, could then mirror perfectly the very nature of things—in short, the Adamic language gave voice to the inarticulate, but divine, language of things inscribed in the Book of Nature.

Postlapsarian man has lost the immediacy of Adam's understanding of things, and through Babel, all trace of the Adamic language as well. What he does have close at hand are things, the only uncorrupted "sound and language" accessible to all mankind.[33] His task, then, is to take the very alphabet of this language into his hands, to learn and master it through laborious searches, and to use it to penetrate to the very core of things themselves. Since that language is structured metonymically, rather than metaphorically and symbolically, mastery must come through careful attention to the contiguity and and spatiotemporal relations of and among things. Hence, man must learn to "read" the Book of Nature first by identifying the fundamental alphabet of nature through the laborious recording of natural histories. Since the Book of Nature is a labyrinth in which man can easily lose his way,[34] and since nature is prone to hide her secrets from him, man must, in addition, learn how to hunt down the traces of things in nature.[35] If "nature free and at large" escapes attempts to catch her in the net of natural history, she must be placed "under constraint and vexed" so as to give up her secrets to active experimentation. Only in this manner will the alphabet of nature coalesce into a meaningful language of things before man's mind.

33. Compare, in addition, Bacon (1857–74:4.327): "for that is the true philosophy which echoes most faithfully the voices of the world itself, and is written as it were at the world's own dictation; being nothing else than the image and reflexion thereof, to which it adds nothing of its own, but only iterates and gives it back."

34. See, for example, Bacon (1857–74:4.18): "But the universe to the eye of the human understanding is framed like a labyrinth; presenting as it does on every side so many ambiguities of way, such deceitful resemblances of objects and signs, natures so irregular in their lines, and so knotted and entangled. . . . No excellence of wit, no repetition of chance experiments, can overcome such difficulties as these. Our steps must be guided by a clue. . . ."

35. See especially recent work by Eamon (1984, 1985, 1990, 1990a, 1994). For related work, see Moran (1981, 1985).

Fundamental to Bacon's turn to things and to his advocacy of natural history and experimentation as the proper way for man to know nature and reclaim his lost Adamic dominion over God's creatures is his understanding of the relationship between the "Word of God" and the "languages of Man." Not only, as we have seen, does Bacon regard human languages as unalterably stamped by the curse of Babel, and hence as thoroughly historical and social artifacts, but he also associates the lure of bookish culture and language with the most burdensome legacies of his famous "Idols of the Mind" that prevent man from apprehending truths of nature.[36] To *overcome* the effects of the Fall, man must reject the way of human languages—the way of philology—as a prideful error. To *repair* the effects of Babel, man must seek Adamic wisdom elsewhere than in language. To *prepare* for the instauration of Adam's dominion over things, Bacon must (paradoxically) use words—writing—as a tool for deconstructing the Babylonic edifice of learning that has obscured and silenced the true voice of things.

For Bacon, Adam himself becomes the model for mankind's instauration. Adam's mastery of nature flowed from his immediate knowledge of the true names of things. As model for postlapsarian man, however, Adam's legacy must reverse this order to take into account man's now imperfect nature and historical condition. If fallen man cannot enjoy direct access to the *verbum Dei*, he can, instead, attempt to mimic, even replicate, Adam's dominion over nature. Where Adam's dominion was immediate, postlapsarian man's mastery over nature must be mediate. As Bacon was so fond of saying, "Truth is the Daughter of Time [, not of Wit]" (1857–74:3.612). Man must work for his knowledge. But, in addition, the key to reclaiming Adam's perfect knowledge of creatures was the divinely established conjunction between knowledge and power in the Garden of Eden. If knowledge enabled Adam to exercise power over nature, postlapsarian man's path to true knowledge lay in refining and cultivating those arts that afforded him some measure of control in his own limited realm. "And so," for Bacon, "those twin objects, human Knowledge and human Power, do really meet in one" (1857–74:4.32). Bacon turned this conjunction of knowledge and power into the basis for man's instauration. To wrest control from nature, man had to learn to subdue her. By constraining nature, forcing

36. Bacon (1857–74:4.60–61): "But the Idols of the Market Place are the most troublesome of all: idols which have crept into the understanding through the alliances of words and names." And Bacon (1857–74:4.62): "But the Idols of the Theater are not innate, nor do they steal into the understanding secretly, but are plainly impressed and received into the mind from the play-books of philosophical systems. . . ."

her to reveal her secrets, and, simultaneously, producing desired effects in nature through a set of procedures or active practices, the art of experimentation reenacted that Adamic coincidence of power and knowledge.

Thus, Bacon's *Instauratio Magna*, his turn to things, and his advocacy of power, knowledge, experiment, and natural history as the proper ends and means of humankind owe their inspiration equally to his radical recasting of both the fundamental narratives of Genesis and the relationship between the "Word of God" and the "languages of man." It is worth noting both the context and the consequences of this latter re-visioning of language. As Martin Elsky has recently argued, Bacon's reform of language represents a sharp reaction against "traditional" Renaissance conceptions of words as inextricably bound up with things. Where scholars have customarily seen Bacon as "seminal" to the emergence of language theories stressing the "correspondence of words and things" during the seventeenth century, Elsky correctly stresses that this *arbitrary* connection of words and things first depended upon Bacon's emphatic rejection of other, intrinsic connections (1984:449, 457).

As we have seen in this book, Bacon had many such models for language to draw upon in his criticism of Renaissance theories. From Ficino to Dee, Neoplatonic, mystical, cabalistic, and Adamic views of language often shared such beliefs in a correspondence between words and things. Elsky points in particular to the Renaissance fascination with hieroglyphs as a context for Bacon's reform. The notion of a language whose characters picture that which they signify and simultaneously suggest both a visible, material connection to things and a deeper spiritual or allegorical affinity proved seductive to many. In particular, this image of the hieroglyph converged nicely with the idea that the *verbum Dei* gave rise to the material universe and its associated spiritual virtues which were parceled into a hierarchically organized array of entities inhabiting distinct, but yet interconnected, strata. More particularly still, Elsky cites the influence of Du Bartas, both in Joshua Sylvester's translation and in Simon Goulart's commentary, and of Alexander Top's *Oliue Leafe*, for a "tradition of alphabetical symbolism . . . current in England" and against which Bacon rebelled (1984:450ff.).

These texts, and numerous others that we could cite from the continent,[37] sketched a view of language as arising from the "Word of

---

37. See discussion and bibliography cited in Chapter 3, above; for the Renaissance fascination with hieroglyphs, emblems, and the like, see Chapter 6 above. In addition, for seventeenth-century England, see Ormsby-Lennon (1988) and Singer (1989).

God" itself, often, as we have noted, identified with an Adamic form of Hebrew. Not only did such a view of language assert an absolute identity between signifiers and signifieds, words and things, but it also gave special religious significance to them. The view of language as hieroglyph reinforced this tendency by adding another allegorical layer to words. In this regard, a proponent of hieroglyphs like Top, as Elsky notes, "suggests that God simultaneously created a Hebrew letter as a hieroglyph of every thing He created. Thus he transfers the allegorical function of Egyptian pictographic hieroglyphs to non-pictographic Hebrew letters" (1984:451). Words, letters, and things are all, in this view, reflections of the "Word of God" itself—indeed, even of the divine mind—and therefore instantiations, however corrupt or veiled, of our Adamic legacy forfeited by the Fall and obscured at Babel.

Bacon, operating within and against this context, rejects not only this "hieroglyphic" theory of language, but also the religious system of meaning associated with the hermeneutic strategies prescribed by such theories. In general, we have seen Bacon adamantly refuse any privileged status to the languages of man. Instead, like all rational instruments of fallen man, words lead man astray insofar as they have been overhastily abstracted from things. The lure of traditional language then constitutes much of the substance and force of those Idols of the Marketplace that our Lord Chancellor so vehemently decries.

But Bacon's criticism of hieroglyphs and of supposed "sacred" languages such as Hebrew (whether or not associated with hieroglyphs, as in Top's text) goes much farther than this general critique of post-Babylonic languages of man. For even the supposed Adamic language, if I understand Bacon correctly, has limitations that man must recognize lest he be led into error. Elsky points to certain passages from book 6, chapter 1 of *The Advancement of Learning* which, he stresses, have been largely ignored by students "of Bacon's linguistic thought" (Elsky 1984:454–58; Bacon 1857–74:4.438–48). Hieroglyphs, according to Bacon, fall under one of two classes of what he calls "the Notes of Things . . . which carry a signification without the help or intervention of words"—namely, those which are *"ex congruo,* where the note has some congruity with the notion." The other category of such notes of things are those which are established *"ad placitum,"* that is, those characters, such as the so-called *"real characters"* of the Chinese language, that directly represent things or concepts, rather than words or letters, but which are nonetheless thoroughly conventional and arbitrary in their assignment of nonnominal characters to their referents. Hieroglyphs, by contrast, are, like gestures, not simply arbitrary, but "always have some similitude to the thing signified, and are a kind of emblem" (Bacon 1857–74:4.439–40).

Even conceding such similitude between hieroglyphs and things, what is striking is that Bacon refuses to attribute any great significance or power to them. Unlike Top, and those in the traditions of Renaissance cabalism and occultism such as Dee, Bacon deems hieroglyphs to be merely accurate pictures of things. Even if, in some sense, their status as accurate pictures confers upon hieroglyphs the distinction of being "natural" representations (and Bacon nowhere leaps to this characterization), they are nevertheless not regarded by Bacon as in any sense sacred or divine. Hieroglyphs, it would appear, do no more than *picture* certain aspects of the observed world; they do not reflect the divine mind or reveal God's intentions, ideas, or nature. Indeed, as pictures, hieroglyphs do not even reveal the true nature, order, or significance of things themselves.[38]

So it is then not surprising that Bacon turns from his discussion of hieroglyphs and mention of the even more ancient writing of the Hebrews to "Grammar." While seeming less exalted than these other dimensions or kinds of language in the pantheon of Renaissance thought and esoterism, grammar, for Bacon, can offer what hieroglyphs and ancient sacred languages cannot. As Bacon asserts, "nor must it [grammar] be esteemed of little dignity, seeing that it serves for an antidote against the curse of the confusion of tongues" (1857–74:4.440–41). What Bacon specifically has in mind, however, is not that grammar that aids learning and stylistic elegance, but rather "a kind of grammar which should inquire, not the analogy of words with one another, but the analogy between words and things, or reason." Bacon calls this "Philosophical Grammar." It interests him precisely because "words are the footsteps of reason, and the footsteps tell something about the body" (1857–74:4.441).

What does Bacon mean here, and how can this philosophical grammar supply him with something more valuable than the hieroglyphs and sacred languages of the Renaissance? Bacon does not mean that the "footsteps" represented by words are the footsteps that God has left upon His creation.[39] Words, for Bacon, remain no substitute for God's works—for things, the language in which God wrote His Book

38. Elsky (1984:456) provides an excellent summary of Bacon's conception of hieroglyphs and discussion of their significance as pictures.

39. For Bacon's reference to God's footsteps, see the second epigraph at the beginning of this chapter.

of Nature. Bacon underlines this point by contrasting his philosophical grammar, and words as "footsteps," to the Neoplatonic-Adamic
theory of language so dominant among Renaissance commentators
and scientists:

> But I must first say that I by no means approve of that curious inquiry,
> which nevertheless so great a man as Plato did not despise; namely con-
> cerning the imposition and original etymology of names; on the supposi-
> tion that they were not arbitrarily fixed at first, but derived and deduced
> by reason and according to significance; a subject elegant indeed, and
> pliant as wax to be shaped and turned, and (as seeming to explore the
> recesses of antiquity) not without a kind of reverence,—but yet sparingly
> true and bearing no fruit. (1857–74:4.441)

Bacon's words permit some latitude of interpretation. He opposes the
belief—commonly, but misleadingly, attributed to Plato's *Cratylus* during the Renaissance[40]—that words and language in general are not arbitrary and conventional. Whether Bacon wants to characterize his opponents' historical argument as suggesting that Adam's "imposition" of
names (and therefore their "original etymology") was "derived and
deduced by reason" is not clear. Nor is it clear what such a mechanism
for the imposition of names might mean; whether, for instance, Bacon
intends to characterize this opposing view as suggesting that, through
the divine light of his pure, unfallen reason, Adam deduced the names
of all creatures from their attributes and imposed the same upon them.
But whatever the specific formulation he wishes to criticize may be,
Bacon clearly wishes to deny the utility of words as repositories of divine truths.

Even, that is, words in the language of Adam! For, unlike the vast
majority of commentators during the Renaissance, Bacon does not see
in the Adamic language any close and necessary link between words
and the divine ideas or mind. While the Adamic language, according to
Valerius Terminus, gave to creatures "their true names," such names are
the product of man's free will and pure, uncorrupted intellect, and not
of the divine will and Word—not, that is, of the "language" spoken by
God. Such names are a mirror of God's works—of nature laid bare
before Adam's gaze—but not of God's nature!

This understanding of the language of Adam enables us finally to
grasp the import of Bacon's *Instauratio magna* and the projects that fos-

---

40. See the discussion of this point, and of Platonic and Neoplatonic theories of language in general, in Chapter 3, above. Note as well relevant works by Coudert, Dubois,
Gombrich, Vickers, and other cited there.

ter it. Philosophical grammar, that "noblest species of grammar," has but a modest purpose, unconnected with any attempts to fathom the mind or nature of God. Rather, it "would be this: if some one well seen in a great number of tongues, learned as well as vulgar, would handle the various properties of *languages;* showing in what points each excelled, in what it failed" (1857–74:4.441). In other words, philosophical grammar aims to improve human languages—the "languages of man"—*not* to uncover an originary language, the "Word of God," assumed by many to be intimately related to the Adamic language. The latter connection between the language of Adam and the *verbum Dei* is at any rate illusory; so that the real aim of philosophical grammar is to proceed through "comparative grammar" to the framing of an improved, or "more serviceable," language.[41]

Here the project of philosophical grammar and that of natural and experimental history converge to form the backbone of Bacon's master project of a great instauration. For, by improving and perfecting language, man comes closer to reconstructing something like the original Adamic language. But such reconstruction requires that humans attend to the mercies that Bacon has uncovered in his new method and that humans can actively instaurate through the operations associated with experimentation and natural history. Only through the latter can man uncover God's inscriptions in His Book of Nature, and hence grasp the language of things. Since, as we have seen above, that language of things is not a spoken language, divine or human, man must frame words to express the intimate nature and order found through his experiential uncovering of that diving language of things. As Elsky so aptly puts it, "Bacon implies that each phenomenon must be studied according to its own natural history, out of which an accurate vocabulary will develop to describe that phenomenon" (1984:453). Bacon ties

41. Elsky (1984:455). Elsky's discussion (esp. 1984:454–55) provides some essential insights into Bacon's linguistic philosophy. But he nonetheless seems to think that there is a tension between Bacon's view of the Adamic language in *Valerius Terminus* and in the *Advancement of Learning,* and that statements in the latter "must severely qualify Bacon's other statements on Adamic naming." What Bacon's position qualifies, I would argue, are Renaissance interpretations of Adam's imposition of names. Bacon himself is not constrained by the religious and metaphysical view that Adam's language is a replica of an alleged divine language that was the blueprint for the Creation. Instead, Bacon, as we shall see below, severs the connection between the divine Word and nature and the nature of created things. This rupture allows Bacon to subscribe to an Adamic language as a mirror of nature, but not of God. As such, Bacon's philosophical speculations open horizons for the play of political motives in the pursuit of power and utility through "philosophical grammar," natural history, and experimentation.

man's instauration to such a strategy, whereby through the work of experimentation and the labor of natural history, man will not so much recover as remake the language of Adam. Bacon's strategy for this great instauration is therefore *prospective* rather than retrospective.

Bacon's *Instauratio Magna* therefore draws much of its inspiration from a radically revised notion of language and, specifically, of the relationship between the "Word of God" and the "languages of man." The very archetype of human languages, the language of Adam, is a creation of man, though it draws inspiration for its choice of words as "true names" of things from the pure light of Adam's divinely created and unfallen intellect. Hence, what is in traditional thought conceived as an originary, primitive tongue derived from the creative "Word of God" becomes, in Bacon's narrative, at best a model for what postlapsarian man can and must recreate in the future. The doubleness of Bacon's *instauratio* may be thought to arise from this dual nature of its model. Man's instauration joins together reform and innovation: Bacon seeks to reform knowledge to return man to his lost Adamic state; but such a return can be effected only by creating knowledge anew. Man cannot recover the whole cloth of a lost Adamic wisdom; he must discover true knowledge, piecing together fragments of hard-won knowledge, to weave, in the end, a new tapestry depicting the natural wisdom inscribed within God's creation.

For Bacon, in fine, the "languages of man"—including both the original, lost language of Adam, and the new, prospective Adamic language, the language of nature described by Baconian natural science— must remain distinct from any "Word of God." This absolute distinction is constitutive of Bacon's discourse and its turn to things. For the *verbum Dei*, to infer from Bacon's text, is *not* generative of the world, of things, in any Neoplatonic sense. The world is not, for Bacon, an emanation or instantiation of the divine ideas that, together, constitute the divine mind and Word. In short, while God created the world through His Word, the world is not itself made in the image of God. As Bacon emphatically declares,

> For as all works show forth the power and skill of the workman, and not his image; so it is of the works of God, which show the omnipotency and wisdom, but do not portray the image of the Maker. And therefore therein the Heathen opinion differs from the sacred truth; for they supposed the world to be the image of God, and man the image of the world; whereas the Scriptures never vouchsafe to attribute to the world such honour as anywhere to call it the image of God, but only the work of his hands. . . . (1857–74:4.341)

Bacon here cuts the knot of necessity that, in the heathen philosophy of the Greeks from Aristotle to Plato to their synthesis in the Neoplatonism that proved so seductive to later Christian philosophers, tied together the essence of the Creator with that which has been created. The world is not God's image, precisely because God, in good medieval voluntarist and Protestant fashion, is not constrained by His nature to create the world according to any one model. Bacon's theory of language and of the relationship of the divine Word to nature and the languages of man is suffused by belief in God's absolute power. Since God's creation is arbitrary, it reflects only His majesty, power, and wisdom. The Word of God expresses itself through nature, but nature does not contain, reflect, or instantiate that Word. Nature, in short, is no image of the Creator.

The consequences of this view for Bacon's *instauratio* are profound. Since nature is no longer an *imago Dei,* man in his strivings to become a new Adam cannot assume that the text of nature presents him with a variety of types, signatures, or symbols that find their unity in a spiritual—whether transcendent or immanent—semiotic system. The materiality and meaning of created things are no longer grounded, as instantiations of the divine mind, in a transcendent and overarching "Word of God." By severing the tie of necessity between things and the divine mind, Bacon underlines the unmooring of language from its safe harbor in the divine Word that is a foundational assumption of his project of instauration. Hence, in seeking to recapture Adam's paradisical dominion over nature, Bacon must replace an originary model of language with an eschatological model. That is to say, rather than attempting to regain knowledge of and power over nature through a retrospective exegesis of human languages, or of the divine symbols imprinted upon nature as traces of a generative *verbum Dei,* Bacon rejects this search for an originative language—a *clavis universalis.*[42] In its stead, Bacon reconceives the Adamic language as a prospective goal: the end, or *eschaton,* to be achieved collectively and cooperatively by mankind through the labor of experimental and natural history. Rejecting etymology, philology, and the deciphering of nature's system of symbols, Bacon looks to things—the divine language of things—as proper founda-

---

42. Rossi (1960). Bacon, in short, turns away from the metaphorically structured, analogical universe and from the vision of God as creating, through His Word, nature as *speculum* so dominant in Western thought from at least the twelfth century. See, for example, Chenu (1968); Leisegang (1937); Bono (1984) with the latter's discussion of "abstract symbolism" versus "symbolic literalism" and the sources cited therein. On Bacon, note Warhaft (1971).

tion for creating words suitable for the true naming of things. Mere words without such foundation in things are but the worthless babblings of mankind.

The metonymy of identity and difference, of contiguity and displacement—ultimately of cause and effect—thus replaces for Bacon all traces of the metaphorics of resemblance, correspondence, and emblematic meaning. The discourses of types—of the recuperation of all variety within the unity of forms, archetypes, ideas, and generative symbols—gives way to a discourse of order and diversity, in which the very multiplicity of things becomes the subject of scientific inquiry, and the ordered table of natural history becomes the objectified artifact of human knowledge of the divine Book of Nature.

In short, for Bacon the various and contending hermeneutics of exegesis give way to a new *de-in-scriptive* hermeneutics of nature. Man's task as Baconian "interpreter of nature" is to de-in-scribe—or, *de-scribe*—what God has inscribed in the very text of nature. Such de-inscription, for Bacon, requires that man interrogate nature relentlessly so that what God has inscribed in the very folds, margins, and unturned leaves of nature's text may be revealed and, once open to the inquirer's gaze, described in fitting, unadorned words.

Afterthoughts

Bacon's transformation of traditional metaphors of God's authorship of nature also deepened and provoked an ongoing early modern transformation of the role of the individual in the making of knowledge about nature. I do not mean by this statement to point out what surely is by now both obvious and unexceptional: that Bacon's natural philosopher is one who acts on nature, who takes it apart through experimentation, and who collaborates with others in the pursuit of physical truths. Rather, for all the incessant activity promoted by the Baconian model of the scientific investigator, it is not the latter, but "methods," "institutions," and especially "experiments" and "facts" that take center stage in the new emplotment of the story of science in the West.

If pre-Baconian—or, more accurately, premodern—science was unabashedly logocentric, and therefore "author"-centric, modern and Baconian science professed to revolve around and upon the "fact." Indeed, there exists a virtual obsession with banishing mere words from the arena of natural knowledge and, therefore, with enforcing a disjunction between words and "facts." Behind words had lain the authorizing presence of the divine Word, a presence that had slowly but dramatically withdrawn itself. Instead, the dazzling presence of "facts" came to etch itself upon the receptive consciousness of the scientist. Never mind that—despite denials and the enforcement of binary oppositions between the likes of words and things, meanings and facts—the authorizing presence of the Word still lay behind these newly constituted facts.

With the triumphant regime of "facts" the role of the individual student of nature changed dramatically. The author, commentator, natural philosopher of the Middle Ages and Renaissance had engaged in a hermeneutics in which all knowledge of nature was, in large measure, a reading and reappropriation of God's text. Origins were all important; originality was not. Yet, at the center of this hermeneutical—often exegetical—enterprise were not facts, but rather the individual investigator's mind and imagination. Thus, while the knowledge produced by the individual author/philosopher was but a deepening and clarifying of a pristine wisdom and not at all an innovation in any modern sense, his actual role as producer of this knowledge was at the forefront of *scientia*.

By contrast, the regime of "facts" continually attempts to efface the role of the individual scientist. It is nature that now, through the vehicles of the experiment and the natural history and through the recording apparatus of the scientific instrument, has a voice. That voice proclaims the necessity, and centrality, of "facts."[43] At the same time, it orders and supervises the activities of scientists as they execute experiments, record results, describe nature, and deploy the instruments and formalisms that generate facts. The scientist may be replaced by others; indeed, the very objectivity of "facts" requires the collective activity of scientists.[44] It is they who are frangible, not the facts.

Yet, paradoxically, this very obsession with "facts" was responsible— to what degree we cannot say—for perhaps the most dramatic of all transformations of the role of the individual in the making of knowledge about nature. I am speaking, of course, of the birth—a birth witnessed by the late sixteenth and seventeenth centuries—of the "scientist" as great discoverer. The paradox lies precisely in the simultaneity of the withdrawal of the individual from the center of inquiry and the emergence of discovery and discoverers as a social and cultural preoccupation. Let me but suggest that it is precisely this paradoxical

43. Bacon himself speaks of "the true philosophy which echoes most faithfully the voices of the world itself"! See note 33 above for full citation.

44. For related discussions of probability, objectivity, and the like in the modern world, see Daston and Galison (1992); Hacking (1975); and Shapiro (1983).

conjunction that *ought* to be expected. For with the emergence of "facts" and their production as the central activity and concern of science, "originality" took precedence over "origins," and innovation came to be valued over understanding and interpretation.[45] Displaced by "facts," no longer at the center of inquiry into nature, the new natural philosopher—"scientist"—had only his claims to innovation to legitimate his practice and authorize the favor of patrons and institutions.[46] With facts and innovations, then, came the age of discoveries and the knotted claims of priority disputes. All was not harmonious within Solomon's House.

45. For the theme of originality in the Renaissance, see Quint (1983).

46. Mario Biagioli is now at work on a project that will address (and reveal the complexities of) legitimating strategies as part of the various "national styles" of early modern scientific culture. Personal communication and unpublished talk, "Civilizing the Scientist's Subjectivity," 1992 Meeting of the History of Science Society, December 30, 1992, Washington, DC.

8

Beyond Babel

Mersenne, Descartes, Language, and the Revolt against Magic

Les pensées qui lui vinrent sur ce sujet lui firent abandonner l'étude particulière de l'Arithmétique & de la Géométrie, pour se donner tout entier à la recherche de cette Science générale, mais vraye & infaillible, que les Grecs ont nommée judicieusement *Mathesis*, & dont toutes les Mathématiques ne sont que des parties. Après avoir solidement considéré toutes les connoissances particulières que l'on qualifie du nom de Mathématiques, il reconnut que pour mériter ce nom, il falloit avoir des rapports, des proportions, & des mesures pour objet. Il jugea delà qu'il y avoit une Science générale destinée à expliquer toutes les questions que l'on pouvoit faire touchant les rapports, les proportions & les mesures, en les considérant comme détachées de toute matière: & que cette Science générale pouvoit à très-juste titre porter le nom de *Mathesis* ou de Mathématique universelle; puis qu'elle renferme tout ce qui peut faire mériter le nom de Science & de Mathématique particulière aux autres connoissances.

—Baillet

The Cartesian project of a universal science—of a "mathesis, ou mathématique universelle"—in many respects stands poles apart from the Baconian project of an experimental and natural history. Where the latter is "laborious" in its attention to the details inscribed by God in the very folds, margins, and unturned leaves of the Book of Nature, the former purports to grasp not fleeting, subjective bits of sensuous impressions, but, rather, the very network of abstract ordered relations within which concrete—"extended"—material things are suspended. Where Bacon places "facts," not the individual scientist, at the center of inquiry, Descartes apotheosizes the autonomous subject—the individual, conscious "I" of the Cartesian *cogito*—as discoverer of clear and distinct ideas and, hence, truths of nature.

247

Despite such fundamental differences, the Baconian and Cartesian projects array themselves on the side of a deinscriptive hermeneutics opposed to the textual, symbolic, "emblematic," and narrative strategies embraced to varying degrees and in changing combinations by the proponents, in all their shades and stripes, of the hermeneutics of exegesis whom we have encountered previously. Both projects, that is, "read" the Book of Nature as a coherent, orderly text produced by an omnipotent author who, nonetheless, remains distinct from, and unmirrored by, nature, His creation. Neither the labyrinth of Bacon's contingent, material nature, nor the geometrical order of Descartes's spaciotemporal extended substance, reflects the divine ideas, the *mens Dei*. The divine author has constituted neither Baconian, nor Cartesian, universe as a *necessary* image of His unchanging nature.

Hence, the knowledge each project purports to uncover is not at all the exegetical recovery of a pristine divine text, but rather the *description* and *discovery* of nature itself—and *only* nature itself—in the language of nature inscribed by God in His very creation of the world. Whether one identifies this divinely inscribed language with the Baconian language of things or the Cartesian language of mathematics (*mathesis universalis*), the text of God's Book of Nature enjoys no privileged relationship to human traditions of learned or esoteric wisdom via, for example, a surviving, post-Babylonic, Adamic link. Rather, the Book of Nature stands apart and independent from the mere vagaries of human traditions, societies, civilizations. Baconian and Cartesian sciences purport, finally, to divorce themselves from the "bookish culture" of the Renaissance; the protoscientific natural philosopher is no longer an exegete. Thus, even Descartes's individual knowing subject, *like* Bacon's pious student of nature, and *unlike* the traditional exegete and the occult naturalist, becomes merely the instrument through which deinscription of nature's text occurs. As the site of an array of material and discursive practices constituting "methodical experience," Baconian and Cartesian natural philosophers encounter the innovative and the original at each significant juncture within the Book of Nature. By contrast, their medieval and Renaissance counterparts—commentators, exegetes, occultists—encountered neither the new nor originality. Rather, they *produced* interpretations and understandings that strove to uncover the originary, divine, pristine, and powerful meanings within the text of nature.

Bacon's reworking of the natural philosopher's role was inseparable, as we saw in the preceding chapter, from his narrative recasting of humankind's Adamic legacy and aspirations. The world of the Cartesian natural philosopher also rests upon a prior reworking of this Adamic heritage and narrative emplotment of the relationship between the "Word of God" and the "languages of man," prompted, in large measure, by the perceived threat associated with the excesses of the occult tradition in the late sixteenth and early seventeenth centuries. This threat was perhaps dramatized by the excitement and furor generated by cryptic announcements of the imminent influx of the Rosicrucian Brotherhood into France in the early 1620s.[1]

While Descartes was allegedly swept up in this furor (Yates 1978:103–17; Baillet 1691), his friend and mentor, Marin Mersenne (1588–1648), by directly confronting the challenge presented by the occult movement, provided an enduring cultural foundation for the post-Renaissance Cartesian natural philosopher.[2] Mersenne's efforts, and hence the cultural underpinning of Cartesian deinscriptive strategies for reading the Book of Nature, gathered force from the antimagical strains of late-sixteenth-century thought and their reconsideration of Renaissance theories of language.

Language and Cultural Narratives: Weyer, Erastus, Del Rio, and Renaissance Magic

Among late-sixteenth-century opponents of magic, Johannes Weyer (1515–88),[3] Thomas Erastus (1523–83),[4] and Martin Del Rio (1551–1608)[5] are among the most vehement and well known. Weyer and Erastus were Protestants; Del Rio, a Catholic. While all are opposed to magic and to belief in special human access to occult virtues through the power of words (which was, in some sense, foundational to magic), Weyer and Erastus adopt rather different strategies than the Catholic Del Rio in their discussions of Renaissance language theory. Thus, we must understand the respective arguments of all three as contributing to larger rhetorical schemes aimed at securing different visions of secular, religious, and interpretive authority.

1. On Rosicrucianism, see Arnold (1970); Peuckert (1928); Waite (1887, 1923); and Yates (1978). For specific Rosicrucian figures, see Craven (1968); Kienast (1970); and Montgomery (1973).

2. On Mersenne and the occult, see Lenoble (1943); Hine (1984); Yates (1964, 1978, 1979); Thorndike (1923–58); and Huffman (1988).

3. On Weyer, see Walker (1958); introduction to Weyer (1991:xxvii–lxxxvii); Thorndike (1923–58:6); and Axenfeld (1866).

4. On Erastus, see Walker (1958:156–66); and also Thorndike (1923–58:5, 652–67).

5. On Del Rio, see Walker (1958:178–85).

The bottom line, from one perspective, is the same among all three: the rejection of any licit, nondiabolic or demonic, dimension of language for the production of magical effects. By telescoping their arguments against magic in such a fashion as to remove from our field of vision any trace of nuance or complexity in the views of their opponents, Weyer and Erastus, however, achieve a powerful rhetorical effect. That effect is to delegitimate not just specific magical practices and beliefs, but also the assumption that such could be grounded in knowledge of the very structure of the created natural cosmos to which humans ordinarily have access through their mental and perceptual powers. The question of humankind's unique history and the access it affords to divine, Adamic wisdom—including wisdom about nature—is left out of the picture.

When Weyer argues, for example, against the notion that words have the power to work physical and psychological changes on men, animals, and nature, he simply dismisses any presumed correspondence between words and celestial powers that might constitute grounds for entertaining even the possibility of such occult, but "natural," effects, concluding only "that this divinely operative power does not exist in words at all" (1991:390–91). Weyer supports this observation by citing the "sacred authority of Chrysostom": "Words," he says, "are brought forth from the mouth of the priest. But they are consecrated by the power and grace of God. If the mutterings of magicians have any efficacy, their hidden power is due to a confident belief in the Devil. In the words, however, there is no efficacy; but, because those who trust in these words impiously rely upon Satan, a most just God often allows them to be deluded by Satan."[6] Clearly this view satisfies a certain strain of Protestant belief that vests ultimate control in nature, as in religion and society, in God's sovereignty and absolute power.[7] While God imparts to secondary causes in nature a certain regular, but contingent, efficacy, the idea that man can somehow harness these divinely originating powers in nature through other mediations that bypass direct physical manipulation of things—God's utterly dependent creatures—strikes Weyer as simply absurd. What Weyer elides is not

6. Weyer (1991:391). Weyer (1583: col. 535): "Verba sacerdotis ore proferuntur, Dei autem virtute consecrantur & gratia: & magici susurri si quid habent efficaciae, id occultae virtutis habent à certa fiducia in diabolum. Nulla vero inest ijs verbis efficacia: sed qui ijs fidunt, hos ob impiam confidentiam saepe illudi à Satana sinit justissimus Deus." See Walker (1958:154 n. 2), from which I have taken this Latin passage; Walker also cites the reference to Chrysostom there.

7. Deason (1986) provides a useful discussion of this Protestant perspective for our purposes. See his bibliography.

just the theoretical justifications for natural magic, but also the very "sacred" narratives that legitimate the idea, for some Renaissance thinkers, that man *can* have access to the divine Adamic language and the dominion over nature that knowledge of the true names of things was thought to bring with it.

Thomas Erastus, in his attack on the foundations of Paracelsian medicine and its occult premises, speaks more directly to the question of the magical power of words,[8] but also ignores those Renaissance cultural narratives that might lend such views plausibility, if not legitimacy. As D. P. Walker rightly noted, Erastus, like Weyer, wished "to achieve a non-magical Christianity" (1958:161). This aspiration was certainly part of Erastus' Protestant desire to defend a "pure" and unadorned Christianity that looked only to the authority of the Bible, literally interpreted (1958:156), and to the awesome power and inscrutable nature of God for understanding spiritual matters and the order of things. Fallen man's distance from God, his utter weakness and fallibility in comparison with God, make a mockery of any deluded Catholic or Platonic attempts to speak of a continuing tradition of esoteric, divine knowledge and to claim continuous access to prelapsarian, Adamic, paradisaical wisdom.

Like Bacon, Erastus conceives of nature as containing secrets and powers that God has, through His absolute power and will, implanted in things and which man can grasp only through direct experience (Walker 1958:157–58). Like Weyer, he condemns any appeal to words as a source of divine wisdom, and especially of occult power, as the workings of demons. Where Bacon explicitly rejects Renaissance myths of the survival of Adamic wisdom in traces left upon the languages of man and creates instead his own revisionist narrative that links mankind's future to reproducing Adam's pious and direct engagement with things in nature (God's works, not words), Erastus consigns such Renaissance myths to an ignominious fate of sheer irrelevance through an effective rhetoric of silence. Like Weyer, then, Erastus simply ignores the possibility of any historic links, of surviving extrabiblical sources for a tradition of divine wisdom, that might underwrite any investment in the "word" as bearer of special power for humankind.

Instead, Erastus chooses to define the terms of the debate differently. The question becomes for Erastus whether or not words have any

8. Erastus (1572). For Erastus' discussion of the power of words, see Erastus (1572:169–81). The *Pars Altera* also devotes a chapter to "De cabala eiusque speciebus seu partibus: & vtrum aliqua licita sit, an non sit." See Erastus (1572: esp, 271–84) for his extensive discussion of Cabala, divine "authorship," medicine, and Paracelsus.

"natural" powers that can explain alleged occult powers attributed to their use. For Erastus words are, in some sense, natural; but, and this is the critical contention for him, such natural dimensions of words do not extend to the attribution of meaning to words—that is, to their status as representations and bearers of meaning. In this one crucial respect, words are thoroughly conventional—based upon human agreement and compact. Only when considering the material basis of words in sounds and, hence, their production by the organs of breath and speech can we truly speak of words as natural.[9]

Having made this critical distinction, Erastus goes on to demonstrate at some length what was already obvious to his contemporaries: namely, that human linguistic conventions differ widely. What a Bovelles, Bibliander, Postel, and Gesner, among others, knew earlier in the century—that the "languages of man" are marked by difference and characterized by a vast array of diversity (see Chapter 3, above)—is used by Erastus as a weapon against the magical power of words (Erastus 1572:170). Words are as various as the peoples that speak them; meanings are therefore conventional, and the ritual invocation and vocalization of mere words can not have the magical effects that are claimed for them, since there is nothing "natural" and bespeaking an inherent natural power about them. Erastus even draws the inference that the deaf ought to be able to speak if words were innate and already present within man from birth, if they were, that is, "natural" as some claims require us to believe.[10] But such is not the case. Hence, Erastus concludes that the so-called magical efficacy of words is no more than illusion: at its root lies the unacknowledged, and dangerous, operation of impious demons.[11]

9. Erastus (1572:169–70): "Multa passim dicit Paracelsus, at nihil probat. Nos igitur omissis nugis, rem ipsam tractabimus. Verba partim sunt naturalia, partim artificialia, sicut Imagines. Equidem materia verborum, vt imaginum, naturalis est, nempe vox: cuius materia est sonus. Forma autem est articulatio certa: quae ab instrumentis animali insitis haud aliter inspiritum expectore & pulmonibus ascendentem imprimitur, quam in exteriorem materiam externis organis insculpitur simulachrum aliquod animo conceptum. Et quemadmodum Imago repraesentat rem aliquam, sic & verba (si modo non sint barbara & ignota) animi sensa significant. Significant autem non natura, sed ex pacto conventoque hominum."

10. Erastus (1572:170): "Adde, quod surdi scirent loqui, tametsi neminem audiussent loquentem, si nobiscum nasceretur sermo, & significatio eius esset naturalis. Ex quibus apertissime videmus, verba esse non a coelo, nec Natura significare, sed esse artificialia, & ex sola constitutione & consensu certorum hominum notas & signa cogitationum nostrarum facta esse."

11. Erastus (1572: e.g., 171): "Nam barbara & Aegyptia, quibus Magi olim quoque vtebantur, & quae nec Graeci nec Aegyptij tum quoque intelligebant, non sunt notae nobis, sed Daemonibus."

Thus, we can see in Erastus a telling example of how modern categories, such as the alleged binary opposition between "natural" and "conventional" theories of language and its alignment with the opposing categories of "magic" and (later) "science," might have arisen out of a highly specific and charged rhetorical context. In order to banish a way of engaging the world that Erastus regards as anti-Christian, distasteful, and dangerous, and in order to bolster an alternative view in which all sorts of "intermediaries"—spiritual, ecclesiastical, and "vitalistic"—are consequently denigrated in favor of a new narrative of an all-powerful God and His directly subordinate creation, Erastus simplifies the very terms of debate over such critical issues in the sixteenth century as magic and the sources of interpretive and intellectual authority. The very evidence he cites—linguistic difference and diversity—was subject, as we have seen, to different interpretations, including interpretations that claimed to recuperate such diversity to a originative unity: one still thought to be present *in potentia*.[12] The entire scheme of Babel and its recuperation in a golden age of renaissance and rebirth is missing, lost, banished from Erastus' account (see Chapter 3, above). With this banished narrative, an alternative to grounding the claims of magic—even of nonmagical exegetical hermeneutics—in a theory of a *natural* language is lost as well. The alternative of, say, a historical, originative, Adamic language to which the conventional languages of man are linked never rises to the surface of Erastus' text. Such telescoping of critical debate and vision was perhaps a significant move opening horizons for other possibilities for reading the Book of Nature in the age of Mersenne and Descartes.[13]

Less strident, and more nuanced in its confrontation with the claims of magic, was the massive examination of all kinds of magical theories and practices by Jesuit Martin Del Rio. While opposed, like Weyer and Erastus, to magic and specifically to the claim that words can have inherent occult powers, Del Rio places great importance upon the existence of a supernatural order pervading our ordinary world. He, in fact, carefully delineates for his reader the characteristics and boundaries of three orders: a natural order (*ordinem naturae*), a supernatural

12. See Chapters 3 and 6 above. For themes of unity and diversity as applied to language and nature in the sixteenth century, see Chapter 6 and Dubois (1970); Céard (1980).

13. See Lloyd (1990) for a detailed and convincing account of how binary oppositions between "magic" (or "myth") and "logos" arose out of specific rhetorical, professional, and social contexts in ancient Greece. Lloyd's example can be seen as a parallel to the rhetorical elision of subtle Renaissance understandings of language and their collapse into the binary opposites "natural" versus "conventional" in the early modern era.

order (*ordinem supernaturalem*), and a human realm (*ordinem homo*) dealing with things produced by human art and industry (*rerum artifi-
cialium*) (1599–1600:51–53). Words, he asserts, have neither natural,
nor supernatural, nor "artificial" powers.[14]

As a Jesuit and Catholic, Del Rio, as Walker again points out, must
delicately balance his opposition to what he sees as the unorthodox
grounds of magical claims against the living, and indeed even quotidian, power of the supernatural, without allowing the charge of magic
and superstition to stick against the practices of the Church.[15] Whatever the configuration of metaphoric and discursive alternatives Del
Rio had to negotiate, however, the effect it had was to provoke in him a
reexamination of the cultural premises of not only magic, but the entire
exegetical hermeneutics adopted by so many natural philosophers,
physicians, natural historians, as well as occultists of the sixteenth century. Hence, Del Rio's rhetorical position as opponent of magic and yet
defender of Church practices had the consequence of opening, rather
than closing, his discourse to the cultural narratives informing his own
and his opponents' views.

What Del Rio accomplishes almost in passing is no less than a complete and radical reversal of the narrative premises of many late Renaissance hermeneutic practices. The crucial narrative link, of course,
had to do with the connection between the *verbum Dei* and the language of Adam. As we have seen, the theoretical possibilities were
many, but the overwhelming consensus, even of those who deny humans postlapsarian or post-Babylonic access to the divine Word, was
to consider Adam's access to the divine, creative, and omniscient
Word—at least as pertains to the earthly creatures and earthly de-
lights of the Garden of Eden—as complete and immediate. However
narrativized—whether as a direct divine imposition, as an essential
and innate presence in man as *imago Dei*, or as the consequence of the
first man's intimate intercourse with God through which he learned
to speak the divine tongue (humankind's first Berlitz course)—Adam
spoke the very Word of God itself and through that singular compe-
tency enjoyed perfect and essential knowledge of all creatures (see
Chapter 3, above).

With several strokes of his pen, Del Rio challenged all that. For, in his

14. Del Rio (1599–1600:1.53): "Verba nullam vim neque artificialem, neque naturalem, neque supernaturalem habent ad huiusmodi effectus."
15. See Walker (1958:178–85) for discussion of Del Rio and various arguments and
implications associated with this balancing act.

recounting of Adam's story, it was Adam alone, and not God through Adam, who gave creatures their names.[16] Unlike the claims of the Platonists (and here we must remember Ficino's adaptation of the Neoplatonic readings of Plato's *Cratylus* and its extraordinary importance for many subsequent thinkers), Del Rio insists that human imposition of words upon things is arbitrary.[17] Indeed, and this I think is where he leaves us in no doubt of the radical nature of his break from previous dominant narratives, Del Rio insists that *only* God knows the proper names of things, no man, not even the first man, Adam. Nor can this originary and divine understanding of the proper and creative language of God's universe come to be known to man through ordinary names in mere human languages or through the alleged insights of magic.[18]

Whatever else it does, Del Rio's critique of Renaissance magical traditions drives a wedge between the languages of man and the Word of God that led to their utter divorce. Yet, and here I think one can see a certain attraction for Father Mersenne, Del Rio's account still retains the central notion of God's creative Word, of His unique and singular knowledge of the true names of things and the knowledge of their essences it implied. Humans can have no claim to access to the hidden mysteries of that Word, but God remains above all an author, one whose Word we can never, it is true, penetrate, but whose plot and action are there to instruct and challenge us. Del Rio, in short, provides both an antidote to the damnable excesses and much-feared claims of the secret brotherhood of early-seventeenth century-occultists, and yet the prospect for rewriting man's role in the uncovering of God's universal plan through the de-in-scription of His Book of Nature.

16. Del Rio (1599–1600:1.46): "Hoc totum est mendacium confictum ad stabiliendam hanc superstitionem. Nomina quidem Deus in scripturis imposuit sanctis quibusdam, mysterij significandi causa, & officio corum, quo isti functuri, conuenientia: sed ipsi nomini Deus vim effectiuam nullam indidit. Animantibus vero cunctis Adam nomina indidit, non Deus."

17. Del Rio (1599–1600:1.56:"Quicquid enim Platonici dicant, nomina sunt hominum arbitratu indita. Nomina quoque nec sunt substantie corporeae aut spirituales, nec sunt accidentia inhaerentia subiecto; quare vitam in se habere non possunt adueniunt quoque seu induntur rebus, quas significant atque denominant, extrinsecus plane: quare necemanantes a rebus radij dicendi."

18. Del Rio (1599–1600:1.56): "*Respondeo* Deum, quidem nosse cunctarum rerum, singulorum etiam angelorum, & stellarum (de quibus eo loco agit propheta) naturas proprias, & sic ea cum lubet proprijs nominibus, hoc est conuenientibus earum naturae, vocare posse: sed ea nomina mortalibus incognita sunt, nec magis reuelata."

Mersenne and Descartes

In 1623, Marin Mersenne, who was to become a celebrated advocate of mathematics and the "new sciences" of the seventeenth century and intellectual mentor to and correspondent of such luminaries as Descartes and Gassendi, published his first major work, significantly entitled *Quaestiones celeberrimae in Genesim*. Mersenne took the occasion of a commentary upon the Book of Genesis to launch a massive attack on Renaissance traditions of magic and the Cabala, singling out the famous Franciscan from Venice Francesco Giorgi (also known as Giorgio, Georgio, or Zorzi), whose 1525 work on the harmony of the world was one of the mainstays of the occult tradition in the sixteenth century. While Mersenne explicitly devoted the last section of his book to a refutation of Giorgi's *In scripturam sacram et philosophos tria millia problemata*, just published in a new edition at Paris in 1622, his Genesis commentary reacted to the threat posed by this recent work by attacking other renowned occultists as well, such as the contemporary Englishman Robert Fludd (Yates 1964; Huffman 1988). More than that, Mersenne sought no less than to undermine the foundations and theoretical premises of magic (Lenoble 1943; Hine 1976, 1984, 1990).

Mersenne's objections to the traditions of Renaissance magic have been surveyed elsewhere, notably by Robert Lenoble (1943) in his fundamental study of Mersenne; Frances Yates (1964) concludes her famous book on Bruno and the hermetic tradition with a reconsideration of the Fludd controversies, in which Mersenne figured prominently. Rather than covering this same ground again, let me simply note that Mersenne was, I believe, rejecting not simply a complex of occult beliefs that he found impious, objectionable, and muddleheaded; he was much more fundamentally rejecting and recasting the entire vision of history and human culture that served to underwrite and legitimize them. This becomes clear when we see that Mersenne, like Del Rio, strikes at the fundamental narrative link in the claims of occultists, like Giorgi and Fludd, who adopt a cabalistic vision of God, man, and the universe: the link, that is, between the sacred "Word of God" that shaped the world and its creatures at the beginning of time and the "languages of man" stemming from our primogenitor, Adam.

Like the Jesuit opponent of magic Del Rio before him, Mersenne—a product of the Jesuit college at La Flèche—had reason to be concerned about the biblical story of the language of Adam. The very exegetical and symbolic hermeneutics embodied in the practices of cabalist and

Renaissance magicians presupposed a special Adamic relationship to such a sacred and power-giving language. Paracelsus' depiction of Adam as the consummate practitioner of the *kunst signata*, for example, took nature on the basis of the biblical authority of Genesis as a vast symbolic text bearing the direct imprint of God's generative Word (see Chapter 5, above). For Paracelsus and other occultists or Neoplatonists of the Renaissance from Ficino to Fludd, the idea and trope of the "Book"—of God's "Book of Nature"—authorized a kind of "bookish" culture (however differently and even contradictorily the term "book" might be interpreted) that Mersenne's work undermined, and that Descartes would altogether reject.

Whether one turned to the books of ancient authors as did Fernel, to the "Book of Nature" alone as did Paracelsus, or to the authority of both as did the Renaissance naturalists Gesner and Aldrovandi, the special status of exegetical practices and the special claims to a deep, often "hidden," and essential knowledge of things in themselves rested ultimately on the primal narrative revelation of an originary Adamic language and its continued accessibility to postlapsarian man. In turning to this story and its implications in his *Quaestiones in Genesim*, and later in his *La vérité des sciences* (1625), Mersenne was thus responding to a legitimate and widely known concern, a concern that might well have been deepened, if not crystallized, by events that erupted in Paris in 1623.

For 1623 was, of course, the year of the so-called "Rosicrucian Scare" in France (Yates 1978:8, 103–17). Whether an elaborate fabrication or not, the prospect of an imminent influx of a secret Rosicrucian brotherhood took hold in parts of Europe in the decade following the publication of the "manifestos" of this esoteric "movement." For some, this prospect brought hope of change, even of a revolutionary millenarian renewal; for others, if we are to believe contemporary accounts, it brought fear and anxiety to a Europe already raised to a feverish pitch of suspicion by the events surrounding and precipitated by the outbreak of the Thirty Years' War (Yates 1978; Wedgwood 1969). The specifics of these events do not concern us now.

What does concern us, however, is what, in the reports of Mersenne's own contemporaries in France, observers believed had come to Paris in 1623. Let me reproduce here two accounts, cited by Yates in her book, of the "placards" allegedly posted in Paris that "announc[ed] the presence in the town of the Brethren of the Rose Cross" (1978:103). The first is from the noted librarian and opponent of magic

Gabriel Naudé; the second, from an anonymous text. Both were published in 1623.

> We, being deputies of the principle College of the Brothers of the Rose
> Cross, are making a visible and invisible stay in this city through the Grace
> of the Most High, towards whom turn the hearts of the Just. *We show and
> teach without books or marks how to speak all languages* of the countries where
> we wish to be, *and to draw men from error and death.*[19]

> We deputies of the College of the Rose Cross, give notice to all those who
> wish to enter our Society and Congregation, that *we will teach them the most
> perfect knowledge of the Most High,* in the name of whom we are today hold-
> ing an assembly, *we will make them from visible, invisible, and from invisible,
> visible.* . . . (Yates 1978:104; emphasis added)

Yates recounts in her book how these "placards" were read in the
context of the times, and claims that they were used to stir up a
"witch-craze," with the Rosicrucians "turned into an organization of
devil worshippers" (1978:104ff.). She further suggests, based chiefly
on his correspondence, that Mersenne himself "believed in the R. C.
Brothers as bogeymen, wicked magicians and subversive agents, whom
he imagined moving invisibly in all countries to spread their evil
doctrines."[20]

Mersenne certainly regarded the claims of magicians as tainted by
demons and even the worship of devils. But we need not believe in
either bogeymen or subversive agents lying all about to imagine
Mersenne reacting sharply to reports of the Rosicrucian placards.
Given the evidence of his own texts, Mersenne would see error and
danger in their purported boasts. Their very claims conjure up a distorted version of apostolic miracle working: the gift of tongues and the
promise of an otherworldly perfection and immortality; the claims to
perfect divine wisdom from those with special access to God and His
secret knowledge of the interpenetrating presence of the invisible in
this our mortal realm. The Pentecostal promises of this secret brotherhood themselves suggest that they hold the key to release from the
curse of Babel. But their claims go beyond those of earlier French luminaries, like Rondelet or Paré, who envisioned in the study and mastery
of languages a kind of universality that held out the secular prospect of
repairing the social and political effects of the Babylonic confusion of

19. Naudé (1623:27; emphasis added) as quoted and translated in Yates (1978:103).
On Naudé, see J. V. N. Rice (1939) and Clarke (1970).
20. Yates (1978:112). Yates specifically cites Mersenne (1932–70:1.37–39, 154, 455;
2.137, 149, 181ff., 496) and Mersenne (1625:566–67).

tongues (see Chapter 6, above; Céard 1980). Coupled as they were with claims of possessing secret and perfect knowledge of a kind that *surpasses* human experience and embraces the divine, the placards likely as not brought Mersenne face-to-face with the very sources—hubristic ones at that—of *magia* and Cabala. That is, they would have confronted him with claims of human access to the primitive, Adamic language and thus to the sort of transcendent archetype of all merely corrupt and fallen human languages that not only granted perfect knowledge of the very natures of all things and the almost Paracelsian ability to grasp the invisible in the visible, and consequently life and death itself, but also enabled one to learn all languages "without books or marks" by grasping wholly how they are but partial, veiled, and distorted reflections of the divine language stamped in nature itself.

In 1623, and again in 1625, Mersenne strikes out against the cultural premises that inform such hubristic claims and that threaten the very fabric of knowledge and society with false hopes, illusions, and delusions of power. In his *Quaestiones in Genesim,* Mersenne allows no more than the possibility that God revealed to Adam the names of things that conform to their natures without granting Adam knowledge of their real essences themselves (Mersenne 1623: cols. 1197–1200; Lenoble 1943:514). As Mersenne puts it in his *La vérité des sciences:*

> On pourroit aussi former autant de dictions diuerses comme il y a de diuers indiuidus au monde, mais on ne peut en inuenter, qui signifient la nature, & l'essence des choses, d'autant que nous ne la cognoissons pas, il n'y à que Dieu qui le puisse faire, ou qui le puisse commander aux Anges: peut estre que les noms qu'Adam imposa, auoient ce priuilege: mais depuis ce temps là les noms se sont tellement éloignez de leur premiere origine, que nous n'en recognoissons plus aucune vestige. (1625:71)

The Adamic language, according to Mersenne, never was the transparent window through which God lay bare for Adam the innermost nature and essences of things. If, in fact, God did reveal to man the "true names" of things (and this is a point on which Mersenne is cautiously skeptical), such names afforded Adam no real, divinelike, insight into their natures. This sort of insight, and the power it carried, belonged solely to God and to the angels acting for Him.

As Robert Lenoble points out, Mersenne goes even further than the cautious, tentative views expressed in his two published works of 1623 and 1625. The full manuscript of the *Quaestiones in Genesim* includes entire sections of Mersenne's commentary that never were published, probably because of their excessive length. At any rate,

Mersenne goes so far as to assert that there is no natural language, certainly no language that man has, or can know, "innately" or by nature ("propterea nullus est sermo innatus hominibus, alioqui et hunc et illum quem a parentibus aut praeceptoribus ediscunt, habebunt"). Moreover, striking at the very "historical" foundations of Renaissance cultural narratives informing various extant strategies for interpreting nature and texts, Mersenne denies the sacred status of Hebrew as constituting a special relationship to a supposed Adamic or originary divine language ("Unde falsum esse constat, quod aliqui sine ratione fingunt de Lingua hebraïca, quam ita hominibus innatum existimant, et hac lingua puer sit locuturus, qui nullam aliam didicerit").[21] Mersenne elaborates on this point, while still further weakening traditional readings of the story of Adam's imposition of names upon the creatures in paradise:

> Les noms qu'Adam a imposé aux animaux sont aussi indifferens de leur nature à signifier les pierres, ou les arbes, que les animaux, comme l'on avoüera si l'on examine iudicieusement les vocables Hebreux ou Chaldeans, que l'on tient avoir esté prononcez par Adam, puis que les lettres, les syllabes, et leur prononciation sont indifferentes, et ne signifient autre chose que ce que nous voulons. . . . (1636:65; Lenoble 1943:516).

In sum, then, there is no primitive, pure, and natural language. Whatever language Adam spoke in the Garden of Eden afforded no special insight into the natures of things. In any case, the Adamic language was irretrievably lost through the march of time and the effects wrought by change, so that the languages of man known to literate men of Mersenne's age bear absolutely no trace of the "mother" tongue. Indeed, for Mersenne, language and words themselves come to bear but a pragmatic utilitarian function related to, but separate from, human thought. That is to say, words are merely signifiers that man attached to his thoughts for the purpose of communication within the context of organized social behavior. The thoughts that words come to signify are prior to and unaffected by the names attached to them: "Au reste les noms ne nous seruent que pour entendre & signifier ce que nous voulons dire, & ce que nous auons dans l'esprit, car si vn homme viuoit tout seul, il n'auroit que faire d'aucun

---

21. Lenoble (1943:514–15). Lenoble cites the *Suite manuscrite des Quaestiones in Genesim*, Paris, BN, f. lat. 17262, p. 511. I have not seen this manuscript. My discussion in this paragraph is evidently dependent upon Lenoble's evidence and comments. As Lenoble (1943:xii–xiv) notes, the published version of the *Quaestiones in Genesim* represents only a portion of the text to the end of chapter 6 of Genesis.

nom, ce seroit assez qu'il eut dans son esprit ce qu'il penseroit, ou ce qu'il voudroit faire" (Mersenne 1625:69). Language is marked, then, not by traces of some lost, ideal coincidence of words and things, but rather by social usage and the concrete effects and limitations of human physiology.

The latter point echoes Erastus' discussion of the close connection between language and sound-producing organs of the human body. Mersenne notes such physiological determinations in various places, but also proposes theories about the diversity of human languages that trace them back to their material conditions as well. He invokes, for example, the entire apparatus of medical temperaments common to Galenic medicine of his day joined to a geographical anthropology of ethnic diversity that is at least as venerable as Hippocratic writings such as *Airs, Waters, Places:*

> Nous voyons neantmoins que les peuples inuentent diuerses langues à cause de leurs diuers temperamens, car ceus qui tirent vers le Nort, meslent plus grand nombre de consonantes, d'autant que le froid rétreint leurs esprits au dedans, & leur faites faire plusieurs repercussions: au contraire ceus du côté du midy, comme les Hespagnols, & les Africains, ont les voyelles plus frequentes, & les dictions plus longues à raison de la chaleur, qui relasche l'autere vocale. Les Venitiens, ont beaucoup de liquides, & de voyelles. . . . (1625:71–72)

For Mersenne, this material explanation of linguistic diversity fits right into his larger retelling of the biblical story of man and his languages. For the geographical dispersion of mankind and the consequent subjection of individual peoples to a variety of climatic conditions only reinforced what God had started at Babel. The net effect of climate and (Babylonic) confusion was diversity. If humans knew but one language, the benefit to the sciences of easy communication would be great, but, as it is, Babel has left humans with "une grande perte des sciences."[22] That is not to say that Mersenne reverts to a nostalgia for a lost linguistic unity of mankind, for as Lenoble sums up his views, "it is not the unity of language, but rather its diversity, which is natural"(1943:516). Nonetheless, we may well detect here in Mersenne a desire for some future state in which man's imperfect grasp of the sci-

---

22. Mersenne (1625:72): "Voyla d'ou sont venües en partie les diuerses langues, ce qui a commencé à la confusion de la tour de Babel auec vne grande perte des sciences, car s'il n'y auoit q'vne langue au monde, on s'entrecommuniqueroit plus facilement les sciences, & on emploiroit tout le temps à les apprendre, qu'on passe à étudier aus langues étrangeres."

ences is made to flourish through the improved communication of scientific knowledge.[23]

Mersenne's views on language, his thoroughgoing recasting of cultural narratives concerning the origins and progress of the languages of man, serve as foundations for his new vision of humans and the sciences. The dream of an originary linguistic essentialism, a dream shared in various forms and to different degrees by those who adopted some variant of the Neoplatonic theory of language in the Renaissance and who saw this myth in the guise of a Christianized quest for the Adamic language, a dream in which there is a perfect, if not presently realized, coincidence between words and things-in-themselves, gives way in Mersenne to a sober search, not for essences, but rather for the structure and dynamics of the world inscribed in God's Book of Nature. Thus, Mersenne sees in language a reflection of the world that parallels his quest, in natural philosophy, for provisional knowledge of the structures and relationships to be found in the world. In both linguistics and the sciences, that is, Mersenne gives up the quest for knowledge of essences as vain and beyond man's capacity as creature of God, in favor of a more modest, but nonetheless powerful, kind of understanding.

As a model for such scientific understanding, language suggests a new kind of hermeneutic strategy for reading the divine text of nature. For Mersenne is adamant that the "Art of Grammar" is not a trivial and inconsequential pursuit, despite the demise of the Adamic quest for essences. Rather, that art "is founded not on air, but on *the nature of things.*"[24] Here, Mersenne does not invoke "nature" as essence, but instead points to other, nonessentialist, modes whereby one may "map" language onto the world:

> Car bien que nos dictions, qu'elle enseigne, ne representent pas l'essence des choses, neantmoins elles nous font ressouuenir de ce que nous auons penetré dans chaque chose par la force de l'esprit: & au lieu que nous ne pouuons fonder les noms sur l'essence, nous les fondons sur les effects sur l'action, sur la resemblance, & sur les autres accidens. . . . (1625:72)

Language provides us, then, with a tableau of the world in its everchanging configuration of relationships. The very grammar of our

23. For such themes as communication, universal languages, and the new science in Mersenne and other seventeenth-century figures, see Aarsleff (1982); Clauss (1982); Cohen (1977); Cornelius (1965); Dear (1988); Formigari (1970); Knowlson (1975); Knox (1990); Robinet (1978); Rossi (1960); Salmon (1979, 1972); Shapiro (1983); and Slaughter (1982). I shall discuss some of these themes in Volume 2.

24. Mersenne (1625:72): "Or de tout ce que dessus vous pouuez conclurre que l'Art de Grammaire n'est pas fondé dans l'air, mais dans la nature des choses. . . ."

chosen tongues constitutes a map within which we can situate the inter-connections, resemblances, differences, movements, actions, conjunc-tions, and separations *among things* in the world. Language captures something real about the world of nature, even if language itself—or, rather, the words framed by man—are thoroughly conventional, the product of history and social use. This reality is not located *in* names, as a reflection of an essence, but can be found instead in the structure of language, in its grammar, and in what grammar and grammatical rela-tions reveal about the world *as a system or network of things*.

Language, and Mersenne's "Art of Grammar," suggest to the phi-losopher who turns his gaze upon nature a new task, one radically different from the exegetical or symbolic hermeneutics of the old, "bookish" culture of textual commentators, occultists, and those en-raptured by the "emblematic world view." For Mersenne's natural phi-losopher, the task is to de-in-scribe the language of things in nature by attending to the grammar, not the semantic content, of that language inscribed by God in His Book of Nature. God has written a text whose structure, relations, and operations bespeak a coherent order. It is this divinely created order—not the essences of things themselves which only God can know—that man can explore and deinscribe. This fun-damentally relational view of nature locates *scientia*—knowledge—not in the grasping of the inherent essences of individual things (as, for example, in "Aristotelianism"), nor in the correspondences and resemblances linking individual essences together into an overarch-ing symbolic, but real, network, but rather, in the system of relations, operations, and actions that constitute the phenomenal world of things. The latter are not, in themselves, knowable as essences—that is, apart from the web of relations in which they are enmeshed. By contrast to the exegetical/symbolic approaches to nature I have de-scribed, this latter view of the language of things seeks knowledge of the "grammatical" or "syntactic" dimensions of the language in which God has inscribed the Book of Nature.[25]

Mersenne's *La vérité des sciences*, in which we have found evidence of

25. I am indebted to Dear (1988:175–76), who puts the point so well: "orthodox ac-counts of the nature of grammar suited Mersenne because they conformed perfectly to his position on what could and could not be known about nature. Man could know the actions, appearances, and phenomenal interactions of bodies, but access to their true essences belonged to God alone. . . . The structure and operations of languages cap-tured the structure and operations of the visible world, but the lexical content of language—words—failed to capture the true contents of the world: that is, the essences instantiated within it."

his profound rewriting of Renaissance cultural narratives and advocacy of a "conventionalist" view of language that nonetheless recuperates the idea of language as a mirror of the Book of Nature, is subtitled
contre les sceptiques ou Pyrrhoniens. Where Mersenne applied skeptical
tools to the "credulity" and "excesses" of the occultists, he also challenged what he saw as the excesses of the Pyrrhonian skeptics. Knowledge was not, according to Mersenne, unattainable; it was not a vain,
prideful dream of weak, fallen man. What proves unattainable is,
rather, only a certain kind of knowledge: an ideal of knowledge that
rests on false premises, on the possibility of obtaining a godlike knowledge of the very essences of things. Mersenne's critique of the cultural
myth that lay behind this overreaching search for essentialist knowledge was central to his critique of magic and of skepticism and is central
to his own project of advocating the "truth of the sciences," *not* as *dogmatic* knowledge, but as reliable, if provisional, knowledge of the relations and operations discoverable in nature. Mersenne therefore caps
his attack on the skeptics by devoting "the last 800 pages of *La vérité des
sciences*" to a catalogue of "the vast body of mathematical and physical
information that is known."[26] Mersenne's deinscriptive hermeneutics
of nature leads then to an effort to describe the very order of God's
nature—its proportions, relations, operations, and the like—in a "language" that can reveal the fundamental relationships mapped upon
nature. This sort of knowledge, Mersenne tells us as part of his story of
Adam and postlapsarian man, is licit, is accessible to man, is empowering, and finally is *possible!* As Popkin says of Mersenne's viewpoint,
"The sciences (considered as the study of phenomenal relationships),
and mathematics (considered as the study of hypothetical relationships), have given us a kind of knowledge that is not really in doubt,
except by madmen" (1968:139).

Mersenne clears a path for Descartes. For Descartes, too, views the
Book of Nature not as a symbolic divine text, but, like Mersenne, as a
text whose intrinsic language bespeaks a grammar of divine order: specifically, of mathematical proportion and relation. Indeed, the key to
unlocking this text for Descartes is "mathesis, ou mathématique universelle." It was precisely this conception of a *universal* and certain
knowledge, however, that marks an important difference between Descartes and Mersenne. While Mersenne clearly advocates a mathematical and mechanical approach to nature, and regards nature as divinely

26. Popkin (1964:139). For skepticism in general, and Mersenne's particular place in
its history, see Popkin's book.

instituted and therefore orderly, he nonetheless insists upon the limits of human cognition. While humans aim to acquire complete knowledge of the relations, operations, and order of nature, human achievements in acquiring such knowledge must always remain provisional. Nature, as God's created order, is rich, diverse, and complex; ultimately, as a result, Mersenne comes to doubt the adequacy of any human explanatory scheme that claims universality. His quest for a mathematical science of nature is, then, tempered by a theological conviction in the contingency of things—a conviction that fostered in Mersenne a respect for the experimental as supplement to the mathematical.

Descartes's turn toward mathematics as a key to unlocking the secrets of nature, while intersecting with Mersenne's inclinations, seems from the start to have been part of a larger and more visionary quest. Having, like Mersenne, attended the Jesuit college of La Flèche, Descartes as a young man left France to begin his travels in Germany. On the now famous night of November 10, 1619, Descartes, according to his biographer Baillet, experienced three successive dreams whose significance haunted him and inspired his lifelong quest for truth. The third dream, in which two books—a "Dictionary" and a "Collection of Poems"—figure prominently, witnesses Descartes struggling from within the logocentric confines of his bookish culture to find his way toward truth. For him, the dictionary—that repository of words!—still symbolizes "toutes les Sciences ramassées ensemble," and the collection of poetry "la Philosophie & la Sagesse jointes ensemble."[27] Together, these images suggest to Descartes the importance of "Revelation and Enthusiasm" and the idea that it was the very "Spirit of Truth that wished to open up to him the treasures of all the sciences through this dream" (Baillet 1691:84).

The young Descartes devoted himself to this quest for truth in the treasure trove of the sciences. Interestingly, however, his biographer next turns from Descartes's famous dream to his infamous search for the secret brotherhood of the Rosy Cross ("*Frères de la Rose-croix*"). Word of this mysterious brotherhood had reached Descartes in Germany and with it came understanding that these invisible figures knew all there was to know and that they could reveal to others a new wisdom, a science as yet undiscovered. ("On luy fit entendre que c'étoient des gens qui sçavoient tout, & qu'ils promettoient aux hommes une nouvelle sagesse, c'est-à-dire, la véritable science qui n'avoit pas en-

---

27. Baillet (1691:83–84). For the dreams as a whole, see Baillet (1691:81–86). The source of the epigraph at the start of the chapter is Baillet (1691:114–15).

core été découverte.")[28] While this search for the Rosicrucians proved futile, Descartes nonetheless came to realize that it was in the application of mathematics and its "rules" to nature that his best hope for unlocking the secrets of the latter lay. Clearly, what Descartes sought was a *universal* science and *sagesse* akin in its range, if not in its occult associations, to what the stories about the followers of the Rosy Cross claimed some young iconoclasts already possessed (Baillet 1691:91).

When Descartes, in 1623, returned to Paris armed with his various revelations, he was, ironically, confronted by the fresh excitement and fear caused by the dramatic rumors of the invisible arrival of the very same Rosicrucian Brotherhood. This time, however, Descartes was alarmed to learn that his own reputation had been tarnished by claims of his association with them while still in Germany. He therefore made every effort, Baillet informs us, to counter such an image by forgoing his reclusive ways and demonstrating, by his active presence in polite society, that he was not one of the "invisible" ones. Mersenne, we learn, who was Descartes's "intimate friend," was much relieved by his behavior. Having just written his *Quaestiones in Genesim*, which took the Rosicrucian sympathizer Robert Fludd as one of his main opponents, Mersenne was convinced that the invisible brotherhood was no mere illusion (Baillet 1691:106–8; see also Yates 1978:114–17).

Descartes's friendship with Mersenne was no accident; both shared, as we have seen, intense interest in the study of nature and in mathematics as one of the keys to such study. Moreover, Descartes's reputation as mathematician and student of nature grew throughout the 1620s. But, close as their friendship, intellectual interests, and even religious ideals were, Descartes and Mersenne did not quite see eye to eye. An episode, sometime in late 1628, is revealing.

Along with Mersenne and others such as the famous Cardinal Bérulle, Descartes attended a meeting of learned men at the home of the papal nuncio, where a certain physician and chemist by the name of Chandoux presented to the assembled luminaries his own rather novel philosophical ideas. Chandoux, we are told, had not less aversion to Aristotelians and Scholastics than did Bacon, Mersenne, Gassendi, or Hobbes.[29] His lecture met with great success, and he received "nearly universal" approval (Baillet 1691:161). However, Cardinal Bérulle, observing that Descartes had not exhibited, like the others, signs of satis-

28. Baillet (1691:87). For Descartes's search for the Rosicrucians, see Baillet (1691:87–91).

29. Baillet (1691:160). The entire story is recounted in Baillet (1691:160–66). Popkin (1964:177–79) discusses this incident.

faction with Chandoux's discourse, pressed him to give his opinion. Having tried unsuccessfully to excuse himself from complying with this request, Descartes was obliged to indulge the curiosity of his host and compatriots lest he risk insulting them (1691:161–62).

Acknowledging Chandoux's eloquent criticisms of Scholastics, Descartes nonetheless confessed his displeasure with philosophical arguments that purport to rest content with merely probable knowledge and, indeed, with the idea of accepting probability in the place of truth ("la force de la vray-semblance qui occupe la place de la Vérité") as a basis for philosophy and scientific knowledge (Baillet 1691:162). Descartes, in what reads as a veritable tour-de-force in Baillet's account, then went on to demonstrate how such a lax standard could easily result in transforming falsehoods into "truths," and vice versa (1691:162–63)! When asked whether there was not some means for avoiding such unacceptable results, Descartes replied

> qu'il n'en connoissoit point de plus infaillible que celuy dont il avoit coutume de se servir, ajoutant qu'il l'avoit tiré du fonds des Mathématiques, & qu'il ne croioit pas qu'il y eut de vérité qu'il ne put démontrer clairement avec ce moien suivant ses propres principes. Ce moien n'étoit autre que sa regle universelle, qu'il appelloit autrement sa Méthode naturelle, sur laquelle il mettoit à l'épreuve toutes sortes de propositions de quelque nature & de quelque espèce qu'elles pussent etre. (1691:163)

Descartes, indeed, went so far as to say that "he did not believe that it was impossible to establish in Philosophy very clear and certain principles through which it would be very easy to give explanations for all the effects of Nature."[30]

Descartes clearly regarded mathematics, and the "méthode naturelle" that could be derived from its principles, as the basis for establishing true and certain knowledge, not merely provisional or probable knowledge. From this moment in 1628 through his mature philosophical works of later decades, Descartes pursued the goal of a *universal* system of knowledge that Mersenne evidently did not share.

Nonetheless, Descartes did share with Mersenne a conception of philosophy and the sciences that abandoned dogmatic claims to knowledge of the innermost essences of individual things. Both rejected a Renaissance Adamic ideal of man seen to be in concert with a nature

---

30. Baillet (1691:164): "il ne croyoit pas qu'il fut impossible d'etablir dans la Philosophie des principes plus clairs & plus certains, par lesquels il seroit plus aisé de rendre raison de tous les effets de la Nature."

that mirrored the divine *mens*. Mersenne's·new narrative of Adam shattered such beliefs in man's occult, microcosmic link with a system of nature and creatures that bore—through a congeries of sympathies and resemblances—deep affinities to man as *imago Dei*. That narrative, in which postlapsarian man was denied any access to essentialist knowledge of things and words that only God Himself enjoyed, envisioned a quite different relationship of God to nature. Nature was the contingent creation of an all-powerful and wise God. As such, nature was no longer thought to be a mirror of God's essence, a text in which the living, creative Word could be read. Instead, the text of nature was a book authored by the divine in a language of things that constituted a world in and of itself, one dependent upon, but radically divergent from, God. In that text, things were defined not as individual essences (these could be known only to the Author Himself), but rather in their configuration, proportion, operations, and activities *in relation to* each other and to the larger systems of which they are collectively a part. For Mersenne and, still more, Descartes, the key to interpreting God's Book of Nature was mathematics itself.

Man, as reader of that text, is for Mersenne fallible: his point of view, and therefore his interpretive stance, is always partial. He lacks a God's eye view; his knowledge, while often exact and precise if using the methods and tools of the sciences, is not certain, and certainly not universal. Descartes envisions man as an ideal reader, as pure thought (the *cogito*), without a limiting *position*, a limiting sight/site.[31] While he does not and cannot know individual essences, he can aspire to a complete, universal, systematic knowledge of things *in* nature: of the very relations and interconnections that constitute the configuration of nature-in-action within the dynamic matrix of *res extensa*.

The affinities and contrasts that mark both Mersenne and Descartes as collaborators in a new deinscriptive hermeneutics of nature can be seen, in germ, in their respective reactions to the proposal of a universal language, concerning which Mersenne sought the opinion of his young friend in 1629. While we have only Descartes's letter in response to Mersenne's report of a new scheme for an artificial language, the general idea of a universal language seems to have held some appeal for Mersenne. As we have seen, Mersenne was, after all, always interested in the benefits that could accrue to science and to mankind

---

31. For a critique of such *unsituated* myths of knowing, see Haraway (1988, 1991). For implications of the *local* nature of all knowing for cultural study of science, see Rouse (1993); Bono (1995).

through the ready communication of scientific knowledge and techniques without the barriers imposed by lack of a common scientific language. But, while Mersenne's mathematical inclinations attracted him to such a systematic possibility, his ingrained sense of the richness and contingency of nature made him come to doubt that a universal language scheme—even assuming that a practicable scheme could be developed—could adequately represent the scheme of nature. As Lenoble concludes, Mersenne's initial "enthusiasm" was "little by little followed by the prudent doubt of the experimenter who knows that one may not do violence to nature" (1943:518). Even more would Mersenne be violently opposed to seventeenth-century proposals for a universal *philosophical* language, that is to say, to schemes that purported to create a language in which words would mirror and capture the very essence of individual things.[32] Such knowledge, as we have seen, Mersenne regards as the province only of God.[33]

In his letter of November 20, 1629, to Mersenne, Descartes quickly dispatches the proposal for a universal language sent to him. For he finds that scheme filled with "inconvenients" that effectively make it impracticable and useless (Mersenne 1932–70:2.323–8; see also Descartes 1969:1.76–82). More interesting is Descartes's own musings about an ideal for such a universal language. He envisions an "invention" for producing "characters" as a means of establishing "un ordre entre tous les pensées" that might enter into the mind of man, and draws an analogy to the order established among numbers. From this

32. Dear (1988:177–200) gives an excellent, comprehensive account of Mersenne's views of language. For the idea of universal and philosophical languages and Mersenne's reactions, see Dear (1988: esp. 185–200). I direct the reader to Dear's account in lieu of providing a comprehensive review here. I should only note that I am inclined to see Descartes not as an advocate of essentialism in the Scholastic sense, in the sense of believing that it is possible for man to know essences of individual things. The kind of true, certain, and universal knowledge Descartes thinks possible is of another sort: that of the proportions or relations obtaining among things in nature.

33. Thus, Mersenne even objects to the thoroughly empirical, deinscriptive practices of Bacon, precisely because he regards Bacon as seeking to establish (at least in some ideal future) the true natures of things: "il ne faut pas penser que nous puissions penétrer la nature des indiuidus, ni ce qui se passe interieurement dans iceus, car nos sens, sans lesquels l'entendement ne peut rien conoître, ne voyent que ce qui est exterieur; qu'on anatomise, & qu'on dissolue les corps tant qu'on voudra soit par le feu, par l'eau, ou par la force de l'esprit, iamais nous n'arriuerons à ce point que de rendre notre intellect pareil à la nature des choses, c'est pourquoy ie croy que le dessein de Verulamius est impossible, & que ces instructions ne seront causes d'autre chose que de quelques nouuelles experiences, lesquelles on pourra facilement expliquer par la Philosophie ordinaire" Mersenne (1625:212–13). Mersenne devotes the final chapter of book 1 to an examination of Bacon's thought (1625:206–24). See also Dear (1988:187–88) and Lenoble (1943:325–35).

foundation Descartes speaks very generally of forming words and a
language (Mersenne 1932–70:2.327). Ultimately, however, Descartes
claims that this language must depend upon "true philosophy":

> l'invention de cette langue depend de la vraye Philosophie; car il est impos-
> sible autrement de denombrer toutes les pensées des hommes, et de les
> mettre par ordre, ny seulement de les distinguer en sorte qu'elles soient
> claires et simples, qui est à mon advis le plus grand secret qu'on puisse
> avoir pour acquerir la bonne science. Et si quelqu'un avoit bien expliqué
> quelles sont les idées simples qui sont en l'imagination des hommes . . .
> j'oserois esperer ensuite une langue universelle fort aisée à aprendre, à
> prononcer et à écrire, et ce qui est le principal, qui aideroit au jugement,
> luy representant si distinctement toutes choses, qu'il luy seroit presque
> impossible de se tromper. (Mersenne 1932–70:2.328)

What purpose could such a language have? Has Descartes here fallen
into the trap of essentialism, of embracing a prior model of words as
essences of things? The answer to the latter question, I believe, is no.
When Descartes speaks of true philosophy, he has at least partially in
mind, I would suggest, something like what has come to be known as
his method. Descartes seeks not a perfect coincidence of word and
things; still less does he think of words as somehow capturing the essence of individual things. Rather, he seeks to establish artificially a
correspondence between words and clear and distinct ideas. That is to
say, words, for Descartes as for Mersenne, are no more than arbitrary
signs for what we have in our minds, for simple ideas. The problem is
that we are led astray in our deliberations and search for "true philosophy" by words that do not denominate clear and distinct ideas. The
problem, then, that a universal language, such as Descartes proposes,
aims to solve is precisely the habit of the mind to wander away from its
focus on clear and distinct ideas, hence producing monstrous, but all
too typical, ideas of a confused or mixed nature. Descartes's proposed
characters and words are not the condition for knowledge, for true
philosophy; they are merely aids for keeping man's mind fixed on the
elements out of which a true picture of the universe can be constructed.

If language included *only* words that denominated clear, distinct, simple ideas, then the possibilities for error would be, as nearly as can be
imagined, eliminated. As the above quotation from Descartes's letter
continues to assert, "au lieu que tout au rebours, les mots que nous
avons n'ont quasi que des significations confuses, ausquelles l'esprit
des hommes s'estant accoutumé de longue main, cela est cause qu'il
n'entend presque rien parfaitement." Confused signification is the ob-

stacle to perfect understanding in Descartes's view. Only the rigorous "mathematical" criteria and habits of thought that will crystallize as Descartes's "method" can eliminate the confused from the simple and "true" ideas man entertains in his mind. And only such a method and "true philosophy" can generate the characters and words of an ideal universal language. But once generated, such a language can work to eliminate sources of false reasoning and error: it can, indeed, turn a mere rustic into a better judge of truth than philosophers (Mersenne 1932–70:2.328).

Descartes is under no illusions. He recognizes that such a language is but a remote possibility, a product of a "paradis terrestre" that has not yet come to be, that indeed is the stuff of myth and the mythic places of "des romans" (Mersenne 1932–70:2.328). Yet the ideal expressed here is but one version of that quest that takes as its mighty steed the hermeneutics of de-in-scription. Descartes's ideal is no less than the complete, perfect, and universal description of the order God inscribed in His Book of Nature. Like Mersenne's vision, that order is a contingent one—dependent upon God's will—and His arbitrary creation, rather than the necessary replication of God's very essence in a world conceived as Platonic mirror of or Neoplatonic emanation from the One. Given that contingent, but orderly and rational, structure to the world, Descartes follows with near obsession his dream of a complete, mathematical description and delineation of the relations that undergird the world's structure. Only clear and distinct ideas can fathom such relations: they become then the "tool" and vehicle of his deinscriptive interpretation of the text of nature. But such perfect, and perfectly certain, understanding as Descartes hopes to generate from his perfect reading of that text, from his "mathesis, ou mathématique universelle," represents what his friend Mersenne declared, in the end, impossible for man. Of such dreams and their diligent inscription are made the sciences and the follies, the noble aspirations and obsessive nightmares, of the modern world.

Epilogue

In *The Art of Describing*, Svetlana Alpers claims that the Dutch and northern art of the seventeenth century "can best be understood as being an art of describing as distinguished from the narrative art of Italy," noting that while "narration has had its defenders and its explicators" among art historians, "the problem remains how to defend and define description" (1983:xx, xxi). Where "true painting" exhibits as its distinguishing mark "surfaces" that give way to "meaningful depth," the "depth" of narrative wholeness and cultural resonance, Alpers argues that "northern images do not disguise meaning or hide it beneath the surface but rather show that meaning by its very nature is lodged in what the eye can take in—however deceptive that might be" (1983:xxiv).

Historians of science and of the Scientific Revolution have not had to defend "description": certainly not as a mark of "true"—or, at least, of "good"—science, and emphatically not against the claims of narrative as source of authority and legitimation for scientific knowledge. Even where the relationship between descriptive facts and theoretical explanations is seen as problematical, description as a mode of investigating nature is itself enshrined as natural. The turn of science during the seventeenth century from texts to things, from language to laboratory, from nature emblematized to nature laid bare—in short, from narrative to description—has not usually been regarded as odd or curious, let alone worthy of serious efforts at historical explanation.

Unlike Alpers, we need to reconstruct the role of narrative in early modern science if we are to understand the substance and significance of late Renaissance science and the nature of the changes produced by the Scientific Revolution. Our critical language as historians of science has been so dominated by models of science that privilege description and employ metaphors evoking the transparency of language that we are apt to misconstrue the turn to description in seventeenth-century science as simply a turn to observation and experiment or to the mapping of nature onto a geometrical grid of mathematicized relations. By reconstructing the role of narrative, I believe that we can better understand the science of the late Renaissance and Scientific Revolution as a whole; we can also begin to offer critical historical analyses of the descriptive turn as a move rooted in language and involving the complex and ideologically situated adoption of interpretive strategies—not the unmotivated, "fortunate" fall back into a "natural" engagement with nature.[1]

I have therefore chosen to call the "descriptive" mode of science emerging during the Scientific Revolution a hermeneutic strategy of *"de-in-scription,"* in order to highlight the narrative foundations and linguistic assumptions legitimating this oft-celebrated "modern" approach to nature. Description *as* de-in-scription assumes nature not as a given, but as a *text* authored, in the idiom of the sixteenth and seventeenth centuries, by God.[2] De-in-scription therefore shares with exegetical and symbolic interpretive strategies a narrative legitimation; it is yet another manifestation and permutation of the Western enthrallment with the "Book." Situating man with respect to the master narratives of Western culture, stories about the Fall, the Tower of Babel, and their aftermath in human cultural history defined the relationship between the "Word of God" and the "languages of man." The net effect of such stories, then, was to provide definition and justification for a variety of different interpretive strategies and practices in natural philosophy, medicine, and the occult and natural sciences.

Like man's descent into history after the Fall, such legitimating stories are never innocent. Rather, they provide, I believe, fertile ground for cultivating a social and cultural interpretation of science and its ideological significance during the "Scientific Revolution." The politically and socially situated and contested nature of narratives of science and language in the culture of Interregnum and Restoration England will provide a focus for exploring such an interpretation in Volume 2 of this study. As we have already seen in this volume, however, by rhetorically presenting itself as no more than a turn toward "things," the descriptive approach to nature attempts to mask its origins in language

1. For a sophisticated analysis of the complexity and sociocultural embeddedness of the experimental enterprise in seventeenth-century science, see Shapin and Schaffer (1985). Latour (1990) offers a stunning analysis of the implications of their book for science studies generally.

2. I do not mean to suggest that "de-in-scription" as a mode of interpretation and scientific practice was universal or monolithic. In fact, the range of "descriptive" strategies in science constituted a wide spectrum; figures like Bacon, Mersenne, Descartes, Boyle and Wilkins define significant differences within this mode. See the essays on Locke, Leibniz, Sprat, and Wilkins in Aarsleff (1982).

and narrative.[3] This attempt becomes reinforced by the very criticisms practitioners of this "new" form of science raise against the self-consciously "narrative" science promoted by exegetical and symbolic interpretations of nature fostered in the late Renaissance. Thus, attacks by such proponents of a descriptive hermeneutics as Galileo, Bacon, and Sprat on, for example, the uses of metaphor and fables in the study of nature implied that, unlike their opponents, they did not surround their descriptions of things with ever-expanding webs of meaning fashioned from words, myth, poetry, and symbols.

While this distinction is in one sense true, it elides the fact that the very discourse of scientific description may itself trace its roots back to the legitimating stories mentioned above. While consciously rejecting overt presentation of nature as symbol or meaningful story, this new science is itself the narrative creation of sixteenth- and seventeenth-century discourses. That is to say, it is the product of discourses that assert postlapsarian man's access to divinely inscribed knowledge of nature through reading, not the hidden symbols of divine wisdom, but the visible, if labyrinthine, language of things constituting the Book of Nature authored by God. In short, although this descriptive science produces a knowledge of things liberated from the dense overlay of stories, it itself is the product of a contingent, local, culturally and ideologically embedded narrative understanding of nature. To describe things, this new science had to construct practices for "de-in-scribing" the text of nature. The absence of narrative explanations of nature depended upon the presence of a not fully acknowledged narrative about nature.

The analysis I have suggested of the Scientific Revolution argues for the centrality of language and interpretation to science and, consequently, to the history of science. It also serves to valorize aspects of science that have previously been consigned to the margins of the Scientific Revolution. More specifically, natural history, which has until recently suffered from a decidedly second-class status in historiographical accounts, can now be seen, thanks to the innovative studies of historians like Ashworth and Findlen, as clearly exhibiting the most telling features of science and scientific change within the cultures of the late Renaissance and the seventeenth century. Symbols, meaning, language theory, and narrative are all forced into the forefront of cultural analysis of science by the attempt to make sense of the signifi-

3. For a similar tendency to mask the origins in language of mathematics, science, and technology, see Knoespel (1987).

cance of natural history during this formative period of Western science. In sum, language theory, assumptions about the relationships between words and things, are not simply the consequence of scientific change and newly established "worldviews"; rather, they help to constitute the very interpretive strategies and concrete practices that science employs to understand and represent nature. And they also become sites of contestation contributing to scientific change. By highlighting such interpretive strategies and practices, and by tracing their legitimation in theories of language and in the cultural narratives that inform the latter, we are enabled to understand science not "in relation to" society or culture, but rather as a practice that simultaneously creates its own contexts.

Bibliography
Index

Bibliography

Aarsleff, Hans. (1964) "Language, Man and Knowledge in the Sixteenth and Seventeenth Centuries." Program in the History and Philosophy of Science, Princeton University. Five lectures.

Aarsleff, Hans. (1982) *From Locke to Saussure: Essays on the Study of Language and Intellectual History.* Minneapolis: University of Minnesota Press.

Aldrovandi, Ulisse. (1599–1603) *Ornithologiae, Hoc est de avibus historiae, libri XII.* Bologna: apud Franciscum de Franciscis Senensem.

Aldrovandi, Ulisse. (1599–1668) *Opera omnia.* Bologna: apud Franciscum de Franciscis Senensem.

Aldrovandi, Ulisse. (1602) *De animalibus insectis libri septem.* Bologna: apud I. B. Bellagambam.

Allen, Don Cameron. (1963) *The Legend of Noah: Renaissance Rationalism in Art, Science, and Letters.* Urbana: University of Illinois Press.

Alpers, Svetlana. (1983) *The Art of Describing: Dutch Art in the Seventeenth Century.* Chicago: University of Chicago Press.

Amrine, Frederick, ed. (1989) *Literature and Science as Modes of Expression.* Boston Studies in the Philosophy of Science, vol. 115. Dordrecht and Boston: Kluwer Academic Publishers.

Anagnine, E. (1937) *Giovanni Pico della Mirandola. Sincretismo Religioso-filosofico.* Bari.

Annius of Viterbo. (1498) *Commentarii Fratris Joannis Annii Viterbiensis super opera diversorum antiquitatibus loquentium.* Rome.

Anon. (1623) *Effroyables pactions faictes entre le diable et les pretendus invisibles, avec leur damnables instructions, perte deplorable de leur escoliers, et leur miserable fin.* n.p.

Apel, Karl-Otto. (1963) *Die Idee der Sprache in der Tradition des Humanismus von Dante bis Vico.* Bonn.

Appleby, Joyce. (1989) "One Good Turn Deserves Another: Moving Beyond the Linguistic; a Response to David Harlan." *American Historical Review* 94:1326–32.

Arens, Hans. (1980) "*Verbum Cordis:* zur Sprachphilosophie des Mittelalters." *Historiographia Linguistica* 7:13–27.

Aristotle. (1963) *Generation of Animals.* Trans. A. L. Peck. Loeb Classical Library. Cambridge/London: Harvard University Press/William Heinemann.

279

Aristotle. (1964) "On Youth and Old Age, on Life and Death." In *Parva Naturalia*. Trans. W. S. Hett, 412–27. Loeb Classical Library. London: William Heinemann.

Arndt, Erwin. (1983) "Sprache und Sprachverstandnis bei Luther." *Zeitschrift für Phonetik, Sprachwissenschaft und Kommunikationsforschung* 36:251–64.

Arnold, Paul. (1970) *La Rose-Croix et ses rapports avec la franc-maçonnerie: essai de synthèse historique*. Paris: Maisonneuve et Larose.

Ashworth, E. J. (1980) "Can I Speak More Clearly Than I Understand? A Problem of Religious Language in Henry of Ghent, Duns Scotus and Ockham." *Historiographia Linguistica* 7:29–38.

Ashworth, E. J. (1988) "Traditional Logic." In Schmitt et al. (1988:143–72).

Ashworth, William B., Jr. (1986) "Catholicism and Early Modern Science." In *God and Nature*, ed. David C. Lindberg and Ronald L. Numbers, 136–66. Berkeley: University of California Press.

Ashworth, William B., Jr. (1989) "Light of Reason, Light of Nature: Catholic and Protestant Metaphors of Scientific Knowledge." *Science in Context* 3:89–107.

Ashworth, William B., Jr. (1990) "Natural History and the Emblematic World View." In Lindberg and Westman (1990:303–32).

Avinyo, J. (1925) *Historia del Lulisme*. Barcelona.

Axenfeld, Alexandre. (1866) *Jean Weir et la sorcellerie*. Paris: G. Balliere.

Bacon, Sir Francis. (1857–74) *The Works of Francis Bacon*. Ed. James Spedding, Robert Leslie Ellis, and Douglas Denon Heath. 7 vols. London: Longman et al.

Bacon, Sir Francis. (1859) *Valerius Terminus. Of the Interpretation of Nature: With the Annotations of Hermes Stella*. In Bacon (1857–74:3.215–52).

Bacon, Sir Francis. (1860) *Parasceve*. In Bacon (1857–74:4.251–63).

Bacon, Sir Francis. (1960) *The New Organon and Related Writings*. Ed. Fulton H. Anderson. Indianapolis:Bobbs-Merrill.

Bacon, Roger. (1928) *De retardatione accidentium senectutis*. Ed. A. G. Little and E. Withington. Oxford.

Baillet, Adrien. (1691) *La Vie de Monsieur Descartes*. Paris.

Baron, Hans. (1927) "Willensfreiheit und Astrologie bei Marsilio Ficino und Pico della Mirandola." In *Kultur- und Universal-geschichte, Festschrift für Walter W. Goetz*, 145–70. Leipzig.

Barone, G. (1949) *L'umanesimo filosofico di G. Pico della Mirandola*. Milan.

Batt, Noelle, and Michel Pierssens, eds. (1993) *Special Issue: Épistémocritique*. *SubStance*, no. 71/72.

Bauhin, Caspar. (1622) *Catalogus plantarum circa basileam sponte nascentium*. Basel.

Bentley, Jerry H. (1983) *Humanists and Holy Writ*. Princeton: Princeton University Press.

Berkhofer, Robert F., Jr. (1988) "The Challenge of Poetics to (Normal) Historical Practice." *Poetics Today* 9:435–52.

Bernal, Martin. (1987) (1991) *Black Athena: The Afroasiatic Roots of Classical Civiliza-tion*. Vols. 1 and 2. New Brunswick, N.J.: Rutgers University Press.

Biagioli, Mario. (1989) "The Social Status of Italian Mathematicians, 1450–1600." *History of Science* 27:41–95.

Biagioli, Mario. (1990) "The Anthropology of Incommensurability." *Studies in History and Philosophy of Science* 21:183–209.

Biagioli, Mario. (1990a) "Galileo the Emblem Maker." *ISIS* 81:230–58.

Biagioli, Mario. (1993) *Galileo, Courtier*. Chicago: University of Chicago Press.

Bianchi, Massimo Luigi. (1987) *Signatura rerum. Segni, magia e conoscenza da Paracelso a Leibniz*. Rome: Edizioni dell'Ateneo.

Bibliander, Theodore. (1548) *De ratione communi omnium linguarum & literarum commentarius Theodori Bibliandri*. Zürich: apud Christop. Frosch.

Boccaccio, Giovanni. (1930) *Boccaccio on Poetry, Being the Preface and Fourteenth and Fifteenth Books of Boccaccio's Genealogia Deorum Gentilium*. Trans. Charles G. Osgood. Princeton: Princeton University Press.

Boehme, Jacob. (1651) *Signatura Rerum: Or, the Signature of All Things Shewing the Sign, and Signification of the Severall Forms and Shapes in the Creation*. Trans. John Elliston. London.

Boehme, Jacob. (1965) *Mysterium Magnum, or an Exposition of the First Book of Moses Called Genesis*. Trans. John Sparrow. Ed. C. J. B. 2 vols. London: John M. Watkins. Reprint of London 1654 edition.

Bono, James J. (1981) "The Languages of Life: Jean Fernel (1497–1558) and *Spiritus* in Pre-Harveian Bio-Medical Thought." Ph.D. diss. Harvard University.

Bono, James J. (1984) "Medical Spirits and the Medieval Language of Life." *Traditio* 40:91–130.

Bono, James J. (1984a) "Review Essay: The Ferment of Van Helmont's Ideas." *Journal of the History of Biology* 17:291–94.

Bono, James J. (1990) "Reform and the Languages of Renaissance Theoretical Medicine: Harvey versus Fernel." *Journal of the History of Biology* 23:341–87.

Bono, James J. (1990a) "Science, Discourse, and Literature: The Role/Rule of Metaphor in Science." In Peterfreund (1990:59–89).

Bono, James J. (1993) Review of Kroll (1991). In *Isis* 84:384–86.

Bono, James J. (1995) "Locating Narratives: Science, Metaphor, Communities, and Epistemic Styles." In Grenzüberschreitungen in der Wissenschaft: *Crossing Boundaries in Science*, ed. Peter Weingart, 115–45. Baden-Baden: Nomos Verlagsgesellschaft.

Bono, James J. (forthcoming) "The Book of Life: Medicine, Aging, and Discourse in the Renaissance." In *Aging and the Life Cycle in the Renaissance*. Newark: University of Delaware Press. [Proceedings of a symposium held at the University of Maryland, April 1988].

Bono, James J. (In progress) *Figuring Science: Metaphor, Narrative, and the Cultural Locations of Scientific Revolutions*.

Bono, James J., and Charles B. Schmitt. (1979) "An Unknown Letter of Jacques Daléchamps to Jean Fernel: Local Autonomy Versus Centralized Government." *Bulletin of the History of Medicine* 53:100–127.

Borst, Arno. (1957–63) *Der Turmbau von Babel: Geschichte der Meinungen über Ursprung und Vielfalt der Sprachen und Volker.* 4 vols. Stuttgart: Anton Hiersemann.

Boulaese, Jean. (1576) . . .*Hebraicum Alphabetum* . . . Paris.

Bouwsma, William J. (1957) *Concordia Mundi: The Career and Thought of Guillaume Postel (1510–1581).* Cambridge: Harvard University Press.

Bovelles, Charles de. (1973) *Sur les langues vulgaires et la variété de la langue fran-çais. Liber de differentia vulgarium linguarium et gallici sermonis varietate (1533).* Trans. Colette Dumont-Demaizière. Strasbourg: Librairie C. Klincksieck.

Boylan, Michael. (1984) "The Galenic and Hippocratic Challenges to Aristotle's Conception Theory." *Journal of the History of Biology* 17:83–112.

Boyle, Marjorie O'Rourke. (1977) *Erasmus on Language and Method in Theology.* Toronto: University of Toronto Press.

Braun, Lucien. (1981) *Paracelse. Nature et philosophie.* Strasbourg: Associations des publications près les Universités de Strasbourg.

Briggs, John C. (1989) *Francis Bacon and the Rhetoric of Nature.* Cambridge: Harvard University Press.

Bullard, Melissa Meriam. (1990) "Marsilio Ficino and the Medici." In *Christianity and the Renaissance,* ed. Timothy Verdon and John Henderson, 467–92. Syracuse.

Burtt, Edwin Arthur. (1954) *The Metaphysical Foundations of Modern Physical Science.* 1924. Rev. ed. Reprint. Garden City, N.Y.: Doubleday Anchor.

Buxtorf, Johann. (1644) *Dissertatio de linguae hebraeae confusione et plurium linguarum origine.* Basel.

Bylebyl, Jerome J. (1971) "Galen on the Non-natural Causes of Variation in the Pulse." *Bulletin of the History of Medicine* 45:482–85.

Bylebyl, Jerome J. (1977) "Nutrition, Quantification and Circulation." *Bulletin of the History of Medicine* 51:369–85.

Bylebyl, Jerome J., and Walter Pagel. (1971) "The Chequered Career of Galen's Doctrine of the Pulmonary Veins." *Medical History* 15:211–29.

Cadden, Joan. (1984) "It Takes All Kinds: Sexuality and Gender Differences in Hildegard of Bingen's Book of Compound Medicine." *Traditio* 40 (1984).

Cadden, Joan. (1992) *The Meaning of Sexual Difference in the Middle Ages: Medicine, Natural Philosophy, and Culture.* Cambridge: Cambridge University Press.

Calvino, Italo. (1985) "Le livre de la nature chez Galilee." In *Exigences et perspectives de la sémiotique: recueil d'hommages pour Algirdas Julien Greimas,* ed. Herman Parret and Hans-George Ruprecht, 683–88. Vol. 2. Amsterdam: John Benjamins.

Camerarius, Joachim. (1590) *Symbolorum & emblematum ex re herbaria desumtorum centuria una collecta.* Nuremberg.

Camerarius, Joachim. (1595) *Symbolorum & emblematum ex animalibus quadru-pedibus desumtorum centuria altera collecta.* Nuremberg.

Camerarius, Joachim. (1596) *Symbolorum & emblematum ex volatilibus ex insectis desumtorum centuria tertia collecta.* Nuremberg.

Camerarius, Joachim. (1604) *Symbolorum & emblematum ex aquatilibus et reptilibus desumptorum centuria quarta.* Nuremberg.

Camporeale, Salvatore I. (1972) *Lorenzo Valla: Umanesimo e teologia.* Florence.

Cardwell, Kenneth William. (1990) "Francis Bacon, Inquisitor." In *Francis Bacon's Legacy of Texts: "The Art of Discovery Grows with Discovery."* Ed. William Sessions. New York: AMS.

Cartari, Vicenzo. (1556) *Le imagini colla sposizione degli dei degli antichi.* Venice.

Casaubon, Meric. (1650) *De quatuor linguis commentationis, pars prior: quae, de lingua hebraica: et de lingua saxonica.* London: J. Flesher, Ric. Mynne.

Cassirer, Ernst. (1942) "Giovanni Pico della Mirandola." *Journal of the History of Ideas* 3:123–44.

Cassirer, Ernst. (1963) *The Individual and the Cosmos in Renaissance Philosophy.* Trans. Mario Domandi. New York: Harper & Row.

Cave, Terence. (1979) *The Cornucopian Text: Problems of Writing in the French Renaissance.* Oxford: Clarendon Press.

Céard, Jean. (1977) *La nature et les prodiges: l'insolite au seizième siècle.* Geneva: Droz.

Céard, Jean. (1980) "De Babel à la Pentecôte: la transformation du mythe de la confusion des langes au XVIe siècle." *Bibliothèque d'Humanisme et Renaissance* 42:577–94.

Céard, Jean. (1988) "Le *De originibus* de Postel et la linguistique de son temps." In *Postello, Venezia, e il suo mondo,* ed. Marion Leathers Kuntz, 19–43. Florence: Olschki.

Certeau, Michel de. (1984) *The Practice of Everyday Life.* Trans. Steven F. Rendall. Berkeley: University of California Press.

Certeau, Michel de. (1986) *Heterologies: Discourse on the Other.* Trans. Brian Massumi. Foreword by Wlad Godzich. Minneapolis: University of Minnesota Press.

Certeau, Michel de. (1988) *The Writing of History.* Trans. Tom Conley. New York: Columbia University Press.

Chenu, M.-D. (1968) *Nature, Man, and Society in the Twelfth Century: Essays on New Theological Perspectives in the Latin West.* Trans. and ed. Jerome Taylor and Lester K. Little. Chicago: University of Chicago Press.

Chéradame, Jean. (1532) *Alphabetvm lingvae sanctae, mystico intellectu refertum.* Paris: apud Aegidium Gormontium.

Chomarat, Jacques. (1981) *Grammaire et rhetorique chez Erasme.* 2 vols. Paris: Société d'edition "Les belles lettres."

Christie, John, and Sally Shuttleworth, eds. (1989) *Nature Transfigured: Science and Literature, 1700–1900.* Manchester and New York: Manchester University Press.

Cigliana, Simona. (1985) "Enrico Cornelio Agrippa e la dignita dell'intellecto: riflesioni quabbalistiche sulla virtu magica delle parole e dei segni." In *Il mago, il cosmo, il teatro degli astri: saggi sulla letteratura esoterica del rinascimento*, ed. Gianfranco Formichetti, 135–57. Rome: Bulzoni.

Clark, John R. (1986) "Roger Bacon and the Composition of Marsilio Ficino's *De Vita Longa*." *Journal of the Warburg and Courtauld Institutes* 49:230–33.

Clarke, Jack A. (1970) *Gabriel Naude, 1600–1653.* Hamden, Conn.: Archon Books.

Clauss, Sidonie. (1982) "John Wilkins' *Essay Toward a Real Character:* Its Place in the Seventeenth Century Episteme." *Journal of the History of Ideas* 42:531–53.

Clulee, Nicholas H. (1977) "Astrology, Magic, and Optics: Facets of John Dee's Early Natural Philosophy." *Renaissance Quarterly* 30:632–80.

Clulee, Nicholas H. (1988) *John Dee's Natural Philosophy: Between Science and Religion.* London: Routledge.

Cohen, Murray. (1977) *Sensible Words: Linguistic Practice in England, 1640–1785.* Baltimore: Johns Hopkins University Press.

Coimbra, University of. (1616) *Commentarii Collegii Conimbricensis, e Societate Iesu, in duos libros De generatione et corruptione, Aristotelis stagiritae.* Venice: Andrea Baba.

Colish, Marcia L. (1968 [1983]) *The Mirror of Language: A Study in the Medieval Theory of Knowledge.* Rev. ed. Reprint. Lincoln: University of Nebraska Press.

Collins, Ardis B. (1974) *The Secular Is Sacred: Platonism and Thomism in Ficino's Platonic Theology.* The Hague.

Colomer, E. (1961) *Nikolaus von Kues und Raimund Lull.* Berlin.

Concasty, Marie-Louise, ed. (1964) *Commentaires de la faculté de médecine de l'université de Paris (1516–1560).* Paris: Imprimerie Nationale.

Conti, Natale. (1551) *Mythologiae.* Venice.

Cook, Harold J. (1990) "The New Philosophy and Medicine in Seventeenth-Century England." In Lindberg and Westman (1990:397–436).

Copenhaver, Brian P. (1977) "Lefèvre d'Étaples, Symphorien Champier, and the Secret Names of God." *Journal of the Warburg and Courtauld Institutes.* 40:189–211.

Copenhaver, Brian P. (1978) *Symphorien Champier and the Reception of the Occultist Tradition in Renaissance France.* The Hague: Mouton.

Copenhaver, Brian P. (1984) "Scholastic Philosophy and Renaissance Magic in the *De Vita* of Marsilio Ficino." *Renaissance Quarterly* 37:523–54.

Copenhaver, Brian P. (1986) "Renaissance Magic and Neoplatonic Philosophy: Ennead 4.3–5 in Ficino's *De vita coelitus comparanda*." In Garfagnini (1986:2.351–69).

Copenhaver, Brian P. (1988) "Hermes Trismegistus, Proclus, and the Question of a Philosophy of Magic in the Renaissance." In *Hermeticism and the Renaissance: Intellectual History and the Occult in Early Modern Europe*, ed. Ingrid

Merkel and Allen G. Debus, 79–110. Washington, D.C.: Folger Shakespeare Library.

Copenhaven, Brian P. (1988a) "Astrology and Magic." In Schmitt et al. (1988: 264–300).

Copenhaver, Brian P. (1990) "Natural Magic, Hermeticism, and Occultism in Early Modern Science." In Lindberg and Westman (1990:261–301).

Coquillette, Daniel R. (1992) *Francis Bacon*. Stanford: Stanford University Press.

Cornelius, Paul (1965) *Language in Seventeenth- and Early-Eighteenth-Century Imaginary Voyages*. Geneva: Droz.

Coudert, Allison. (1978) "Some Theories of a Natural Language from the Renaissance to the Seventeenth Century." In *Magia Naturalis und die Entstehung der Modernen Naturwissenschaften*, ed. Albert Heinekamp and Dieter Mettler, 56–114. Wiesbaden: Franz Steiner Verlag.

Couliano, Ioan P. (1987) *Eros and Magic in the Renaissance*. Trans. Margaret Cook. Chicago: University of Chicago Press.

Courtine, Jean-François. (1980) "Leibniz et la langue adamique." *Revue des sciences philosophiques et théologiques* 64:373–91.

Craven, J. B. (1968) *Count Michael Maier, Doctor of Philosophy and of Medicine, Alchemist, Rosicrucian, Mystic, 1568–1622: Life and Writings*. London: Dawsons.

Croll, Oswald. (1609) *Basilica Chymica, continens philosophicam propria laborum experientia confirmatam discriptionem & usum remediorum chymicorum selectissimorum e lumine gratiae et naturae desumptorum. In fine libri additus est eiusdem autoris tractatus novus de signaturis rerum internis*. Frankfurt.

Croll, Oswald. (1657) *Philosophy Reformed and Improved in Four Profound Tractates. The I. Discovering the Great and Deep Mysteries of Nature: By That Learned Chymist & Physitian Osw. Crollius*. Trans. H[enry] Pinnell. London: Printed by M.S. for Lodowick Lloyd.

Curtius, Ernst Robert. (1963) *European Literature and the Latin Middle Ages*. Trans. Willard R. Trask. New York: Harper & Row.

Dan, R. (1977) "The Age of Reformation Versus 'Linguam Sanctam Hebraicam'— a Survey." *Annales Universitatis Scientiarum Budapestinensis de Rolando Eotvos Nominatae Sectio Linguistica* 8:131–44.

Dannenfeldt, Karl H. (1959) "Egypt and Egyptian Antiquities in the Renaissance." *Studies in the Renaissance* 6:7–27.

Daston, Lorraine, and Peter Galison. (1992) "The Image of Objectivity." *Representations* 40:81–128.

Deacon, Richard. (1968) *John Dee: Scientist, Geographer, Astrologer, and Secret Agent to Elizabeth I*. London: Frederick Muller.

Dean-Jones, Lesley. (1992) "The Politics of Pleasure: Female Sexual Appetite in the Hippocratic Corpus." *Helios* 19:72–91.

Dear, Peter. (1987) "Jesuit Mathematical Science and the Reconstitution of Experience in the Early Seventeenth Century." *Studies in History and Philosophy of Science* 18:133–75.

Dear, Peter. (1988) *Mersenne and the Learning of the Schools*. Ithaca: Cornell University Press.

Deason, Gary B. (1986) "Reformation Theology and the Mechanistic Conception of Nature." In *God and Nature: Historical Essays on the Encounter between Christianity and Science*, ed. David C. Lindberg and Ronald L. Numbers, 167–91. Berkeley: University of California Press.

Dee, John. (1564) *Monas Hieroglyphica*. Antwerp: Guliel. Silvius Typog. Regius.

Deer, Linda Allen. (1980) "Academic Theories of Generation in the Renaissance: The Contemporaries and Successors of Jean Fernel (1497–1558)." Diss. University of London.

De Grazia, Margreta. (1978) "Shakespeare's View of Language: An Historical Perspective." *Shakespeare Quarterly* 29:374–88.

Del Rio, Martino. (1599–1600) *Disqvisitionvm magicarvm libri sex, in tres tomos partiti*. Louvain: ex officina Gerardi Rivii.

Demonet, Marie-Luce. (1992) *Les voix du signe: nature et origine du langage à la renaissance (1480–1580)*. Paris: Librairie Honoré Champion.

Derrida, Jacques. (1982) "White Mythology: Metaphor in the Text of Philosophy." In *Margins of Philosophy*, trans. Alan Bass, 207–71. Chicago: University of Chicago Press.

Descartes, René. (1969) *Oeuvres de Descartes*. Ed. Charles Adam and Paul Tannery. Vol. 1. Paris: Librairie Philosophique J. Vrin.

Devereux, Daniel, and Pierre Pellegrin, eds. (1990) *Biologie, logique et métaphysique chez Aristote*. Paris: Centre National de la Recherche Scientifique.

Dilg-Frank, Rosemarie, ed. (1981) *Kreatur and Kosmos. Internationale Beitrage zur Paracelsusforschung*. Stuttgart: Gustav Fischer.

Du Bellay, Joachim. (1948) *La deffence et illustration de la langue françoyse*. Ed. H. Chamard. Paris.

Dubois, Claude-Gilbert. (1970) *Mythe et langage au seizième siècle*. Lyons: Editions Ducros.

Dubois, Claude-Gilbert. (1977) *La conception de l'histoire en France au XVIe siècle*. Paris: A. G. Nizet.

Dulles, A. (1941) *Princeps Concordiae. Pico della Mirandola and the Scholastic Tradition*. Cambridge: Harvard University Press.

Duret, Claude. (1605) *Histoire admirable des plantes et herbes*. Paris.

Duret, Claude. (1613) *Thresor de l'histoire des langues de cest univers contenant les origines, beautés, perfections, decadences, mutations, changemens, conversions et ruines des langues (1613)*. Cologny. Reprint Geneva: Statkine Reprints, 1972.

Eamon, William. (1984) "Arcana Disclosed: The Advent of Printing, the Books of Secrets Tradition and the Development of Experimental Science in the Sixteenth Century." *History of Science* 22:111—50.

Eamon, William. (1985) "Books of Secrets in Medieval and Early Modern Science." *Sudhoffs Archiv* 69:26–49.

Eamon, William. (1990) "From the Secrets of Nature to Public Knowledge." In Lindberg and Westman (1990:333–65).

Eamon, William. (1990a). "Science as a Hunt, and Some Other Metaphors of Discovery in the Renaissance." Paper presented at the annual meeting of the Society for Literature and Science in Portland, Ore.

Eamon, William. (1994) *Science and the Secrets of Nature: Books of Secrets in Medieval and Early Modern Culture.* Princeton: Princeton University Press.

Egli, E. (1901) "Theodor Biblianders Leben und Schriften." *Analecta Reformatoria* 2:1–143.

Eisenstein, Elizabeth L. (1979) *The Printing Press as an Agent of Change: Communciations and Cultural Transformations in Early-Modern Europe.* 2 vols. Cambridge: Cambridge University Press.

Elert, Claes-Christian. (1978) "Andreas Kempe (1622–1689) and the Languages Spoken in Paradise." *Historiographia Linguistica* 5:221–26.

Elsky, Martin. (1984) "Bacon's Hieroglyphs and the Separation of Word and Thing." *Philological Quarterly* 63:449–60.

Elsky, Martin. (1989) *Authorizing Words: Speech, Writing, and Print in the English Renaissance.* Ithaca: Cornell University Press.

Erastus, Thomas. (1572) *Disputationum de medicina nova Philippi Paracelsi.* Basel: apud Petrum Pernam.

Eros, John Francis. (1972) "Diachronic Linguistics in Seventeenth-Century England, with Special Attention to the Theories of Meric Casaubon." Diss. University of Wisconsin.

Estienne, Henri. (1569) *Traicté de la conformité du langage françois avec le Grec.* Paris.

Farrington, Benjamin. (1966) *The Philosophy of Francis Bacon.* Chicago: University of Chicago Press.

Feldhay, Rivka. (1987) "Knowledge and Salvation in Jesuit Culture." *Science in Context* 1:195–213.

Fernel, Jean. (1542) *De naturali parte medicinae.* Paris: Simon de Colines.

Fernel, Jean. (1548) *De abditis rerum causis.* Paris: A. Wechel.

Fernel, Jean. (1550) *De abditis rerum causis.* Venice: Andrea Arrivabene.

Fernel, Jean. (1554) *Physiologia.* In Fernel (1554a).

Fernel, Jean. (1554a) *Medicina.* Paris: A. Wechel.

Festugière, A.-J. (1949–54) *La révélation d'Hermès Trismégiste.* 4 vols. Paris.

Ficino, Marsilio. (1489) *De vita libri tres.* Florence.

Ficino, Marsilio. (1576) *Opera Omnia.* Ed. M. Sancipriano and P. O. Kristeller. 4 vols. Basel. Reprint Turin, 1959.

Ficino, Marsilio. (1975) *Marsilio Ficino: The Philebus Commentary.* Ed. and trans. Michael J. B. Allen. Berkeley: University of California Press.

Ficino, Marsilio. (1989) *Three Books on Life: A Critical Edition and Translation with Introduction and Notes.* Trans. and ed. Carol V. Kaske and John R. Clark. Binghamton, N.Y.: Medieval and Renaissance Texts and Studies.

Field, Arthur. (1988) *The Origins of the Platonic Academy of Florence.* Princeton: Princeton University Press.

Findlen, Paula. (1989) "Museums, Collecting and Scientific Culture in Early Modern Italy." Diss. University of California, Berkeley.

Findlen, Paula. (1989a) "The Museum: Its Classical Etymology and Renaissance Genealogy." *Journal of the History of Collections* 1:59–78.

Findlen, Paula. (1990) "Jokes of Nature and Jokes of Knowledge: The Playfulness of Scientific Discourse in Early Modern Europe." *Renaissance Quarterly* 43:292–331.

Findlen, Paula. (1990a) "Empty Signs? Reading the Book of Nature in Renaissance Science." *Studies in the History and Philosophy of Science* 21:511–18.

Findlen, Paula. (1991) "The Economy of Scientific Exchange in Early Modern Italy." In *Patronage and Institutions*, ed. Bruce T. Moran, 5–24. Rochester: Boydell Press.

Fingesten, Peter. (1970) *The Eclipse of Symbolism*. Columbia: University of South Carolina Press.

Finocchiaro, Maurice A. (1980) *Galileo and the Art of Reasoning*. Boston Studies in the Philosophy of Science, vol. 61. Dordrecht: Reidel.

Finocchiaro, Maurice A., ed. and trans. (1989) *The Galileo Affair: A Documentary History*. Berkeley: University of California Press.

Fludd, Robert. (1659) *Mosaicall Philosophy: Grounded Upon the Essential Truth or Eternal Sapience*. London: Printed for Humphrey Moseley.

Formigari, Lia. (1970) *Linguistica ed empiricismo nel seicento inglese*. Bari: Laterza.

Foucault, Michel. (1966) *Les mots et les choses*. Paris: Gallimard.

Foucault, Michel. (1973) *The Order of Things: An Archaeology of the Human Sciences*. New York: Vintage.

Fraser, Russell. (1977) *The Language of Adam: On the Limits and Systems of Discourse*. New York: Columbia University Press.

French, Peter J. (1972) *John Dee: The World of an Elizabethan Magus*. London: Routledge & Kegan Paul.

Funkenstein, Amos. (1986) *Theology and the Scientific Imagination from the Middle Ages to the Seventeenth Century*. Princeton: Princeton University Press.

Fussler, Jean-Pierre. (1986) *Les idées éthiques, sociales et politiques de Paracelse (1493–1541) et leur fondement*. Strasbourg: Associations des publications près les Universités de Strasbourg.

Gabbey, Alan. (1977) "Anne Conway et Henri More, lettres sur Descartes." *Archives de philosophie* 40:379–404.

Galen, Claudius. (1821–33) *De marcore liber*. In *Opera omnia Galeni*, ed. C. G. Kuhn, 7.666–704. Leipzig.

Galen, Claudius. (1971) *De Marasmo*. Trans. Theoharis C. Theoharides. In *Journal of the History of Medicine and Allied Sciences* 26:371–90.

Galilei, Galileo. (1895) *Lettera a Madama Cristina di Lorena Granduchessa di Toscana [1615]*. In *Le opere di Galileo Galilei*, ed. Antonio Favaro, 5.309–48. Florence: G. Barbera.

Galilei, Galileo. (1896) *Il Saggiatore*. In *Le opere di Galileo Galilei*, ed. Antonio Favaro, 6.199–372. Florence: G. Barbera.

Galilei, Galileo. (1957) *Discoveries and Opinions of Galileo*. Trans. Stillman Drake. Garden City, N.Y.: Doubleday Anchor.

Galison, Peter. (1988) "History, Philosophy, and the Central Metaphor." *Science in Context* 2:197–212.

Gandillac, Maurice de. (1960) "Astres, anges et génies chez Marsile Ficin." In *Umanesimo e esoterismo*, ed. Enrico Castelli, 85–109. Padua.

Garfagnini, Giancarlo, ed. (1986) *Marsilio Ficino e il ritorno di Platone.* 2 vols. Florence.

Garin, Eugenio. (1937) *Giovanni Pico della Mirandola. Vita e dottrina.* Florence.

Garin, Eugenio. (1961) "La nuova scienza e il simbolo del 'libro.' " In his *La cultura filosofica del rinascimento italiano*, 451–65. Florence: G. C. Sansoni.

Garin, Eugenio. (1969) *Science and Civic Life in the Italian Renaissance.* Trans. Peter Munz. New York.

Garin, Eugenio. (1972) *Portraits from the Quattrocento.* Trans. Victor A. Velen and Elizabeth Velen. New York: Harper & Row.

Garin, Eugenio. (1976) *Astrology in the Renaissance: The Zodiac of Life.* Trans. Carolyn Jackson and June Allen. London: Routledge & Kegan Paul.

Geertz, Clifford. (forthcoming) "The Strange Estrangement: Charles Taylor and the Natural Sciences." In *Philosophy in an Age of Pluralism*, ed. James Tully. Cambridge: Cambridge University Press.

Gehl, Paul F. (1984) "Mystical Language Models in Monastic Educational Psychology." *Journal of Medieval and Renaissance Studies* 14:219–43.

Geiger, Ludwig. (1871) *Johann Reuchlin: Sein Leben und seine Werke.* Leipzig: Verlag von Duncker & Humblot.

Gellrich, Jesse M. (1985) *The Idea of the Book in the Middle Ages: Language Theory, Mythology, and Fiction.* Ithaca: Cornell University Press.

Gesner, Conrad. (1551–58) *Historiae animalium lib. I–IIII.* Zürich: apud C. Froschoverum.

Gesner, Conrad. (1555) *Historica animalium lib. III: De avium.* Zurich: Froschover.

Gesner, Conrad. (1974) *Mithridates, De differentiis linguarum.* Ed. Manfred Peters. Aalen: Scientia Verlag. Reprint of 1555 Zurich edition.

Giambullari, M. Pierfrancesco. (1546) *Il Gello.* Florence.

Giambullari, M. Pierfrancesco. (1549) *Origine della lingua fiorentina altrimenti il gello di M. Pierfrancesco Giambullari.* Florence.

Giambullari, M. Pierfrancesco. (1551) *De la lingua che si parla e scrive in Firenze, et uno dialogo di Giovan Batista Gelli sopra la difficulta dello ordinare detta lingua.* Florence.

Giorgi, Francesco. (1622) *In scripturam sacram et philosophos tria millia problemata.* Paris.

Giorgi, Francesco. (1625) *De harmonia mundi.* Venice.

Giraldi, Lilio. (1548) *De deis gentium varia et multiplex historia.* Basel.

Goldammer, Kurt. (1948–1952) "Paracelsische Eschatologie: zum Verstandnis derAnthropologie und Kosmologie Hohenheims." *Nova Acta Paracelsica* 5:45–85, 6:68–102.

Goldammer, Kurt. (1953) *Paracelsus: Natur und Offenbarung.* Hanover: Theodor Oppermann.

Goldammer, Kurt. (1973) "La contribution de Paracelse à la nouvelle méthod-
ologie scientifique et à la théorie de la connaissance." In *Sciences de la renaissance, VIIIe Congrès international de Tours*, 229–43. Paris: Vrin.

Goldammer, Kurt. (1973a). "La conception Paracelsienne de l'homme entre
la tradition théologique, la mythologie et la science de la nature." In
Sciences de la renaissance, VIIIe Congrès international de Tours, 245–59.

Golinski, Jan. (1990) "The Theory of Practice and the Practice of Theory: Socio-
logical Approaches in the History of Science." *ISIS* 81:492–505.

Goltz, Dietlinde. (1972) "Die Paracelsisten und die Sprache." *Sudhoffs Archiv*
56:337–52.

Gombrich, Ernst H. (1978) "Icones Symbolicae: Philosophies of Symbolism and
Their Bearing on Art." In *Symbolic Images: Studies in the Art of the Renaissance
II*, 123–95, 228–35. Oxford: Oxford University Press.

Goropius Becanus, Johannes. (1569) *Origines Antwerpianae, sive cimmeriorum
becceselana novem libros complexa*. Antwerp: Christophorus Plantinus.

Goropius Becanus, Johannes. (1580) *Opera . . . hactenus in lucem non edita*. Ed.
L. Torrentius. Antwerp: Plantinus.

Gould, Stephen Jay. (1991) "The Birth of the Two-Sex World." *New York Review of
Books*, June 13, 11–13.

Grafton, Anthony. (1979) "Rhetoric, Philology and Egyptomania in the 1570s: J.
J. Scaliger's Invective against M. Guilandinus's *Papyrus*." *Journal of the Warburg and Courtauld Institutes* 42:167–94.

Grafton, Anthony. (1983) *Joseph Scaliger: A Study in the History of Classical Scholarship*. Oxford: Clarendon Press; New York: Oxford University Press.

Grafton, Anthony. (1991) *Defenders of the Text: The Traditions of Scholarship in an
Age of Science, 1450–1800*. Cambridge: Harvard University Press.

Grant, Douglass. (1957) *Margaret the First: A Biography of Margaret Cavendish,
Duchess of Newcastle, 1623–1673*. London: Hart-Davis.

Gravelle, Sarah Stever. (1988) "The Latin-Vernacular Question and Humanist Theory of Language and Culture." *Journal of the History of Ideas*
49:367–86.

Gray, Floyd. (1990) "Montaigne et le langage des animaux." In *Le signe et le texte:
études sur l'écriture au XVIe siècle en France*, ed. Lawrence D. Kritzman, 149–
59. Lexington, Ky.: French Forum.

Greenblatt, Stephen. (1980) *Renaissance Self-Fashioning: From More to Shakespeare*. Chicago: University of Chicago Press.

Gruman, Gerald J. (1966) *A History of Ideas about the Prolongation of Life: The
Evolution of Prolongevity Hypotheses to 1800*. Transactions of the American
Philosophical Society, vol. n.s. 56, no. 9. Philadelphia: American Philo-
sophical Society.

Guerlac, Rita. (1979) *Jean Luis Vives against the Pseudodialecticians. A Humanist
Attack on Medieval Logic*. Dordrecht: Reidel.

Guibbory, Achsah. (1975) "Francis Bacon's View of History: The Cycles of
Error and the Progress of Truth." *Journal of English and Germanic Philology*
74:336–50.

Hacking, Ian. (1975) *The Emergence of Probability.* New York.

Hall, Thomas S. (1971) "Life, Death and the Radical Moisture: A Study of Thematic Patterns in Medieval Medical Theory." *Clio Medica* 6:3–23.

Hallyn, Fernand. (1990) *The Poetic Structure of the World: Copernicus and Kepler.* Trans. Donald M. Leslie. New York: Zone Books.

Hankins, James. (1990) *Plato in the Italian Renaissance.* Columbia Studies in the Classical Tradition, vol. 17. Leiden.

Hankins, James. (1990a) "Cosimo de' Medici and the 'Platonic Academy.' " *Journal of the Warburg and Courtauld Institutes* 53:144–62.

Hankins, James. (1991) "The Myth of the Platonic Academy of Florence." *Renaissance Quarterly* 44:429–75.

Hankins, James. (1991a) "Cosimo de' Medici as Patron of Humanistic Literature." In *Cosimo "Il Vecchio" de' Medici, 1389–1464,* ed. Francis Ames-Lewis, 69–94. Oxford.

Hannaway, Owen. (1975) *The Chemists and the Word: The Didactic Origins of Chemistry.* Baltimore: Johns Hopkins University Press.

Hanson, Ann Ellis. (1992) "Conception, Gestation, and the Origin of Female Nature in the *Corpus Hippocraticum.*" *Helios* 19:31–71.

Haraway, Donna. (1988) "Situated Knowledges: The Science Question in Feminism and the Privilege of Partial Perspective." *Feminist Studies* 14:575–99.

Haraway, Donna. (1991) *Simians, Cyborgs, and Women: The Reinvention of Nature.* New York: Routledge.

Harlan, David. (1989) "Intellectual History and the Return of Literature." *American Historical Review* 94:581–609.

Harms, Wolfgang. (1985) "On Natural History and Emblematics in the Sixteenth Century." In *The Natural Sciences and the Arts,* ed. Allan Ellenius, 67–83. Acta Universitatis Upsaliensis, Figura Nova, vol. 22. Uppsala: Almqvist & Wiksell.

Harms, Wolfgang. (1989) "Bedeutung als Tiel der Sache in zoologischen Standardwerken der frühen Neuzeit (Konrad Gesner, Ulisse Aldrovandi)." In *Lebenslehren und Weltenwürfe in Übergang von Mittelalter zur Neuzeit,* 352–69. Gottengen: Vandenhoeck & Ruprecht.

Harris, Steven J. (1989) "Transposing the Merton Thesis: Apostolic Spirituality and the Establishment of the Jesuit Scientific Tradition." *Science in Context* 3:29–65.

Harth, Dietrich. (1970) *Philologie und praktische Philosophie: Untersuchungen zum Sprach- und Traditionsverstandnis des Erasmus von Rotterdam.* Munich.

Harvey, William. (1649) *Exercitatio anatomica de circulatione sanguinis. Ad Joannem Riolanum Filium Parisiensem.* Cambridge: Roger Daniels.

Harvey, William. (1651) *Exercitationes de generatione animalium.* London: O. Pulleyn.

Harvey, William. (1653) *Anatomical Exercitations Concerning the Generation of Living Creatures.* London: Printed by James Young, for Octavian Pulleyn.

Harvey, William. (1766) *Exercitatio altera, ad J. Riolanum. In Guilielmi Harveii opera omnia: a Collegio Medicorum Londinensi edita.* London.

Harvey, William. (1847) *The Works of William Harvey, M.D.* Trans. Robert Willis. London: Printed for the Sydenham Society. Reprint, New York: Johnson Reprint Corporation, 1965.

Harvey, William. (1928) *Exercitatio anatomica de motu cordis et sanguinis in animalibus.* Ed. Chauncey D. Leake. Springfield, Ill./Baltimore: Charles C. Thomas. Facsimile of the first edition, Frankfurt: Wilhelm Fitzer, 1628.

Harvey, William. (1963) *The Circulation of the Blood and Other Writings.* Trans. Kenneth J. Franklin. London/New York: Dent/Dutton.

Hattaway, Michael. (1978) "Bacon and 'Knowledge Broken': Limits for Scientific Method." *Journal of the History of Ideas* 39:183–97.

Heilbron, John L. (1978) "Introductory Essay." In *John Dee on Astronomy,* ed. Wayne Shumaker. Berkeley: University of California Press.

Henderson, John B. (1991) *Scripture, Canon, and Commentary: A Comparison of Confucian and Western Exegesis.* Princeton: Princeton University Press.

Henry, John, and Sarah Hutton, eds. (1990) *New Perspectives on Renaissance Thought: Essays in Memory of Charles B. Schmitt.* London: Duckworth.

Hermes Trismegistus. (1924–36) *Hermetica.* Ed. Walter Scott. 4 vols. Oxford.

Hermes Trismegistus. (1945–54) *Corpus Hermeticum.* Trans. and ed. A. D. Nock and A.-J. Festugière. 4 vols. Paris.

Hillgarth, J. N. (1972) *Ramon Lull and Lullism in Fourteenth Century France.* Oxford: Oxford University Press.

Hine, William L. (1976) "Mersenne and Vanini." *Renaissance Quarterly* 29:52–65.

Hine, William L. (1984) "Marin Mersenne: Renaissance Naturalism and Renaissance Magic." In Vickers (1984:165–76).

Hine, William L. (1990) "Mersenne and Alchemy." In *Alchemy Revisited,* ed. Z. von Martels, 188–91. Leiden: Brill.

Hippocrates. (1923) *Hippocrates.* With English Translation by W. H. S. Jones. Vol. 2. Cambridge, Mass.: Harvard University Press; London: William Heinemann.

Hollinger, David A. (1989) "The Return of the Prodigal: The Persistence of Historical Knowing." *American Historical Review* 94:610–21.

Horton, Robin. (1967) "African Traditional Thought and Modern Science." *Africa* 37:50–71, 155–87.

Horton, Robin. (1982) "Tradition and Modernity Revisited." In *Rationality and Relativism,* ed. Martin Hollis and Steven Lukes, 201–60. Cambridge: MIT Press.

Howell, W. S. (1961) *Logic and Rhetoric in England, 1500–1700.* New York: Russell & Russell.

Huffman, William H. (1988) *Robert Fludd and the End of the Renaissance.* London: Routledge.

Hunt, Lynn, ed. (1989) *The New Cultural History.* Berkeley: University of California Press.

Hunter, Michael. (1981) *Science and Society in Restoration England.* Cambridge: Cambridge University Press.

Hunter, Michael. (1990) "Science and Heterodoxy: An Early Modern Problem Reconsidered." In Lindberg and Westman (1990:437–60).

Huppert, George. (1974) " 'Divinatio et Eruditio': Thoughts on Foucault." *History and Theory* 13:191–207.

Hutin, Serge. (1960) *Les disciples anglais de Jacob Boehme*. Paris.

Idel, Moshe. (1988) *Kabbalah: New Perspectives*. New Haven: Yale University Press.

Idel, Moshe. (1989) *Language, Torah, and Hermeneutics in Abraham Abulafia*. Trans. Menahem Kallus. Albany: State University of New York Press.

Impey, Oliver, and Arthur MacGregor, eds. (1985) *The Origins of Museums: The Cabinet of Curiosities in Sixteenth- and Seventeenth-Century Europe*. Oxford: Clarendon Press.

Iversen, E. (1961) *The Myth of Egypt and Its Hieroglyphs in European Tradition*. Copenhagen.

Jacquart, Danielle, and Claude Thomasset. (1988) *Sexuality and Medicine in the Middle Ages*. Princeton: Princeton University Press.

Jaeckle, Erwin. (1945) "Paracelsus und Agrippa von Nettesheim." *Nova Acta Paracelsica* 2:83–109.

Jarcho, Saul. (1970) "Galen's Six Non-naturals: A Bibliographical Note and Translation." *Bulletin of the History of Medicine* 44:372–77.

Jardine, Lisa. (1974) *Francis Bacon: Discovery and the Art of Discourse*. Cambridge: Cambridge University Press.

Jardine, Lisa. (1977) "Lorenzo Valla and the Origins of Humanist Dialectic." *Journal of the History of Philosophy* 15:143–64.

Jardine, Nicholas. (1979) "The Forging of Modern Realism: Clavius and Kepler against the Sceptics." *Studies in the History and Philosophy of Science* 10:141–73.

Jardine, Nicholas. (1984) *The Birth of History and Philosophy of Science: Kepler's A Defence of Tycho against Ursus with Essays on Its Provenance and Significance*. Cambridge: Cambridge University Press.

Jensen, Kristian. (1985) "Julius Caesar Scaliger's Concept of Language: A Case Study in Sixteenth Century Aristotelianism." Ph.D. diss. European University Institute, Florence.

Jensen, Kristian. (1990) *Rhetorical Philosophy and Philosophical Grammar: Julius Caesar Scaliger's Theory of Language*. Humanistische Bibliothek, Texte und Abhandlungen. Munich: Fink.

Jonston, Joannes. (1650–53) *Historia Naturalis*. 6 vols. Frankfurt.

Josten, C. H. (1964) "A Translation of John Dee's 'Monas Hieroglyphica' (Antwerp, 1564), with an Introduction and Annotations." *Ambix* 12:84–221.

Joubert, Laurent. (1579) *Erreurs populaires et propos vulgaires touchant la médecine et le régime de santé*. Bordeaux.

Kammerer, Ernst Wilhelm. (1971) *Das Leib-Seele-Geist-Problem bei Paracelsus und einigen Autoren des 17. Jahrhunderts*. Weisbaden: Steiner.

Kammerer, Ernst Wilhelm. (1980) *Le problème du corps, de l'âme et de l'esprit chez*

Paracelse et chez quelques auteurs du XVIIe siècle. Trans. P. Kessler. In *Cahiers de l'hermetisme. Paracelse,* 89–231. Paris: Aubin Michel.

Kaske, Carol V. (1986) "Ficino's Shifting Attitude toward Astrology in the *De vita coelitus comparanda,* the Letter to Poliziano, and the *Apologia* to the Cardinals." In Garfagnini (1986:371–80).

Keefer, Michael H. (1988) "Agrippa's Dilemma: Hermetic 'Rebirth' and the Ambivalences of *De vanitate* and *De occulta philosophia.*" *Renaissance Quarterly* 41:614–53.

Kelley, Donald R. (1970) *Foundations of Modern Historical Scholarship: Language, Law, and History in the French Renaissance.* New York: Columbia University Press.

Kessler, Eckhard; Charles H. Lohr; and Walter Sparn, eds. (1988) *Aristotelismus und Renaissance: In Memoriam Charles B. Schmitt.* Weisbaden: Harrassowitz.

Keynes, Sir Geoffrey. (1966) *The Life of William Harvey.* Oxford: Clarendon Press.

Kienast, Richard. (1970) *Johann Valentin Andreae und die Vier echten Rosenkreutzer Schriften.* New York: Johnson Reprint Corporation.

Kircher, Athanasius. (1679) *Turris Babel, sive archontologica qua primo priscorum post diluvium hominum vita, mores rerumque gestarum magnitudo, secundo turris fabrica civitatumque ex structio, confusio linguarum, & inde gentium transmigrationis, cum principalium inde enatorum idiomatum historia, multiplici eruditione describuntur & explicantur.* Amsterdam: ex officina Janssonio-Waesbergiana.

Klibansky, R. (1939) *The Continuity of the Platonic Tradition during the Middle Ages.* London.

Klibansky, R.; E. Panofsky; F. Saxl. (1964) *Saturn and Melancholy.* London.

Kloppenberg, James T. (1987) "Deconstructive and Hermeneutic Strategies for Intellectual History; The Recent Work of Dominick LaCapra and David Hollinger." *Intellectual History Newsletter* 9:3–22.

Knoespel, Kenneth J. (1987) "The Narrative Matter of Mathematics: John Dee's Preface to the *Elements* of Euclid of Megara (1570)." *Philological Quarterly* 66:27–46.

Knowlson, J. (1975) *Universal Language Schemes in England and France, 1600–1800.* Toronto: University of Toronto Press.

Knox, Dilwyn. (1990) "Ideas on Gesture and Universal Languages c. 1550–1650." In *New Perspectives on Renaissance Thought: Essays in the History of Science, Education, and Philosophy in Memory of Charles B. Schmitt,* ed. John Henry and Sarah Hutton, 101–36. London: Duckworth.

Koerner, Konrad. (1980) "Medieval Linguistic Thought: A Comprehensive Bibliography." *Historiographia Linguistica* 7:265–99.

Konopacki, Steven A. (1979) *The Descent into Words: Jakob Boehme's Transcendental Linguistics.* Ann Arbor: Karoma Publishers.

Koyré, Alexandre. (1929) *La philosophie de Jacob Boehme.* Paris.

Koyré, Alexandre. (1943) "Galileo and Plato" *Journal of the History of Ideas* 4:400–428.

Koyré, Alexandre. (1966) *Etudes Galiléennes*. Paris: Hermann.

Kretzmann, Norman; Anthony Kenny; Jan Pinborg; and Eleanor Stump, eds. (1982) *The Cambridge History of Later Medieval Philosophy*. Cambridge: Cambridge University Press.

Kristeller, Paul Oskar. (1943) *The Philosophy of Marsilio Ficino*. Trans. Virginia Conant. New York.

Kristeller, Paul Oskar. (1956) *Studies in Renaissance Thought and Letters*. Roma: Edizioni di Storia e Letteratura.

Kristeller, Paul Oskar. (1964) *Eight Philosophers of the Italian Renaissance*. Stanford: Stanford University Press.

Kristeller, Paul Oskar. (1965) "Giovanni Pico della Mirandola and His Sources," In *L'opera e il pensiero di Giovanni Pico della Mirandola nella storia dell'umanesimo*, 35–142. Florence.

Kristeller, Paul Oskar. (1976) "L'état présent des études sur Marsile Ficin." In *Platone et Aristote à la Renaissance*. XVIe Colloque International de Tours. Paris.

Kristeller, Paul Oskar. (1987) *Marsilio Ficino and His Work after Five Hundred Years*. Istituto Nazionale di Studi sul Rinascimento, Quaderni di "Rinascimento," vol. 7. Florence.

Kroll, Richard W. F. (1991) *The Material Word: Literate Culture in the Restoration and Early Eighteenth Century*. Baltimore: Johns Hopkins University Press.

LaCapra, Dominick (1983) *Rethinking Intellectual History: Texts, Contexts, Language*. Ithaca: Cornell University Press.

LaCapra, Dominick. (1988) "Of Lumpers and Readers." *Intellectual History Newsletter* 10:3–10.

Ladner, Gerhard B. (1967) *The Idea of Reform: Its Impact on Christian Thought and Action in the Age of the Fathers*. New York.

Lambin, D. (1572) *De utilitate linguae graecae*. Paris.

Laqueur, Thomas. (1990) *Making Sex: Body and Gender from the Greeks to Freud*. Cambridge: Harvard University Press.

Latour, Bruno. (1987) *Science in Action*. Cambridge: Harvard University Press.

Latour, Bruno. (1990) "Postmodern? No, Simply Amodern! Steps towards an Anthropology of Science." Essay Review. *Studies in History and Philosophy of Science* 21:145–71.

Latour, Bruno. (1993) "Pasteur on Lactic Acid Yeast: A Partial Semiotic Analysis." *Configurations* 1:129–45.

Lehoux, Françoise. (1976) *Le cadre de la vie des médecins parisiens aux XVIe et XVIIe siècles*. Paris: A. & J. Picard.

Leigh, Edward. (1656) *A Treatise of Religion & Learning, and of Religious and Learned Men . . . a Work Seasonable for These Times, Wherein Religion and Learning Have So Many Enemies*. London: Printed by A. M. for Charles Adams.

Leisegang, Hans. (1937) "La connaissance de Dieu au miroir de l'âme et de la nature." *Revue d'histoire et de philosophie religieuses* 17:145–71.

Lenoble, Robert. (1943) *Mersenne, ou la naissance du mécanisme*. Paris: J. Vrin.

Le Roy, Loys. (1577) *De la vicissitude ou variete des choses en l'univers*. Paris: chez Pierre L'Hullier.

Levine, George, ed. (1987) *One Culture: Essays in Science and Literature*. Madison: University of Wisconsin Press.

Lindberg, David C., and Robert S. Westman, eds. (1990) *Reappraisals of the Scientific Revolution*. Cambridge: Cambridge University Press.

Lloyd, G. E. R. (1983) *Science, Folklore and Ideology: Studies in the Life Sciences in Ancient Greece*. Cambridge: Cambridge University Press.

Lloyd, G. E. R. (1990) *Demystifying Mentalities*. Cambridge: Cambridge University Press.

Lohr, Charles H. (1968) "Ramon Llull: Liber Alquindi and Liber Telif." *Estudios Lulianos* 12:145–60.

Lohr, Charles H. (1990) "Raimondo Lullo: l'azione e il pensiero." In *Conciliarismo, stati nazionali, inizi dell'umanesimo*, 235–43. Spoleto: Centro Italiano di Studi sull'Alto Medioevo.

Lugli, Adalgisa. (1983) *Naturalia et mirablia. Il collezionsimo enciclopedico nelle wunderkammern d'Europa*. Milan.

Luther, Martin. (1524) "An die Ratherren aller Stadte deutsches Lands (1524)." In *Werke* [Kritische Gesamtausgabe]. Weimar: Hermann Böhlaus.

Luther, Martin. (1958) *Lectures on Genesis: Chapters 1–5*. Vol. 1 of *Luther's Works*, ed. Jaroslav Pelikan. Saint Louis: Concordia Publishing House.

Luther, Martin. (1960) *Lectures on Genesis: Chapters 6–14*. Vol. 2 of *Luther's Works*, ed. Jaroslav Pelikan and Daniel E. Poellot. Saint Louis: Concordia Publishing House.

Lynch, Michael. (1991) "Laboratory Space and the Technological Complex: An Investigation of Topical Contextures." *Science in Content* 4:51–78.

Lyotard, Jean-François. (1984) *The Postmodern Condition: A Report on Knowledge*. Trans. Geoff Bennington and Brian Massumi. Minneapolis: University of Minnesota Press.

Maclean, Ian. (1980) *The Renaissance Notion of Woman: A Study in the Fortunes of Scholasticism and Medical Science in European Intellectual Life*. Cambridge: Cambridge University Press.

Maclean, Ian. (1984) "The Place of Interpretation: Montaigne and Humanist Jurists on Words, Intention and Meaning." In *Neo-Latin and the Vernacular in Renaissance France*, ed. Grahame Castor and Terence Cave, 252–72. Oxford: Clarendon Press.

Maier, Anneliese. (1949) *Der Vorlaufer Galileis im 14. Jahrhundert*. Rome: Edizioni di storia e litteratura.

Marcel, Raymond. (1958) *Marsile Ficin, 1433–1499*. Paris.

Margolin, Jean-Claude. (1971) "Erasme et le verbe: de la rhetorique à l'hermeneutique." In *Erasme, l'Alsace, et son temps. Catalogue de l'exposition realisée à la bibliothèque nationale et universitaire de Strasbourg*. Strasbourg: Palais de l'université.

Margolin, Jean-Claude. (1985) "Science et nationalisme linguistiques ou la bataille pour l'etymologie au XVIe siècle: Bovelles et sa posterité critique." In *The Fairest Flower: The Emergence of Linguistic National Consciousness in Renaissance Europe*, 139–65. Florence: Accademia della Crusca.

Markley, Robert. (1985) "Robert Boyle on Language: Some Considerations Touching the Style of the Holy Scriptures." *Studies in Eighteenth Century Culture* 14:159–71.

Markley, Robert. (1993) *Fallen Languages: Crises of Representation in Newtonian England, 1600–1740*. Ithaca: Cornell University Press.

Martin, R. J. J. (1988) " 'Knowledge Is Power': Francis Bacon, the State, and the Reform of Natural Philosophy." Ph.D. diss. Cambridge University.

McMullin, Erwin, ed. (1967) *Galileo: Man of Science*. New York: Basic Books.

McNamee, Maurice. (1971) "Bacon's Inductive Method and Humanistic Grammar." *Studies in the Literary Imagination* 4:81–106.

McVaugh, Michael. (1974) "The 'Humidum Radicale' in Thirteenth-Century Medicine." *Traditio* 30:259–83.

Meinhold, Peter. (1958) *Luthers Sprachphilosophie*. Berlin: Lutherisches Verlagshaus.

Merchant, Carolyn. (1979) "The Vitalism of Anne Conway: Its Impact on Leibniz's Concept of the Monad." *Journal of the History of Philosophy* 17:255–69.

Merchant, Carolyn. (1980) *The Death of Nature: Women, Ecology, and the Scientific Revolution*. New York: Harper & Row.

Mersenne, Marin. (1623) *Quaestiones Celeberrimae in Genesim*. Paris: Sebastian Cramoisy.

Mersenne, Marin. (1625) *La vérité des sciences: contre les sceptiques ou pyrrhoniens*. Facsimile ed., Stuttgart-Bad Cannstatt: Frommann, 1969.

Mersenne, Marin. (1636) *Harmonie universelle contenant la theorie et la pratique de la musique*. Paris: chez Sebastien Cramoisy.

Mersenne, Marin. (1932–70) *Correspondance du P. Marin Mersenne*. Ed. Cornelius De Waard and René Pintard. Paris: Presses Universitaires de France.

Metcalf, George J. (1963) "The Views of Konrad Gesner on Language." In *Studies in Germanic Languages and Literatures in Memory of Fred O. Nolte*, ed. Erich Hofacker and Liselotte Dieckmann, 15–26. St. Louis: Washington University Press.

Metcalf, George J. (1963a) "Konrad Gesner's Views on the Germanic Languages." *Monatshefte für deutschen Unterricht, deutsche Sprache und Literatur* 55:149–56.

Metcalf, George J. (1974) "The Indo-European Hypothesis in the Sixteenth and Seventeenth Centuries." In *Studies in the History of Linguistics: Traditions and Paradigms*, ed. Dell Hymes, 233–57. Bloomington: Indiana University Press.

Metcalf, George J. (1980) "Theodore Bibliander (1504–1564) and the Languages of Japheth's Progeny." *Historiographia Linguistica* 7:323–33.

Monnerjahn, E. (1960) *Giovanni Pico della Mirandola. Ein Beitrag zur philosophischen Theologie des italienischen Humanismus*. Wiesbaden.

Montgomery, John Warwick. (1973) *Cross and Crucible: Johann Valentin Andreae (1586–1654), Phoenix of the Theologians.* The Hague: M. Nijhoff.

Moody, E. A. (1951) "Galileo and Avempace: The Dynamics of the Leaning Tower Experiment." *Journal of the History of Ideas* 12:163–93, 375–422.

Moran, Bruce T. (1981) "German Prince-Practitioners: Aspects in the Development of Courtly Science, Technology, and Procedures in the Renaissance." *Technology and Culture* 22:253–74.

Moran, Bruce T. (1985) "Privilege, Communication, and Chemiatry: The Hermetic-Alchemical Circle of Moritz of Hessen-Kassel." *Ambix* 32:110–26.

Moran, Bruce T., ed. (1991) *Patronage and Institutions: Science, Technology, and Medicine at the European Court, 1500–1750.* Rochester: Boydell Press.

Morinus, Stephanus. (1694) *Exercitationes de lingua primaeva.* Utrecht: apud Gulielmum Broedelet.

Mungello, David E. (1985) *Curious Land: Jesuit Accommodation and the Origins of Sinology.* Stuttgart: Franz Steiner Verlag.

Naudé, Gabriel. (1623) *Instruction à la France sur la vérité de l'histoire des freres de la Roze-Croix.* Paris: François Julliot.

Naudé, Gabriel. (1625) *Apologie pour tous les grands personnages qui ont este faussement soupçonnez de magie.* Paris: François Targa.

Nauert, Charles G., Jr. (1965) *Agrippa and the Crisis of Renaissance Thought.* Urbana: University of Illinois Press.

Nicolson, Marjorie. (1930) *Conway Letters: The Correspondence of Anne, Viscountess Conway, Henry More and Their Friends.* New Haven: Yale University Press.

Niebyl, Peter H. (1971) "Old Age, Fever, and the Lamp Metaphor." *Journal of the History of Medicine and Allied Sciences* 26:351–68.

Norena, Carlos G. (1970) *Juan Luis Vives.* The Hague: Martinus Nijhoff.

Novick, Peter. (1988) *That Noble Dream: The "Objectivity Question" and the American Historical Profession.* Cambridge: Cambridge University Press.

Nutton, Vivian. (1983) "The Seeds of Disease: An Explanation of Contagion and Infection from the Greeks to the Renaissance." *Medical History* 27:1–34.

Olmi, Giuseppe. (1976) *Ulisse Aldrovandi: scienze e natura nel secondo cinquecento.* Quaderni di storia e filosofia della scienze, vol. 4. Trento: University of Trento.

Omont, H. (1917) "Un humaniste du seizième siècle: excellence de l'affinité de la langue grecque avec la françoise par Blasset." *Revue des études grecques.* Paris.

Ormsby-Lennon, Hugh. (1988) "Rosicrucian Linguistics: Twilight of a Renaissance Tradition." In *Hermeticism and the Renaissance: Intellectual History and the Occult in Early-Modern Europe,* ed. Allen G. Debus and Ingrid Merkel, 311–41. Washington, D.C.: Folger Shakespeare Library.

Orr, Linda. (1986) "The Revenge of Literature: A History of History." *New Literary History* 18:1–22.

Ozment, Steven E. (1973) *Mysticism and Dissent: Religious Ideology and Social Protest in the Sixteenth Century.* New Haven: Yale University Press.

Padley, G. Arthur. (1976) *Grammatical Theory in Western Europe, 1500–1700: The Latin Tradition.* Cambridge: Cambridge University Press.

Pagel, Walter. (1935) "ReligiousMotivesintheMedicalBiologyoftheSeventeenth Century." *Bulletin of the History of Medicine* 3:97–128, 213–31, 265–312.

Pagel, Walter. (1967) *William Harvey's Biological Ideas.* Basel/New York: Karger.

Pagel, Walter. (1969–70) "William Harvey Revisited." *History of Science* 8:1–31, 9:1–41.

Pagel, Walter. (1976) *New Light on William Harvey.* Basel/New York: Karger.

Pagel, Walter. (1982) *Paracelsus: An Introduction to Philosophical Medicine in the Era of the Renaissance.* 2d rev. ed. Basel and New York: Karger.

Pagel, Walter. (1982a) *Joan Baptista Van Helmont: Reformer of Science and Medicine.* Cambridge: Cambridge University Press.

Pagel, Walter. (1984) *The Smiling Spleen: Paracelsianism in Storm and Stress.* Basel: Karger.

Paracelsus, Theophrast von Hohenheim. (1520) *Elf Traktat von Ursprung, Ursachen, Zeichen und Kur einzelner Krankheiten.* In *Samtliche Werke,* vol. 1, ed. Karl Suchoff, 1–161. Munich and Berlin: R. Oldenbourg, 1929.

Paracelsus, Theophrast von Hohenheim. (1525) *Von den naturlichen Dingen.* In *Samtliche Werke,* vol. 2, ed. Karl Sudhoff, 59–175. Munich and Berlin: R. Oldenbourg, 1930.

Paracelsus, Theophrast von Hohenheim. (1529–30) *Paragranum.* In *Samtliche Werke,* vol. 8, ed. Karl Sudhoff, 31–113. Munich: Otto Wilhelm Barth, 1924.

Paracelsus, Theophrast von Hohenheim. (1537) *Die 9 Bucher De natura rerum.* In *Samtliche Werke,* vol. 11, ed. Karl Sudhoff, 307–403. Munich and Berlin: R. Oldenbourg, 1928.

Paracelsus, Theophrast von Hohenheim. (1537–38) *Astronomia Magna oder die ganze Philosophia Sagax der grossen und kleinen Welt.* In *Samtliche Werke,* vol. 12, ed. Karl Sudhoff, 1–444. Munich and Berlin: R. Oldenbourg, 1929.

Paracelsus, Theophrastus von Hohenheim. (1537–38a) *Labyrinthus medicorum errantium.* In *Samtliche Werke,* vol. 11, ed. Karl Sudhoff, 161–221. Munich and Berlin: R. Oldenbourg, 1928.

Paracelsus, Theophrast von Hohenheim. (1922–33) *Theophrast von Hohenheim, genannt Paracelsus, Samtliche Werke. I. Abteilung: Medizinische, naturwis-senschaftliche und philosophisce Schriften.* Ed. Karl Sudhoff. 14 vols. Munich: Oldenbourg.

Paracelsus, Theophrast von Hohenheim. (1976) *The Hermetic and Alchemical Writings of Aureolus Philippus Theophrastus Bombast of Hohenheim, Called Paracelsus the Great.* Ed. Arthur Edward Waite. 2 vols. Boulder: Shambhala.

Paré, Ambroise. (1841) *Le livre des animavx et de l'excellence de l'homme.* Ed. J. F. Malgaigne. In vol. 3 of *Oeuvres complètes d'Ambroise Paré,* 735–69. Paris: J. B. Ballière.

Paré, Ambroise. (1971) *Des monstres et prodiges. Ed. critique et commentée.* Ed. Jean Céard. Geneva: Droz.

Paré, Ambroise. (1982) *On Monsters and Marvels.* Trans. Janis L. Pallister. Chicago: University of Chicago Press.

Park, Katharine, and Lorraine J. Daston. (1981) "Unnatural Conceptions: The Study of Monsters in Sixteenth- and Seventeenth-Century France and England." *Past and Present* 92:20–54.

Pattaro, Sandra Tugnoli. (1981) *Metodo e sistema delle scienze nel pensiero de Ulisse Aldrovandi.* Bologna: CLUEB.

Payne, J. B. (1969) "Toward the Hermeneutics of Erasmus." In *Scrinium Erasmianum II*, ed. J. Coppens, 13–49. Leiden.

Peers, E. A. (1929) *Ramon Lull: A Biography.* London.

Pererius, Benedictus. (1589) *Prior tomus commentariorum & disputationum in Genesim: continens historiam Mosis ad exordio mundi usque ad noeticum diluvium, septem libris explanatum.* Rome: apud Georgium Ferrarium.

Pereius, Benedictus. (1592) *Commentariorum et disputationum in Genesim. Tomus secundus. Continens novem libros circa historiam Mosis de dilvvio, arca Noe, aedificatione turris Babel; confusione linguarum, aliique usque ad vocationem Abrahae, id est, a capite quinto usque ad duodecimum.* Rome: ex typographia Aloysii Zannetti.

Perez-Ramos, Antonio. (1988) *Francis Bacon's Idea of Science and the Maker's Knowledge Tradition.* Oxford/New York: Clarendon Press/Oxford University Press.

Perion, Joachim. (1554) *Dialogorum de linguae gallicae origine ejusque cum graeca cognatione libri quatuor.* Paris.

Peterfreund, Stuart, ed. (1990) *Literature and Science: Theory and Practice.* Boston: Northeastern University Press.

Peters, Manfred. (1971) "Conrad Gesner als Linguist und Germanist." *Gesnerus* 28: 115–45.

Peters, Manfred. (1984) "Theodore Bibliander, *De ratione communi omnium linguarum et literarum commentarius*, Zurich 1548." *Archiv dür das Studium der neueren Sprachen und Literaturen* 221:1–18.

Peuckert, Will-Erich. (1928) *Die Rosenkreutzer.* Jena.

Picard, Jean. (1556) *De prisca celtopaedia libri quinque.* Paris.

Pickering, Andrew, ed. (1992) *Science as Practice and Culture.* Chicago: University of Chicago Press.

Pico della Mirandola, Giovanni, and Gian Francesco Pico. (1969) *Opera omnia (1557–1573).* Hildesheim: Georg Olms. Reprint of Basel 1557 edition.

Pico della Mirandola, Giovanni. (1965) *On the Dignity of Man.* Trans. Charles Glenn Wallis, 3–34. Indianapolis: Bobbs-Merrill.

Plancy, Guillaume. (1656) "Vita Fernelii." In Jean Fernel, *Universa Medicina.* Utrecht: G. à Zijll & T. ab Ackersdijck.

Plancy, Guillaume. (1946) "Life of Fernel." In Sherrington (1946).

Platter, Felix. (1976) *Tagebuch (Lebensbeschreibung) 1536–1567.* Ed. Valentin Lötscher. Basel/Stuttgart: Schwabe.

Popkin, Richard H. (1968) *The History of Scepticism from Erasmus to Descartes.* New York: Harper & Row.

Postel, Guillaume. (1538) *De originibus de hebraicae linguae & gentis antiquitate, deque variarum linguarum affinitate, liber.* Paris: apud Dionysium Lescuier.

Postel, Guillaume. (1547) *Clavis absconditorum a constitutione mundi, qua mens humana tam in divinis quam humanis pertingit ad interiora velaminis aeternae veritatis.* Basel.

Praz, Mario. (1939) *Studies in Seventeenth Century Imagery.* Vol. 1. London: Warburg Institute.

Preus, Anthony. (1970) "Science and Philosophy in Aristotle's *Generation of Animals.*" *Journal of the History of Biology* 3:1–52.

Preus, Anthony. (1977) "Galen's Criticism of Aristotle's Conception Theory." *Journal of the History of Biology* 10:65–85.

Pring-Mill, R. D. F. (1955–56) "The Trinitarian World Picture of Ramon Lull." *Romanistisches Jahrbuch* 7:229–56.

Quecke, K. (1955) "Die Signaturlehren im Schriften der Paracelsus." *Beitrage zur Geschichte der Pharmazie* 2:41–55.

Quicherat, J. E. J. (1860) *Histoire de Sainte-Barbe: Collège, Communauté, Institution.* Vol. 1. Paris.

Quint, David. (1983) *Origin and Originality in Renaissance Literature: Versions of the Source.* New Haven: Yale University Press.

Randall, John Herman, Jr. (1961) *The School of Padua and the Emergence of Modern Science.* Padua: Antenore.

Rather, Leslie J. (1968) "The 'Six Things Non-natural': A Note on the Origins and Fate of a Doctrine and a Phrase." *Clio medica* 3:333–47.

Rattansi, P. M. (1964) "The Helmontian-Galenist Controversy in Restoration England." *Ambix* 12:1–23.

Redondi, Pietro. (1987) *Galileo Heretic.* Trans. R. Rosenthal. Princeton: Princeton University Press.

Reeds, Karen. (1975) "Botany in Medieval and Renaissance Universities." Ph.D. Diss. Harvard University.

Rees, Graham. (1975) "Francis Bacon's Semi-paracelsian Cosmology." *Ambix* 22:81–101.

Rees, Graham. (1975a) "Francis Bacon's Semi-paracelsian Cosmology and the Great Instauration." *Ambix* 22:161–73.

Rees, Graham. (1977) "The Fate of Bacon's Cosmology in the Seventeenth Century." *Ambix* 24:27–38.

Rees, Graham. (1977a) "Matter Theory: A Unifying Factor in Bacon's Natural Philosophy?" *Ambix* 24:110–25.

Rees, Graham, assisted by Christopher Upton. (1984) *Francis Bacon's Natural Philosophy: A New Source.* A transcription of manuscript Hardwick 72A with translation and commentary. Chalfont St. Giles: British Society for the History of Science.

Reuchlin, Johannes. (1964) *De verbo mirifico (1494). De arte cabalistica (1517).* Stuttgart-Bad Cannstatt: Friedrich Frommann Verlag.

Rice, Eugene F., Jr. (1970) "Humanist Aristotelianism in France. Jacques Lefèvre d'Étaples and His Circle." In *Humanism in France at the End of the Middle Ages and in the Early Renaissance*, ed. H. T. Levi, 132–49. Manchester: Manchester University Press.

Rice, Eugene F., Jr. (1971) "Jacques Lefèvre d'Étaples and the Medieval Christian Mystics." In *Florilegium Historiale: Essays Presented to Wallace K. Ferguson*. Ed. J. G. Rowe and W. H. Stockdale, 89–124. Toronto: University of Toronto Press.

Rice, Eugene F., Jr., ed. (1972) *The Prefatory Epistles of Jacques Lefèvre D'Étaples and Related Texts*. New York: Columbia University Press.

Rice, Eugene F., Jr. (1975) "The *De Magia Naturali* of Jacques Lefèvre D'Etaples." In *Philosophy and Humanism: Renaissance Essays in Honor of Paul Oskar Kristeller*, ed. E. P. Mahoney, 19–29. New York: Columbia University Press.

Rice, James Van Nostran. (1939) *Gabriel Naudé, 1600–1653*. Baltimore: Johns Hopkins University Press.

Richardson, Linda Deer. (1985) "The Generation of Disease: Occult Causes and Diseases of the Total Substance." In *The Medical Renaissance of the Sixteenth Century*, ed. A. Wear, R. K. French, and I. M. Lonie, 175–94. Cambridge: Cambridge University Press.

Ripa, Cesare. (1593) *Iconologia*. Rome.

Robinet, A. (1978) *Le langage à l'âge classique*. Paris.

Rorty, Richard. (1991) "Is Natural Science a Natural Kind." In *Objectivity, Relativism, and Truth: Philosophical Papers*, 1.42–62. Cambridge:Cambridge University Press.

Rossi, Paolo. (1960) *Clavis universalis: arti mnemoniche e logica combinatoria da Lullo a Leibniz*. Milan/Naples: Ricciardi.

Rossi, Paolo. (1961) "The Legacy of Ramon Lull in Sixteenth Century Thought." *Medieval and Renaissance Studies* 5:182–213.

Rossi, Paolo. (1968) *Francis Bacon: From Magic to Science*. Trans. S. Rabinovitch. London: Routledge & Kegan Paul.

Rossi, Paolo. (1970) *Philosophy, Technology and the Arts in the Early Modern Era*. Trans. Salvator Attanasio. Ed. Benjamin Nelson. New York: Harper & Row.

Rossi, Paolo. (1971) *Aspetti della rivoluzione scientifica*. Naples: Morano Editore.

Rothacker, E. (1979) *Das "Buch der Natur": Materialien und Grundsatzliches zur Metapherngeschichte*. Bonn: Bouvier.

Rothstein, Marian. (1990) "Etymology, Genealogy, and the Immutability of Origins." *Renaissance Quarterly* 43:332–47.

Rouse, Joseph. (1990) "The Narrative Reconstruction of Science." *Inquiry* 33:179–96.

Rouse, Joseph. (1993) "What Are Cultural Studies of Scientific Knowledge?" *Configurations* 1:1–22.

Rousseau, G. S., ed. (1986) *Science and the Imagination*. Annals of Scholarship, vol. 4.

Rudolph, Hartmut. (1980) "Kosmosspekulation und Trinitätslehre, Weltbild und Theologie bei Paracelsus." In *Paracelsus in der Tradition*, ed. S. Domandl. *Salzb. Beitr. Paracelsus-Forschung* 21:32–47.

Rudolph, Hartmut. (1981) "Einige Gesichtspunkte zum Thema 'Paracelsus und Luther' " *Archiv für Reformationsgeschichte* 72:34–54.

Saitta, Giuseppe. (1954) *Marsilio Ficino e la filosofia dell'umanesimo*. 3d ed. Bologna.

Salmon, Vivian. (1972) *The Works of Francis Lodwick: A Study of His Writings in the Intellectual Context of the Seventeenth Century*. London: Longman.

Salmon, Vivian. (1979) *The Study of Language in Seventeenth Century England*. Studies in the History of Linguistics, vol. 17. Amsterdam.

Sanches, Francisco. (1988) *That Nothing is Known (Quod nihil scitur)*. Trans. Douglas F. S. Thomson. Ed. Elaine Limbrick. Latin text from 2d 1581 edition. Cambridge: Cambridge University Press.

Sarasohn, Lisa. (1984) "A Science Turned Upside Down: Feminism and the Natural Philosophy of Margaret Cavendish." *Huntington Library Quarterly* 47:289–307.

Sargent, Rose-Mary. (1989) "Scientific Experiment and Legal Expertise: The Way of Experience in Seventeenth-Century England." *Studies in the History and Philosophy of Science* 20:19–45.

Saulnier, V. L. (1953) *François Rabelais, ouvrage publié pour le quatrième centenaire de sa mort*. Geneva: Droz; Lille: Giard.

Scaliger, Joseph Justus. (1610) *Opuscula varia antehac non edita*. 119–22. Paris: apud Hieronymum Drouart.

Scaliger, Julius Caesar. (1540) *De causis linguae latinae*. Lyons: Seb. Gryphium.

Schiebinger, Londa. (1989) *The Mind Has No Sex? Women in the Origins of Modern Science*. Cambridge: Harvard University Press.

Schipperges, Heinrich. (1988) *Die Entienlehre des Paracelsus: Aufbau und Umriss seiner theoretischen Pathologie*. Berlin: Springer.

Schmitt, Charles B. (1969) "Experience and Experiment: A Comparison of Zabarella's Views with Galileo's in *De motu*." *Studies in the Renaissance* 16:80–138.

Schmitt, Charles B. (1970) "Prisca theologia e philosophia perennis: due temi del rinascimento italiano e la loro fortuna." In *Il pensiero italiano del rinascimento e il tempo nostro*, 211–36. Florence: Leo S. Olschki.

Schmitt, Charles B. (1972) *Cicero Scepticus: A Study of the Influence of the Academica in the Renaissance*. The Hague: Martinus Nijhoff.

Schmitt, Charles B. (1978) "Reappraisals in Renaissance Science." *History of Science* 16:200–214.

Schmitt, Charles B. (1981) *Studies in Renaissance Philosophy and Science*. London: Variorum.

Schmitt, Charles B. (1983) *Aristotle and the Renaissance*. Cambridge: Harvard University Press.

Schmitt, Charles B. (1983a) *John Case and Aristotelianism in Renaissance England*. Kingston: McGill-Queens University Press.

Schmitt, Charles B. (1984) *The Aristotelian Tradition and Renaissance Universities*. London: Variorum.

Schmitt, Charles B. (1984a) "William Harvey and Renaissance Aristotelianism: A Consideration of the *Praefatio* to *'De generatione animalium'* (1651)." In *Humanismus und Medizin*, ed. Rudolf Schmitz and Gundolf Keil, 117–38. Weinheim: Acta Humaniora.

Schmitt, Charles B.; Quentin Skinner; Eckhard Kessler; and Jill Kraye, eds. (1988) *The Cambridge History of Renaissance Philosophy*. Cambridge: Cambridge University Press.

Scholem, Gershom G. (1946) *Major Trends in Jewish Mysticism*. New York: Schocken Books.

Schwarz, W. (1955) *Principles and Problems of Biblical Translation: Some Reformation Controversies and Their Background*. Cambridge: Cambridge University Press.

Scott, Joan W. (1986) "Gender: A Useful Category of Historical Analysis." *American Historical Review* 91:1053–75.

Secret, François. (1964) *Les kabbalistes chrétiens de la renaissance*. Paris.

Seznec, Jean. (1953) *The Survival of the Pagan Gods: The Mythological Tradition and Its Place in Renaissance Humanism and Art*. Trans. Barbara F. Sessions. N.Y.: Pantheon Books.

Shapin, Steven, and Simon Schaffer. (1985) *Leviathan and the Air-pump: Hobbes, Boyle, and the Experimental Life*. Princeton: Princeton University Press.

Shapiro, Barbara J. (1983) *Probability and Certainty in Seventeenth-Century England: A Study of the Relationships between Natural Science, Religion, History, Law, and Literature*. Princeton: Princeton University Press.

Shapiro, Barbara J. (1991) *"Beyond Reasonable Doubt" and "Probable Cause": Historical Perspectives on the Anglo-American Law of Evidence*. Berkeley: University of California Press.

Shea, William R. (1977) *Galileo's Intellectual Revolution: Middle Period, 1610–1632*. 2d ed. New York: Science History Publications.

Sherrington, Sir Charles. (1946) *The Endeavour of Jean Fernel*. Cambridge: Cambridge University Press.

Simoncelli, Paolo. (1984) *La lingua di Adamo: Guillaume Postel tra accademici e fuoriusciti fiorentini*. Florence: Olschki.

Singer, Thomas C. (1989) "Hieroglyphs, Real Characters, and the Idea of Natural Language in English Seventeenth-Century Thought." *Journal of the History of Ideas* 50:49–70.

Slaughter, Mary M. (1982) *Universal Language and Scientific Taxonomy in the Seventeenth Century*. Cambridge: Cambridge University Press.

Smocovitis, V. B. (1992) "Unifying Biology: The Evolutionary Synthesis and Evolutionary Biology." *Journal of the History of Biology* 25:1–65.

Speroni, Sperone. (1978) *Dialogo delle lingue*. Trattatisti del cinquecento. Milan/Naples. [1542].

Spitz, Lewis W. (1963) *The Religious Renaissance of the German Humanists*. Cambridge: Harvard University Press.

Staden, Heinrich von. (1991) "Matière et signification." *L'Antiquité classique* 60:42–61.

Staden, Heinrich von. (1992) "Women and Dirt." *Helios* 19:7–30.

Staden, Heinrich von. (1993) "Spiderwoman and the Chaste Tree: The Semantics of Matter." *Configurations* 1:23–56.

Stevin, Simon. (1955) "Discourse on the Worth of the Dutch Language." In *The Principal Works of Simon Stevin*, ed. E. J. Dijksterhuis, 58–93. Amsterdam.

Stillingfleet, Edward. (1662) *Origines Sacrae, or a Rational Account of the Grounds of Christian Faith, as to the Truth and Divine Authority of the Scriptures, and the Matters Therein Contained*. London: Printed by P. W. for Henry Mortlock.

Stock, Brian. (1983) *The Implications of Literacy: Written Language and Models of Interpretation in the Eleventh and Twelfth Centuries*. Princeton: Princeton University Press.

Temkin, Owsei. (1951) "On Galen's Pneumatology." *Gesnerus* 8:180–89.

Temkin, Owsei. (1977) *The Double Face of Janus*. Baltimore: Johns Hopkins University Press.

Thomas, Keith. (1971) *Religion and the Decline of Magic*. Harmondsworth: Penguin.

Thorndike, Lynn. (1923–58) *A History of Magic and Experimental Science*. 8 vols. New York: Columbia University Press.

Toews, John. (1987) "Intellectual History after the Linguistic Turn: The Autonomy of Meaning and the Irreducibility of Experience." *American Historical Review* 92:879–907.

Topsell, Edward. (1607) *A Historie of Four-footed Beastes*. London.

Traweek, Sharon. (1988) *Beamtimes and Lifetimes*. Cambridge: Harvard University Press.

Trinkaus, Charles. (1970) *In Our Image and Likeness: Humanity and Divinity in Italian Humanist Thought*. 2 vols. Chicago: University of Chicago Press.

Trippault, Leon. (1580) *Celt' hellenisme ou etymologie des mots françois tires du grec plus preuve en general de la descente de notre langue*. Orléans.

Tyard, Pontus de. (1587) "Le second curieux, ou second discours de la nature du monde, et de ses parties, traitant des choses intellectuelles." *Les discours philosophiques de Pontus de Tyard*. Paris: chez Abel L'Angelier.

Tyard, Pontus de. (1603) *De recta nominum impositione*. Lyons: apud Iacobvm Rovssin.

Tymme, Thomas. (1963) *A Light in Darkness. Which Illumineth for All the Monas Hieroglyphica of the Famous and Profound Dr. John Dee, Discovering Natures Closet and Revealing the True Christian Secrets of Alchimy*. From MS Ashmole 1459. Oxford: Printed at the New Bodleian Library.

Valeriano, Pierio. (1556) *Hieroglyphica*. Basel.

Vasoli, Cesare. (1980) "L'analogie dans le langage de la magie à la renaissance." Trans. D. Lesur. In *La magie et ses langages*, ed. Margaret Jones-Davies. Travaux et Recherches, Université de Lille, vol. 3. Lille.

Vernant, Jean-Pierre. (1982) *The Origins of Greek Thought*. Ithaca: Cornell University Press.

Vickers, Brian, ed. (1984) *Occult and Scientific Mentalities in the Renaissance*. Cambridge: Cambridge University Press.

Vickers, Brian. (1984a) "Analogy versus Identity: The Rejection of Occult Symbolism, 1580–1680." In Vickers (1984:95–163).

Victor, Joseph M. (1975) "The Revival of Lullism at Paris, 1499–1516." *Renaissance Quarterly* 28:504–34.

Victor, Joseph M. (1978) *Charles de Bovelles, 1479–1553: An Intellectual Biography.* Geneva: Droz.

Waite, Arthur Edward. (1887) *The Real History of the Rosicrucians.* London.

Waite, Arthur Edward. (1923) *The Brotherhood of the Rosy Cross.* London.

Walker, D. P. (1954) "The *Prisca Theologia* in France." *Journal of the Warburg and Courtauld Institutes* 17:204–59.

Walker, D. P. (1958) *Spiritual and Demonic Magic from Ficino to Campanella.* London: Warburg Institute, University of London.

Walker, D. P. (1958a) "The Astral Body in Renaissance Medicine." *Journal of the Warburg and Courtauld Institutes* 21:119–33.

Walker, D. P. (1972) *The Ancient Theology.* Ithaca: Cornell University Press.

Walker, D. P. (1972a) "Leibniz and Language." *Journal of the Warburg and Courtauld Institutes* 35:294–307. Reprinted in Walker (1985).

Walker, D. P. (1985) *Music, Spirit and Language in the Renaissance.* Ed. Penelope Gouk. London: Variorum.

Walker, D. P. (1985a) "Francis Bacon and *Spiritus.*" In Walker (1985) 121–130.

Walker, D. P. (1986) "Marsilio Ficino and Astrology." In Garfagnini (1986:341–49).

Wallace, William A. (1981) *Prelude to Galileo: Essays on Medieval and Sixteenth-Century Sources of Galileo's Thought.* Dordrecht: Reidel.

Wallace, William A. (1984) *Galileo and His Sources: The Heritage of the Collegio Romano in Galileo's Science.* Princeton: Princeton University Press.

Wallace, William A., ed. (1986) *Reinterpreting Galileo.* Washington, D.C.: Catholic University of America Press.

Walton, Michael T. (1976) "John Dee's *Monas Hieroglyphica:* Geometrical Cabala." *Ambix* 23:116–23.

Walton, Michael T. (1981) "Hermetic Cabala in the *Monas Hieroglyphica* and the *Mosaicall Philosophy.*" *Essentia* (Summer): 7–18.

Warhaft, Sidney. (1971) "The Providential Order in Bacon's New Philosophy." *Studies in the Literary Imagination* 4:49–64.

Waswo, Richard. (1979) "The 'Ordinary Language Philosophy' of Lorenzo Valla." *Bibliothèque d'Humanism et Renaissance* 41:255–71.

Waswo, Richard. (1987) *Language and Meaning in the Renaissance.* Princeton: Princeton University Press.

Watts, Pauline Moffitt. (1982) *Nicolaus Cusanus: A Fifteenth Century Vision of Man.* Leiden: Brill.

Wear, Andrew. (1983) "William Harvey and the 'Way of the Anatomists.' " *History of Science* 21:223–49.

Webb, John. (1678) *The Antiquity of China, or an Historical Essay, Endeavouring a Probability That the Language of the Empire of China is the Primitive Language*

Spoken Through the Whole World Before the Confusion of Babel. London: Printed for Obadiah Blagrave.

Webster, Charles. (1967) "The College of Physicians: 'Solomon's House' in Commonwealth England." *Bulletin of the History of Medicine* 41:393–412.

Webster, Charles. (1967a) "Harvey's *De generatione:* Its Origin and Relevance to the Theory of Circulation." *British Journal for the History of Science* 3:262–74.

Webster, Charles. (1975) *The Great Instauration: Science, Medicine and Reform, 1626–1660.* London: Duckworth.

Webster, Charles. (1982) *From Paracelsus to Newton: Magic and the Making of Modern Science.* Cambridge: Cambridge University Press.

Wedgwood, C. V. (1969) *The Thirty Years War.* Gloucester, Mass.: P. Smith.

Westman, Robert S. (1972) "Kepler's Theory of Hypothesis and the 'Realist Dilemma.' " *Studies in History and Philosophy of Science* 3:233–64.

Westman, Robert S. (1977) "Magical Reform and Astronomical Reform: The Yates Thesis Reconsidered." In *Hermeticism and the Scientific Revolution,* ed. Robert S. Westman and J. E. McGuire, 1–91. Los Angeles: Clark Library.

Westman, Robert S. (1980) "The Astronomer's Role in the Sixteenth Century: A Preliminary Study." *History of Science* 18:105–47.

Westman, Robert S. (1990) "Proof, Poetics, and Patronage: Copernicus's Preface to *De revolutionibus.*" In Lindberg and Westman (1990:167–205).

Weyer [Weir], Johann. (1583) *De praestigiis daemonum, & incantationibus, ac veneficiis libri sex.* Basel.

Weyer, Johann. (1991) *Witches, Devils, and Doctors in the Renaissance: Johann Weyer, De praestigiis daemonum.* Trans. John Shea. Ed. George Mora and Benjamin Kohl. Binghamton, N.Y.: Medieval and Renaissance Texts and Studies.

Wheeler, Harvey. (1990) "Francis Bacon's *New Atlantis:* The 'Mould' of a Lawfinding Commonwealth." In *Francis Bacon's Legacy of Texts: "The Art of Discovery Grows with Discovery,"* 291–310. New York: AMS Press.

White, John S. (1986) "William Harvey and the Primacy of the Blood." *Annals of Science* 43:239–55.

Whitney, Charles. (1986) *Francis Bacon and Modernity.* New Haven: Yale University Press.

Wilcox, Donald J. (1987) *The Measure of Times Past: Pre-Newtonian Chronologies and the Rhetoric of Relative Time.* Chicago: University of Chicago Press.

Wilkins, John. (1668) *An Essay Towards a Real Character, and a Philosophical Language.* Facsimile of original London edition, Menston, England: Scolar Press, 1968.

Williams, Arnold. (1948) *The Common Expositor: An Account of the Commentaries on Genesis, 1527–1633.* Chapel Hill: University of North Carolina Press.

Williams, George H. (1962) *The Radical Reformation.* London.

Wilson, Leonard G. (1959) "Erasistratus, Galen, and the *Pneuma.*" *Bulletin of the History of Medicine* 33:293–314.

Wirszubski, Chaim. (1989) *Pico della Mirandola's Encounter with Jewish Mysticism.* Cambridge: Harvard University Press.

Wojciehowski, Dolora Ann. (1990) "The Will to Read: Galileo and the Book of Nature." Unpublished essay.

Woolgar, Steve. (1986) "On the Alleged Distinction between Discourse and *Praxis." Social Studies of Science* 16:309–17.

Wormius, Ole. (1636) *Runica, seu Danica literatura antiquissima.* Copenhaven: M. Martzan.

Yates, Frances A. (1964) *Giordano Bruno and the Hermetic Tradition.* London.

Yates, Frances A. (1978) *The Rosicrucian Enlightenment.* Boulder: Shambhala.

Yates, Frances A. (1979) *The Occult Philosophy in the Elizabethan Age.* London: Routledge & Kegan Paul.

Yates, Frances A. (1982) *Lull and Bruno.* Vol. 1 of *Collected Essays.* London/ Boston: Routledge & Kegan Paul.

Zambelli, Paola. (1966) "*Humanae litterae, verbum Divinum, docta ignorantia* negli ultimi scritti di Enrico Cornelio Agrippa." *Giornale critico della filosofia italiana* 20:187–217.

Zambelli, Paola. (1973) "Platone, Ficino, e la magia." In *Studia Humanitatis: Ernesto Grassi zum 70. Geburtstag,* ed. E. Hora and E. Kessler, 121–42. Munich.

Zambelli, Paola. (1976) "Magic and Radical Reformation in Agrippa of Nettesheim." *Journal of the Warburg and Courtauld Institutes* 39:69–103.

Zanier, Giancarlo. (1977) *La medicina astrologica e la sua teoria: Marsilio Ficino e i suoi critici contemporanei.* Rome.

Zika, Charles. (1976) "Reuchlin's *De verbo mirifico* and the Magic Debate of the Late Fifteenth Century." *Journal of the Warburg and Courtauld Institutes* 39:104–38.

Zika, Charles. (1976–77) "Reuchlin and Erasmus: Humanism and Occult Philosophy." *Journal of Religious History* 9:223–46.

Index